ATLAS OF
FUNCTIONAL
NEUROANATOMY

SECOND EDITION

ATLAS OF
FUNCTIONAL
NEUROANATOMY
SECOND EDITION

WALTER J. HENDELMAN, M.D., C.M.

Taylor & Francis
Taylor & Francis Group
Boca Raton   London   New York

A CRC title, part of the Taylor & Francis imprint, a member of the
Taylor & Francis Group, the academic division of T&F Informa plc.

Published in 2006 by
CRC Press
Taylor & Francis Group
6000 Broken Sound Parkway NW, Suite 300
Boca Raton, FL 33487-2742

International Standard Book Number-10: 0-8493-3084-X (Softcover)
International Standard Book Number-13: 978-0-8493-3084-1 (Softcover)
Library of Congress Card Number 2005049418

Library of Congress Cataloging-in-Publication Data

Hendelman, Walter.
    Atlas of functional neuroanatomy / Walter Hendelman.-- 2nd ed.
        p. ; cm.
    Includes bibliographical references and index.
    ISBN 0-8493-3084-X
    1. Neuroanatomy--Atlases. I. Title: Functional neuroanatomy. II. Title.
    [DNLM: 1. Central Nervous System--anatomy & histology--Atlases. WL 17 H495a 2005]

QM451.H347 2005
611.8'022'2--dc22

                                                                       2005049418

Taylor & Francis Group is the Academic Division of Informa plc.

Visit the Taylor & Francis Web site at
http://www.taylorandfrancis.com

and the CRC Press Web site at
http://www.crcpress.com

DEDICATION

I wish to dedicate this book to people who have made a meaningful impact on my life
as a professional, both teacher and scientist, and as a person.

To my wife and life partner, Teena
and to our daughter, Lisanne
and sadly now to the memory of our daughter, Devra

To the many teachers and mentors and colleagues in my career as a neuroscientist,
and particularly with respect and gratitude to

Dr. Donald Hebb
Dr. Richard Bunge
Dr. Malcolm Carpenter

To all those students, staff, and colleagues who have assisted me in this endeavor
and to all the students who have inspired me in this learning partnership.

PREFACE

This atlas grew out of the seeds of discontent of a teacher attempting to enable medical students to understand the neuroanatomical framework of the human brain, the central nervous system. As a teacher, it is my conviction that each slide or picture that is shown to students should be accompanied by an explanation; these explanations formed the basis of an atlas. Diagrams were created to help students understand the structures and pathways of the nervous system and each illustration was accompanied by explanatory text, so that the student could study both together.

The pedagogical perspective has not changed over the various editions of the atlas as it expanded in content, but the illustrations have evolved markedly. They changed from simple artwork to computer-based graphics, from no color to 2-color, to the present edition in full color. The illustrations now include digital photographs, using carefully selected and dissected specimens.

Most of the diagrams in the atlas were created by medical students, with artistic and/or technological ability, who could visualize the structural aspects of the nervous system. These students, who had completed the basic neuroanatomy course, collaborated with the author to create the diagrams intended to assist the next generation of students to learn the material more easily and with better understanding. I sincerely thank each of them for their effort and dedication and for their frequent, intense discussions about the material (please see the acknowledgements). They helped decide which aspects should be included in an atlas intended for use by students early in their career with limited time allotted for this course of study during their medical studies.

This atlas has benefited from the help of colleagues and staff in the department of which I have been a member for over 30 years, and from professional colleagues who have contributed histological and radiological enhancements, as well as advice. Their assistance is sincerely appreciated.

The previous edition of this atlas included a CD ROM containing all the images in full color. At that time, few texts had such a learning companion. It is to the credit of CRC Press that they were willing to accept the idea of this visual enhancement as an aid to student learning. The CD-ROM accompanying this new edition of the atlas, thanks to another student, employs newer software that allows the creative use of "rollover" labeling, and also adds animation to some of the illustrations (please see the User's Guide).

A final comment about the word "functional" in the title is appropriate. The central nervous system, the CNS, is a vast, continually active set of connections, ever-changing and capable of alteration throughout life. The orientation of the written text is to describe both the structural aspects of the CNS and the connections between the parts, and to explain the way those structures of the brain operate as a functional unit. In addition, there are clinically relevant comments included in the descriptive text, where there is a clear relation between the structures being described and neurological disease.

No book could be completed without the support and encouragement of the people who are part of the process of transforming a manuscript to a published work, from the publisher and the project editor, to the technical staff that handles the illustrations, to the proofreaders and copyeditors who work to improve and clarify the text. Each individual is an important contributor to the final product, and I wish to thank them all.

I sincerely hope that you, the learner, enjoy studying from the *Atlas of Funtional Neuroanatomy* and its accompanying CD-ROM, and that the text and illustrations, along with the dynamic images, help you to gain a firm understanding of this fascinating, complex organ—the brain.

Walter J. Hendelman, M.D., C.M.
Ottawa, Canada

AUTHOR BIOGRAPHY

Dr. Walter Hendelman, M.D.,C.M., is a Canadian, born and raised in Montreal. He did his undergraduate studies at McGill University in science with honors in psychology. As part of his courses in physiological psychology, he assisted in an experimental study of rats with lesions of the hippocampus, which was then a little known area of the brain. At that time, Professor Donald Hebb was the chair of the Psychology Department and was gaining prominence for his theory known as "cell assembly," explaining how the brain functions.

Dr. Hendelman proceeded to do his medical studies at McGill. The medical building is situated in the shadow of the world-famous Montreal Neurological Institute (MNI) where Dr. Wilder Penfield and colleagues were forging a new frontier in the understanding of the brain. Subsequently, Dr. Hendelman completed an internship and a year of pediatric medicine, both in Montreal.

Having chosen the brain as his lifelong field of study and work, the next decision involved the choice of either clinical neurology or brain research—Dr. Hendelman chose the latter, with the help of Dr. Francis McNaughton, a senior neurologist at the MNI. Postgraduate studies continued for 4 years in the United States, in the emerging field of developmental neuroscience, using the "new" techniques of nerve tissue culture and electron microscopy. Dr. Richard Bunge was his research mentor at Columbia University Medical Center in New York City, while his neuroanatomy mentor was Dr. Malcolm Carpenter, author of the well-known textbook Human Neuroanatomy.

Dr. Hendelman returned to Canada and has made Ottawa his home for his academic career at the Faculty of Medicine of the University of Ottawa, in the Department of Anatomy, now merged with Physiology and Pharmacology into the Department of Cellular and Molecular Medicine. He began his teaching in gross anatomy and neuroanatomy, and in recent years has focused on the latter. His research continued, with support from Canadian granting agencies, using nerve tissue culture to examine the development of the cerebellum; more recently he has been involved in studies on the development of the cerebral cortex. Several investigations were carried out in collaboration with summer and graduate students and with other scientists. He has been a member of various neuroscience and anatomy professional organizations, has attended and presented at their meetings, and has numerous publications on his research findings.

In addition to research and teaching and the usual academic "duties," Dr. Hendelman was involved with the faculty and university community, including a committee on research ethics. He has also been very active in curriculum planning and teaching matters in the faculty. During the 1990s, when digital technology became available, Dr. Hendelman recognized its potential to assist student learning, particularly in the anatomical subjects and helped bring the new technology into the learning environment of the faculty. Recently, he organized a teaching symposium for the Canadian Association of Anatomy, Neurobiology and Cell Biology on the use of technology for learning the anatomical sciences.

In 2002, Dr. Hendelman completed a program in medical education and received a Master's degree in Education from the Ontario Institute of Studies in Education (OISE), affiliated with the University of Toronto. In the same year, following retirement, he began a new stage of his career, with the responsibility for the development of a professionalism program for medical students at the University of Ottawa.

As a student of the brain, Dr. Hendelman has been deeply engaged as a teacher of the subject throughout his career. Dedicated to assisting those who wish to learn functional neuroanatomy, he has produced teaching videotapes and four previous editions of this atlas. As part of this commitment he has collaborated in the creation of two computer-based learning modules, one on the spinal cord based upon the disease syringomyelia and the other on voluntary motor pathways; both contain original graphics to assist in the learning of the challenging and fascinating subject matter, the human brain.

In his nonprofessional life, Walter Hendelman is a husband, a father, an active member of the community, a choir member, a commuter cyclist, and an avid skier.

ACKNOWLEDGMENTS

This atlas has been a cumulative "work-in-progress," adding and altering and deleting material over time. The illustrations have been created by talented and dedicated individuals—artists, photographers, and students, and with the help of staff and colleagues—whom the author has had the pleasure of working with over these many years.

PREVIOUS EDITIONS

The atlas was originally published with the title of *Student's Atlas of Neuroanatomy*. The diagrams in the first editions were created by Mr. Jean-Pierre Morrissey, a medical student at the time he did the work. To these were added photographs of brain specimens taken by Mr. Stanley Klosevych, who was then the director of the Health Sciences Communication Services, University of Ottawa. Mr. Emil Purgina, a medical artist with the same unit, assisted in these early editions and added his own illustration. Dr. Andrei Rosen subsequently created the airbrush diagrams (note particularly the basal ganglia, thalamus, and limbic system) and expanded the pool of illustrations. For the previous edition of the atlas under its new title The Atlas of Functional Neuroanatomy many of the earlier illustrations were replaced by computer-generated diagrams done by Mr. Gordon Wright, a medical illustrator. Mr. Wright also put together the CD-ROM for the previous edition, which contained all the illustrations in this atlas. The efforts of the staff of the University of Ottawa Press and of W.B. Saunders, who published the previous editions, are very much appreciated and acknowledged.

PRESENT EDITION

ILLUSTRATIONS AND PHOTOGRAPHS

Dr. Tim Willett, a medical student during the preparation of the atlas, created many new illustrations and retouched several others. In addition, all the photographs were redone, using original dissections and digital photography, with the assistance of Dr. Willett.

CD-ROM

Mr. Patrick O'Byrne, a doctoral candidate in the nursing program at the Faculty of Health Sciences, University of Ottawa, has put together the present CD-ROM, using Macromedia Flash software to create "rollover" labeling and animated illustrations.

MEDICAL ARTIST

Mr. Mohammad Dayfallah created the overview diagrams and those of the ventricular system.

RADIOGRAPHS

Colleagues at the Ottawa Hospital contributed the radiographs to the previous edition, and all have been replaced with new images, using the upgraded capability of the newer machines and accompanying software.

HISTOLOGICAL SECTIONS

Colleagues and staff of the Department of Pathology, Children's Hospital of Eastern Ontario, are responsible for preparing the histological sections of the human brainstem, added to in the present edition by sections of the human spinal cord.

SUPPORT

The previous editions were supported, in part, by grants from Teaching Resources Services of the University of Ottawa. The present edition received support from CRC Press.

The support of my home department at the Faculty of Medicine of the University of Ottawa, initially the Department of Anatomy and now called the Department of Cellular and Molecular Medicine, including colleagues, secretaries, and other support staff in the gross anatomy laboratory, is gratefully acknowledged.

Finally, thanks to the many classes of students, who have provided inspiration, as well as comments, suggestions and feedback.

With thanks to all

Dr. Walter J. Hendelman

CONTENTS

LIST OF ILLUSTRATIONS

Part II: Reticular Formation

Part III: Motor Systems

Section C: Neurological Neuroanatomy

Section D: The Limbic System

Sensory:

Dorsal columns & medial lemniscus
(fine touch, vibration & proprioception from the body)

Anterolateral system
(pain, temperature & crude touch from the body)

Trigeminal system
(touch, pain, temperature & proprioception from the head)

Special senses
(vision, audition & taste)

Reticular Formation
(arousal & regulation of muscle t one and reflexes)

Motor:

Voluntary
(movement of body and face)

Parasympathetic
("rest & digest")

Other
(non-voluntary motor & visual coordination)

Vestibular nuclei & tracts
(balance & gravity adjustments)

Cerebellum & associated tracts
(motor coordination)

Special Nuclei:

Substantia nigra
(motor initiation)

Red nucleus & tract
(non-voluntary motor)

Other
(miscellaneous)

USER'S GUIDE

COLOR CODING

Color adds a significant beneficial dimension to the learning of neuroanatomy. The colors have a functional role in this atlas, in that they are used consistently for the presentation of sensory, motor, and other components. The following is the color coding used in this atlas, as shown on the opposite page:

Sensory (nuclei and tracts)

Dorsal Column – Medial Lemniscus	Cobalt Blue
Anterolateral System (Pain and Temperature)	Deep Blue
Trigeminal Pathways	Purple
Special Senses (Audition, Vision, Taste)	Violet
Reticular Formation	Yellow

Motor (nuclei and tracts)

Voluntary	Cadmium Orange
Parasympathetic	Orange
Other Motor (e.g. visual motor)	Light Red
Vestibular (nuclei and tracts)	Lime Green
Cerebellum (nuclei and tracts)	Turquoise

Special Nuclei:

Substantia Nigra	Brown
Red Nucleus (and tract)	Red
Other (e.g., area postrema)	Peach

For students who enjoy a different learning approach, a black and white photocopy of the illustration can be made and then the color added, promoting active learning.

Some students may wish to add color to some of the airbrush diagrams, including the basal ganglia, thalamus, and limbic system.

REFERENCE TO OTHER FIGURES

Reference is made throughout the atlas to other illustrations that contain material relevant to the subject matter or structure being discussed. Although this may be somewhat disruptive to the learner reading a page of text, the author recommends looking at the illustration and the accompanying text being referenced, in order to clarify or enhance the learning of the subject matter or structure.

CLINICAL ASPECT

Various clinical entities are mentioned where there is a clear connection between the structures being discussed and a clinical disease, for example, Parkinson's disease and the substantia nigra. In Section C, the vascular territories are discussed and the deficits associated with occlusion of these vessels is reviewed. Textbooks of neurology should be consulted for a detailed review of clinical diseases (see the Annotated Bibliography). Management of the disease and specific drug therapies are not part of the subject matter of this atlas.

ADDITIONAL DETAIL

On occasion, a structure is described that has some importance but may be beyond what is necessary, at this stage, for an understanding of the system or pathway under discussion. In other cases, a structure is labeled in an illustration but is discussed at another point in the atlas.

DEVELOPMENTAL ASPECT

For certain parts of the nervous system, knowledge of the development contributes to an understanding of the structure seen in the adult. This is particularly so for the spinal cord, as well as for the ventricular system. Knowledge of development is also relevant for the cerebral hemispheres, and for the limbic system (i.e., the hippocampal formation).

NOTE TO THE LEARNER

This notation is added at certain points in the text when, in the author's experience, it might be beneficial for a student learning the matter to review a certain topic; in other cases there is a recommendation to return to the section at a later stage. Sometimes, consulting other texts is suggested. Of course, this is advice only, and each student will approach the learning task in his or her own way.

THE CD-ROM

The CD-ROM adds another dimension to the learning process. Ideally, the student is advised to read the text, using *both* the text illustration and the illustration on the CD. In addition, animation has been added to certain illustrations, such as the pathways, where understanding and seeing the tract that is being described, along with the relays and crossing (decussation), can hopefully assist the student in developing a 3-dimensional understanding of the nervous system.

Labeling of structures on the CD-ROM has been accomplished using "rollover" technology, so that the name of the structure is seen when the cursor is on the area, or when the cursor is over the label, the named structure is highlighted in the illustration.

FOREWORD

We are about to embark on an amazing and challenging journey — an exploration of the human brain. The complexity of the brain has not yet been adequately described in words. The analogies to switchboards or computers, although in some ways appropriate to describe some aspect of brain function, do not do the least bit of justice to the totality. The brain functioning as a whole is infinitely more than its parts. Our brains encompass and create a vast universe.

In the past decade we have come to appreciate that our brains are in a dynamic state of change in all stages of life. We knew that brain function was developing throughout childhood and this has been extended into the teen years, and even into early adulthood. We now are beginning to understand that the brain has the potential to change throughout life, in reaction to the way we live and our personal experiences in this world. The generic term for this is *plasticity*, and the changes may significantly alter the connections of the brain and its pattern of "processing" information, whether from the external world, from our internal environment, or from the brain itself as it generates thoughts and feelings.

ORGANIZATION

The *Atlas* is divided into four sections, each with an introductory text. The focus is on the illustrations, photographs, diagrams, radiographs, and histological material, accompanied by explanatory text on the opposite page.

Section A: The Atlas starts with an **Overview** of the various parts of the central nervous system, the CNS. Then we embark on an **Orientation** to the structural components of the CNS, and this is presented from the spinal cord upward to "the brain"; additional material on the spinal cord is added in other parts of the Atlas. Radiographic images have been included, because that is how the CNS will be viewed and investigated in the clinical setting.

Section B: The second section, **Functional Systems**, uses these structural components to study the sensory ascending pathways (Part I), and the various motor descending tracts (Part III), from origin to termination. Interspersed between them is a discussion of the Reticular Formation (Part II), which has both sensory and motor aspects. Included as part of the motor systems are the major contributors to motor function, the basal ganglia and the cerebellum.

Section C: The third section, **Neurological Neuroanatomy**, includes a neurological orientation and detailed neuroanatomical information, to allow the student to work through *the* neurological question: *Where* is the disease process occurring (i.e., neurological localization)? Because vascular lesions are still most common and relate closely to the functional neuroanatomy, the blood supply to the brain is presented in some detail, using photographs with overlays. The emphasis in this section is on the brainstem, including a select series of histological cross-sections of the human brainstem. In addition, there is a summary of the spinal cord nuclei and tracts, along with a histological view of levels of the human cord.

Section D: The section on the **Limbic System** has once again been revised. New photographs of limbic structures enhance the presentation. This material is sometimes taught within the context of other systems in the curriculum.

ANNOTATED BIBLIOGRAPHY

Students may wish to consult more complete texts on the anatomy and physiology of the nervous system, and certainly some neurology books concerning diseases of the nervous system. A guide to this reference material is included, with commentary, as an annotated bibliography, with an emphasis on recent publications. Added are suggestions for material available on CD-ROM, as well as the Internet. Students are encouraged to search out additional (reliable) resources of this nature.

GLOSSARY

Much of the difficulty of the subject matter is the terminology — complex, difficult to spell, sometimes inconsistent, with a Latin remnant, and sometimes with names of individuals who have described or discovered structures or disease entities, used often by neurologists, neurosurgeons, and neuroradiologists. A Glossary of terms is appended to help the student through this task.

Section A

ORIENTATION

INTRODUCTION

An understanding of the *central nervous system* — the CNS — and how it functions requires knowing its component parts and their specialized operations, and the contribution of each of the parts to the function of the whole. The first section of this atlas introduces the student to the CNS from an anatomical and functional viewpoint. The subsequent section (Section B) will use these components to build the various systems, such as the sensory and motor systems. The blood supply and the detailed anatomical organization are found in Section C. Emotional behavior is discussed in Section D.

FUNCTIONAL NEUROHISTOLOGY

The major cell of the CNS is the **neuron**. Human brains have billions of neurons. A neuron has a cell body (also called soma, or **perikaryon**); **dendrites**, which extend a short distance from the soma; and an **axon**, which connects one neuron with others. Neuronal membranes are specialized for electro-chemical events, which allow these cells to receive and transmit messages to other neurons. The dendrites and cell bodies of the neurons receive information, and the axons transmit the firing pattern of the cell to the next neuron. Generally, each neuron receives synaptic input from hundreds or perhaps thousands of neurons, and its axon distributes this information via collaterals (branches) to hundreds of neurons.

Within the CNS, neurons that share a common function are usually grouped together; such groupings are called **nuclei** (singular **nucleus**, which is somewhat confusing as it does not refer to the part of a cell). In other parts of the brain, the neurons are grouped at the surface, forming a **cortex**. In a cortical organization, neurons are arranged in layers and the neurons in each layer are functionally alike and different from those in other layers. Older cortical areas have three layers (e.g., the cerebellum); more recently evolved cortices have six layers (the cerebral cortex) and sometimes sublayers.

Some neurons in the nervous system are directly linked to sensory (afferent) or motor (efferent) functions. In the CNS, the overwhelming majority of neurons interconnect, that is, form circuits that participate in the processing of information. These neurons are called **interneurons**, and more complex information processing, such as occurs in the human brain, is correlated with the dramatic increase in the number of interneurons in our brains.

Communication between neurons occurs almost exclusively at specialized junctions called **synapses**, using biological molecules called **neurotransmitters**. These modify ion movements across the neuronal membranes of the synapse and alter neurotransmission — they can be excitatory or inhibitory in their action, or modulate synaptic excitability. The post-synaptic neuron will modify its firing pattern depending on the summative effect of all the synapses acting upon it at any moment in time. The action of neurotransmitters depends also on the specific receptor type; there is an ever increasing number of receptor subtypes allowing for even more complexity of information processing within the CNS. Drugs are being designed to act on those receptors for therapeutic purposes.

Much of the substance of the brain consists of **axons**, also called **fibers**, which connect one part of the brain with other areas. These fibers function so that the various parts of the brain communicate with each other, some going a short distance linking neurons locally and others traveling a long distance connecting different areas of the brain and spinal cord. Many of the axons are myelinated, an "insulation," which serves to increase the speed of axonal conduction; the thicker the **myelin sheath**, the faster the conduction. Axons originating from one area (cortex or nucleus) and destined for another area usually group together and form a **tract**, also called a **pathway** (or fasciculus).

The other major cells of the CNS are **glia**; there are more glia than neurons. There are two types of glial cells:

- **Astrocytes**, which are involved in supportive structural and metabolic events
- **Oligodendrocytes**, which are responsible for the formation and maintenance of the myelin that ensheaths the axons

Some of the early maturation that we see in infants and children can be accounted for by the progressive myelination of the various pathways within the CNS throughout childhood.

FUNCTIONAL NEUROANATOMY
OF THE CNS

One approach to an understanding of the nervous system is to conceptualize that it is composed of a number of functional modules, starting with simpler ones and evolving in higher primates and humans to a more complex organizational network of cells and connections. The function of each part is dependent upon and linked to the function of all the modules acting in concert.

The basic unit of the CNS is the **spinal cord** (see Figure 1 and Figure 2), which connects the CNS with the skin and muscles of the body. Simple and complex reflex circuits are located within the spinal cord. It receives sensory information (**afferents**) from the skin and body wall, which are then transmitted to higher centers of the brain. The spinal cord receives movement instructions from the higher centers and sends motor commands (**efferents**) to the muscles. Certain motor patterns are organized in the spinal cord, and these are under the influence of motor areas in other parts of the brain. The autonomic nervous system, which supplies the internal organs and the glands, is also found within the spinal cord.

As the functional systems of the brain become more complex, new control "centers" have evolved. These are often spoken of as higher centers. The first set of these is located in the **brainstem**, which is situated above the spinal cord and within the skull (in humans). The brainstem includes three distinct areas — the **medulla**, **pons**, and **midbrain** (see Figure 0A, Figure 0L, Figure 6, and Figure 7). Some nuclei within the brainstem are concerned with essential functions such as pulse, respiration, and the regulation of blood pressure. Other nuclei within the brainstem are involved in setting our level of arousal and play an important role in maintaining our state of consciousness. Special nuclei in the brainstem are responsible for some basic types of movements in response to gravity or sound. In addition, most of the **cranial nerves** and their nuclei, which supply the structures of the head, are anchored in the brainstem (see Figure 8A and Figure 8B). Many nuclei in the brainstem are related to the cerebellum.

The **cerebellum** has strong connections with the brainstem and is situated behind the brainstem (inside the skull) in humans (see Figure 0A, Figure 0L, and Figure 9A). The cerebellum has a simpler form of cortex, which consists of only three layers. Parts of the cerebellum are quite old in the evolutionary sense, and parts are relatively newer. This "little brain" is involved in motor coordination and also in the planning of movements. How this is accomplished will be understood once the input/output connections of the various parts of the cerebellum are studied.

Next in the hierarchy of the development of the CNS is the area of the brain called the **diencephalon** (see Figure 0A, Figure 0L, and Figure 11). Its largest part, the **thalamus**, develops in conjunction with the cerebral hemispheres and acts as the gateway to the cerebral cortex. The thalamus consists of several nuclei, each of which projects to a part of the cerebral cortex and receives reciprocal connections from the cortex. The **hypothalamus**, a much smaller part of the diencephalon, serves mostly to control the neuroendocrine system via the pituitary gland, and also organizes the activity of the autonomic nervous system. Parts of the hypothalamus are intimately connected with the expression of basic drives (e.g., hunger and thirst), with the regulation of water in our bodies, and with the manifestations of "emotional" behavior as part of the limbic system (see below).

With the continued evolution of the brain, the part of the brain called the forebrain undergoes increased development, a process called encephalization. This has culminated in the development of the **cerebral hemispheres**, which dominate the brains of higher mammals, reaching its zenith (so we think) in humans. The neurons of the cerebral hemispheres are found at the surface, the **cerebral cortex** (see Figure 13 and Figure 14A), most of which is six-layered (also called the **neocortex**). In humans, the cerebral cortex is thrown into ridges (gyri, singular **gyrus**) and valleys (sulci, singular **sulcus**). The enormous expansion of the cerebral cortex in the human, both in terms of size and complexity, has resulted in this part of the brain becoming the dominant controller of the CNS, capable, so it seems, of overriding most of the other regulatory systems. We need our cerebral cortex for almost all interpretations and actions related to the functioning of the sensory and motor systems, for consciousness, language, and thinking.

Buried within the cerebral hemispheres are the **basal ganglia**, large collections of neurons (see Figure 0A, Figure 0L, and Figure 22) that are involved mainly in the initiation and organization of motor movements. These neurons affect motor activity through their influence on the cerebral cortex.

A number of areas of the brain are involved in behavior, which is characterized by the reaction of the animal or person to situations. This reaction is often termed "emotional" and, in humans, consists of both psychological and physiological changes. Various parts of the brain are involved with these activities, and collectively they have been named the **limbic system**. This network includes the cortex, various subcortical areas, parts of the basal ganglia, the hypothalamus and parts of the brainstem. (The limbic system is described in Section D of this atlas.)

In summary, the nervous system has evolved so that its various parts have "assigned tasks." In order for the nervous system to function properly, there must be communication between the various parts. Some of these links are the major sensory and motor pathways, called **tracts** (or fascicles). Much of the mass of tissue in our hemispheres is made up of these **pathways** (e.g., see Figure 33 and Figure 45).

Within all parts of the CNS there are the remnants of the neural tube from which the brain developed; these spaces are filled with **cerebrospinal fluid (CSF)**. The spaces in the cerebral hemispheres are actually quite large and are called **ventricles** (see Figure OA, Figure OL, Figure 20A, Figure 20B, and Figure 21).

The CNS is laced with blood vessels as neurons depend upon a continuous supply of oxygen and glucose. This aspect will be discussed further with the section on vasculature (e.g., see Figure 58).

STUDY OF THE CNS

Early studies of the normal brain were generally descriptive. Brain tissue does not have a firm consistency, and the brain needs to be fixed for gross and microscopic examination. One of the most common fixatives used to preserve the brain for study is formalin, after which it can be handled and sectioned. Areas containing predominantly neuronal cell bodies (and their dendrites and synapses) become grayish in appearance after formalin fixation, and this is traditionally called **gray matter**. Tracts containing myelinated axons become white in color with formalin fixation, and such areas are likewise simply called the **white matter** (see Figure 27 and Figure 29).

We have learned much about the normal function of the human CNS through diseases and injuries to the nervous system. Diseases of the nervous system can involve the neurons, either directly (e.g., metabolic disease) or by reducing the blood supply, which is critical for the viability of nerve cells. Some degenerative diseases affect a particular group of neurons. Other diseases can affect the cells supporting the myelin sheath, thereby disrupting neurotransmission. Biochemical disturbances may disrupt the balance of neurotransmitters and cause functional disease states.

The recent introduction of functional imaging of the nervous system is revealing fascinating information about the functional organization of the CNS. We are slowly beginning to piece together an understanding of what is considered by many as the last and most important frontier of human knowledge, an understanding of the brain.

CLINICAL ASPECT

Certain aspects of clinical neurology will be included in this atlas, both to amplify the text and to indicate the importance of knowing the functional anatomy of the CNS. Knowing where a lesion is located (the localization) often indicates the nature of the disease (the diagnosis), leading to treatment and allowing the physician to discuss the prognosis with the patient.

FIGURE 0A

OVERVIEW — ANTERIOR VIEW

Constructing a three-dimensional visualization of the brain and its various parts is a challenging task for most people, and this diagram and its companion (the next illustration) are designed to assist the learner in this task.

This is a semi-anatomic representation of the brain and the parts of the CNS. This general diagrammatic view should be consulted as the learner is orienting to the placement of the structures within the brain. These same structures are viewed from the lateral perspective with the next illustration.

The cerebral hemispheres: The large cerebral hemispheres, with its extensive **cerebral cortex**, is by far the most impressive structure of the CNS and the one that most are referring to when speaking about "the brain." In fact there are two cerebral hemispheres that are connected across the midline by a massive communication link called the **corpus callosum** (see Figure 16 and Figure 19A). The hemispheres are discussed with Figure 13–Figure 19 of the Orientation section.

Many parts of the brain are found deep inside the hemispheres. This illustration is done so that these structures should be visualized "within" the hemispheres. Included are:

- **Basal ganglia:** These large neuronal areas are found within the brain; its three parts are shown — the **caudate** nucleus (head and tail), the **putamen**, and the **globus pallidus**. The basal ganglia are discussed with Figure 22–Figure 30 of the Orientation section.
- **Ventricles of the brain:** Each hemisphere has within it a space remaining from the neural tube, from which the brain developed, called a ventricle — the **lateral ventricle** (also called ventricles 1 and 2). The ventricles are presented in this anterior perspective with Figure 20B.

The massive cerebral hemispheres hide the other parts of the brain from view, when looking from the anterior perspective, although some of these parts can be seen if the brain is viewed from below (see Figure 15A and Figure 15B). These structures include:

- **Diencephalon:** The largest part of the diencephalon is the **thalamus**; in fact, this is a paired structure. The unpaired third ventricle should be noted between the thalamus of each side. The thalamus is discussed with Figure 11 and Figure 12 of the Orientation section.
- **Brainstem:** By definition, the brainstem consists of the **midbrain, pons,** and **medulla**; the cranial nerves are attached to the brainstem. The brainstem and cranial nerves are considered in Figure 6–Figure 10 of the Orientation section. The ventricular space within the brainstem is the fourth ventricle.
- **Cerebellum:** Part of the **cerebellum** can be seen from this perspective. This "little brain" is usually considered with the brainstem and is discussed with Figure 9A and Figure 9B of the Orientation section.
- **Spinal cord:** This long extension of the CNS continues from the medulla and is found in the vertebral canal. The **spinal cord** is discussed with Figure 1–Figure 5 of the Orientation section.

Note on the safe handling of brain tissue: Current guidelines recommend the use of disposable gloves when handling any brain tissue, to avoid possible contamination with infectious agents, particularly the "slow" viruses. In addition, formalin is a harsh fixative and can cause irritation of the skin. Many individuals can react to the smell of the formalin and may develop an asthmatic reaction. People who handle formalin-fixed tissue must take extra precautions to avoid these problems. In most labs, the brains are soaked in water before being put out for study.

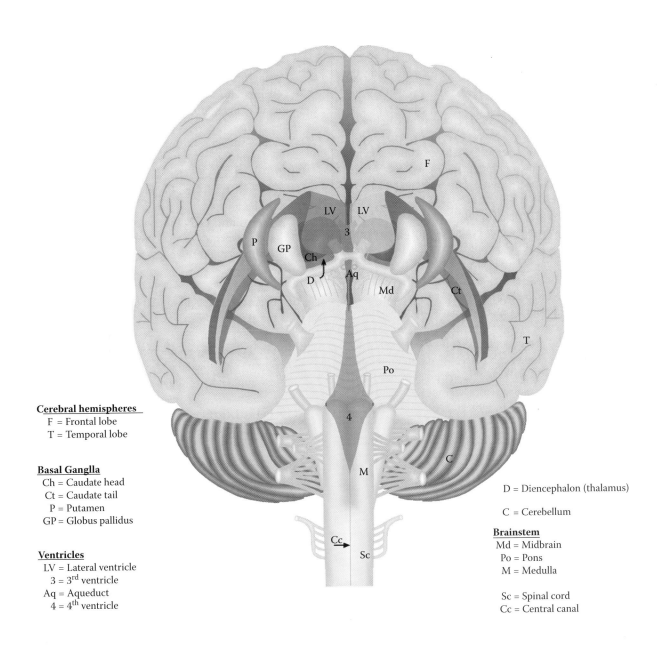

Cerebral hemispheres
 F = Frontal lobe
 T = Temporal lobe

Basal Ganglla
 Ch = Caudate head
 Ct = Caudate tail
 P = Putamen
 GP = Globus pallidus

Ventricles
 LV = Lateral ventricle
 3 = 3rd ventricle
 Aq = Aqueduct
 4 = 4th ventricle

D = Diencephalon (thalamus)

C = Cerebellum

Brainstem
Md = Midbrain
Po = Pons
M = Medulla

Sc = Spinal cord
Cc = Central canal

FIGURE OA: Overview Diagram — Anterior View

FIGURE OL

OVERVIEW — LATERAL VIEW

This is the companion diagram to the previous illustration, created to assist the learner in placing the brain and its various divisions in a three-dimensional construct.

This is a semi-anatomic view of the brain from the lateral perspective. The front pole of the brain is on the left side of this illustration; the posterior pole is on the right side. The structures included are:

- **Cerebral hemispheres**: The extensive cerebral hemisphere of one side is seen, with the top edge of the other hemisphere in view (this same view is presented in Figure 14). The lower part of the hemisphere seen on this view is the temporal lobe.
- **Lateral ventricles**: The shape of the ventricles within the hemispheres is now clearly seen (like a reversed letter C), with its continuation into the temporal lobe. The ventricle of the other hemisphere is seen as a "shadow." (A similar view is presented in Figure 20B.)
- **Basal ganglia**: The three parts of the basal ganglia are represented in this view. The caudate (head, body, and tail) follows the ventricle. The putamen can be seen from the lateral perspective, but the globus pallidus is hidden from view because it lies medial to the putamen; its position is indicated by the dashed ellipse. (A similar view is presented in Figure 25.) The two nuclei together are called the **lentiform** or **lenticular nucleus**.

One additional nucleus belonging, by definition, with the basal ganglia is seen within the temporal lobe — the amygdala. It will be discussed with the limbic system (in Section D).

- **Diencephalon**: The thalamus of one side can be visualized from this perspective, almost completely hidden from view by the putamen and the globus pallidus, the lentiform nucleus. The third ventricle is seen just behind it, occupying the midline (see Figure 25).
- **Brainstem**: The upper parts of the brainstem, namely the midbrain and upper pons, cannot be seen from this view of the brain, but their position is shown as if one could "see through" the temporal lobe. The lower part of the pons and the medulla may be seen. The shape of the fourth ventricle within the brainstem should also be noted.
- **Cerebellum**: Only the lower portion of one of the hemispheres of the cerebellum can be seen from this lateral perspective, below the cerebral hemispheres.

The brainstem and cerebellum occupy the posterior cranial fossa of the skull.

- **Spinal cord**: The spinal cord continues from the bottom of the medulla. A view similar to this is seen in a neuroradiologic image in Figure 3.

Note to the Learner: These overview illustrations are only sometimes referred to in this atlas but should be consulted as often as necessary while developing a three-dimensional understanding of the various parts of the brain.

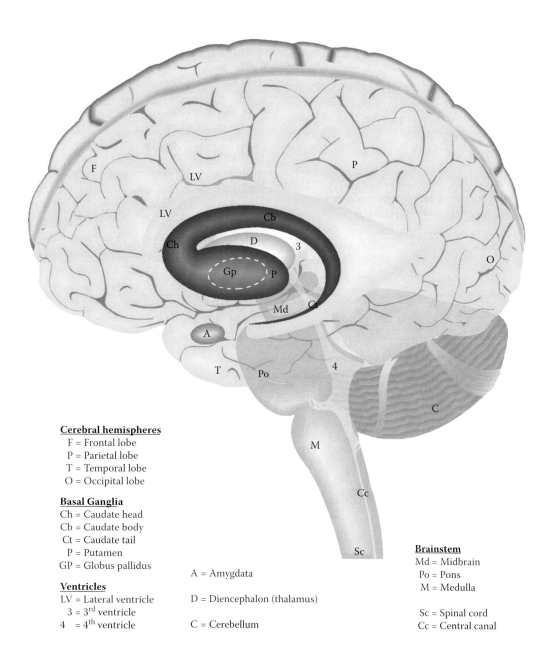

Cerebral hemispheres
F = Frontal lobe
P = Parietal lobe
T = Temporal lobe
O = Occipital lobe

Basal Ganglia
Ch = Caudate head
Cb = Caudate body
Ct = Caudate tail
P = Putamen
GP = Globus pallidus

Ventricles
LV = Lateral ventricle
3 = 3rd ventricle
4 = 4th ventricle

A = Amygdata

D = Diencephalon (thalamus)

C = Cerebellum

Brainstem
Md = Midbrain
Po = Pons
M = Medulla

Sc = Spinal cord
Cc = Central canal

FIGURE OL: Overview Diagram — Lateral View

FIGURE 1
SPINAL CORD 1

SPINAL CORD: LONGITUDINAL VIEW

The spinal cord is the extension of the CNS below the level of the skull. It is an elongated structure that is located in the vertebral canal, covered with the **meninges** — dura, arachnoid, and pia — and surrounded by the **subarachnoid space** containing cerebrospinal fluid (CSF) (see Figure 21). There is also a space between the dura and vertebra, known as the **epidural space**. Both of these spaces have important clinical implications (see Figure 2C and Figure 3).

The spinal cord, notwithstanding its relatively small size compared with the rest of the brain, is absolutely essential for our normal function. It is the connector between the central nervous system and our body (other than the head). On the sensory (afferent) side, the information arriving from the skin, muscles, and viscera informs the CNS about what is occurring in the periphery; this information then "ascends" to higher centers in the brain.

On the motor (efferent) side, the nerves leave the spinal cord to control our muscles. Although the spinal cord has a functional organization within itself, these neurons of the spinal cord receive their "instructions" from higher centers, including the cerebral cortex, via several descending tracts. This enables us to carry out normal movements, including normal walking and voluntary activities. The spinal cord also has a motor output to the viscera and glands, part of the autonomic nervous system (see Figure 4).

UPPER INSET: CERVICAL SPINAL CORD
CROSS-SECTION

The neurons of the spinal cord are organized as nuclei, the **gray matter**, and the various pathways are known as **white matter**. In the spinal cord, the gray matter is found on the inside, with the white matter all around. The divisions of the gray matter are introduced with Figure 4; the functional aspects will be described with the sensory (see Figure 32) and motor (see Figure 44) systems. The tracts of the spinal cord are described with the pathways in Section B (e.g., see Figure 33 and Figure 45). All the pathways are summarized in one cross-section (see Figure 68). Histological cross-sections of the spinal cord are also presented (see Figure 69).

LOWER INSET: NERVE ROOTS

The **dorsal root** (sensory) and **ventral root** (motor) unite within the intervertebral foramina to form the (mixed) **spinal nerve** (see also Figure 5). The nerve cell bodies for the dorsal root are located in the **dorsal root ganglion (DRG)**. Both the roots and the dorsal root ganglion belong to the peripheral nervous system (PNS) (where the Schwann cell forms and maintains the myelin).

DEVELOPMENTAL PERSPECTIVE

During early development, the spinal cord is the same length as the vertebral canal and the entering/exiting nerve roots correspond to the spinal cord vertebral levels. During the second part of fetal development, the body and the bony spine continue to grow, but the spinal cord does not. After birth, the spinal cord only fills the vertebral canal to the level of L2, the second lumbar vertebra (see also Figure 3). The space below the termination of the spinal cord is the **lumbar cistern**, filled with cerebrospinal fluid.

Therefore, as the spinal cord segments do not correspond to the vertebral segments, the nerve roots must travel in a downward direction to reach their proper entry/exit level between the vertebra, more so for the lower spinal cord roots (see the photographic view in Figure 2A and Figure 2C). These nerve roots are collectively called the **cauda equina**, and they are found in the lumbar cistern (see Figure 2A, Figure 2C, and Figure 3).

CLINICAL ASPECT

The four vertebral levels — cervical, thoracic, lumbar, and sacral — are indicated on the left side of the illustration. The spinal cord levels are indicated on the right side. One must be very aware of which reference point — the vertebral or spinal — is being used when discussing spinal cord injuries.

Nerve roots can be anesthetized by injection of a local anesthetic into their immediate vicinity. One of the locations for this is in the epidural space. The sensory nerve roots to the perineal region, which enter the cord at the sacral level, are often anesthetized in their epidural location during childbirth. This procedure requires a skilled anesthetist.

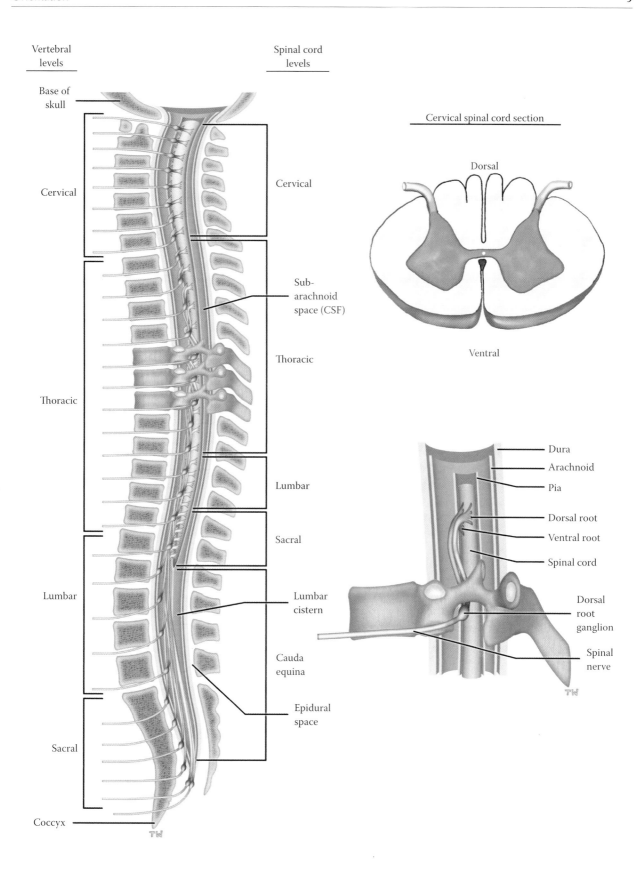

Vertebral
levels

Spinal cord
levels

Base of
skull

Cervical

Cervical spinal cord section

Dorsal

Sub-
arachnoid
space (CSF)

Ventral

Thoracic

Thoracic

Lumbar

Dura

Arachnoid

Pia

Sacral

Dorsal root

Ventral root

Spinal cord

Lumbar

Lumbar
cistern

Dorsal
root
ganglion

Cauda
equina

Spinal
nerve

Sacral

Epidural
space

Coccyx

FIGURE 1: Spinal Cord 1 — Longitudinal (Vertebral) View

FIGURE 2A
SPINAL CORD 2

SPINAL CORD: LONGITUDINAL VIEW (PHOTOGRAPH)

This is a photographic image of the spinal cord removed from the vertebral canal. The dura-arachnoid has been opened and the anterior aspect of the cord is seen, with the attached spinal roots; from this anterior perspective, most of the roots seen are the ventral (i.e., motor) roots.

The spinal cord is divided into parts according to the region innervated: cervical (8 spinal roots), thoracic (12 spinal roots), lumbar (5 spinal roots), sacral (5 spinal roots), and coccygeal (1 root).

The nerve roots attached to the spinal cord, connecting the spinal cord with the skin and muscles of the body, give the cord a segmented appearance. This segmental organization is reflected onto the body in accordance with embryological development. Areas of skin are supplied by certain nerve segments — each area is called a **dermatome** (e.g., inner aspect of the arm and hand = C8; umbilical region = T10), with overlap from adjacent segments. The muscles are supplied usually by two adjacent segments, called **myotomes** (e.g., biceps of the upper limb = C5 and C6; quadriceps of the lower limb = L3 and L4). This known pattern is very important in the clinical setting (see below).

There are two enlargements of the cord: at the cervical level for the upper limb (seen at greater magnification in Figure 2B), the roots of which will form the **brachial plexus**, and at the lumbosacral level for the lower limb, the roots of which form the **lumbar and sacral plexuses**. The cord tapers at its ending, and this lowermost portion is called the **conus medullaris**. Below the vertebral level of L2 in the adult, inside the vertebral canal, are numerous nerve roots, both ventral and dorsal, collectively called the **cauda equina**; these are found within the **lumbar cistern**, an expansion of the subarachnoid space, a space containing CSF (see Figure 1, and shown at a greater magnification and discussed in Figure 2C; also shown in the MRI in Figure 3).

CLINICAL ASPECT

The segmental organization of the spinal cord and the known pattern of innervation to areas of skin and to muscles allows a knowledgeable practitioner, after performing a detailed neurological examination, to develop an accurate localization of the injury or disease (called the lesion) at the spinal cord (segmental) level.

The spinal cord can be affected by tumors, either within the cord (intramedullary), or outside the cord (extramedullary). There is a large plexus of veins on the outside of the dura of the spinal cord (see Figure 1), and this is a site for metastases from pelvic (including prostate) tumors. These press upon the spinal cord as they grow and cause symptoms as they compress and interfere with the various pathways (see Section B).

Traumatic lesions of the spinal cord occur following car and bicycle accidents and still occur because of diving accidents into shallow water (swimming pools). Protruding discs can impinge upon the spinal cord. Other traumatic lesions involve gunshot and knife wounds. If the spinal cord is completely transected (i.e., cut through completely), all the tracts are interrupted. For the ascending pathways, this means that sensory information from the periphery is no longer available to the brain. On the motor side, all the motor commands cannot be transmitted to the anterior horn cells, the final common pathway for the motor system. The person therefore is completely cut off on the sensory side and loses all voluntary control, below the level of the lesion. Bowel and bladder control are also lost.

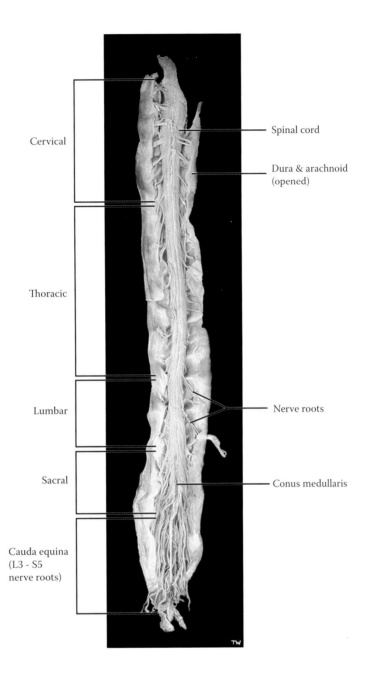

Cervical

Thoracic

Lumbar

Sacral

Cauda equina
(L3 - S5
nerve roots)

Spinal cord

Dura & arachnoid
(opened)

Nerve roots

Conus medullaris

FIGURE 2A: Spinal Cord 2 — Longitudinal View (photograph)

FIGURE 2B
SPINAL CORD 3

SPINAL CORD: CERVICAL REGION (PHOTOGRAPH)

This is a higher magnification photographic image of the cervical region of the spinal cord. Most of the attached roots are the motor/ventral roots, coming from the ventral horn of the spinal cord (discussed with Figure 4); a few of the dorsal/sensory roots can be seen, which enter the cord in the dorsal horn. These roots exit the vertebral canal and carry a sleeve of arachnoid-dura with them for a very short distance, as they head for the intervertebral spaces (see Figure 1).

The somewhat tortuous artery running down the midline of the cord is the **anterior spinal artery**. This artery, which is the major blood supply to the ventral portion of the upper part of the cord, is formed by a branch from each of the vertebral arteries (see Figure 58). This artery receives supplementary branches from the aorta along its way, called radicular arteries, which follow the nerve roots. There are two very small posterior spinal arteries. The most vulnerable area of the spinal cord blood supply is around the mid-thoracic level. There is a particularly important branch off the aorta that supplies this critical region of the spinal cord. This is important clinically (see below).

The pia is attached directly to the spinal cord. Sheets of pia are found in the subarachnoid space, between the ventral and dorsal roots, and can be seen attaching to the inner aspect of the arachnoid — these pial extensions are called **denticulate ligaments**. These ligaments, which are located at intervals along the cord, are thought to tether the cord, perhaps to minimize movement of the cord.

CLINICAL ASPECT

Because of its tenuous blood supply, the spinal cord is most vulnerable in the mid-thoracic portion. A dramatic drop in blood pressure, such as occurs with a cardiac arrest or excessive blood loss, may lead to an infarction of the spinal cord. The result can be just as severe as if the spinal cord was severed by a knife. The most serious consequence of this would be the loss of voluntary motor control of the lower limbs, known as paraplegia. The clinical picture will be understood once the sensory and motor tracts of the spinal cord have been explained (in Section B).

Surgeons who operate on the abdominal aorta, for example, for aortic aneurysm, must make every effort to preserve the small branches coming off the aorta as these are critical for the vascular supply of the spinal cord. One would not want the end result of an aneurysmal repair to be a paraplegic patient.

DEVELOPMENTAL ASPECT

Embryologically, the spinal cord commences as a tube of uniform size. In those segments that innervate the limbs (muscles and skin), all the neurons reach maturity. However, in the intervening portions, there is massive programmed cell death during development because there is less peripheral tissue to be supplied. In the adult, therefore, the spinal cord has two "enlargements": the cervical for the upper limb, and the lumbosacral for the lower limb, each giving rise to the nerve plexus for the upper and lower limbs, respectively.

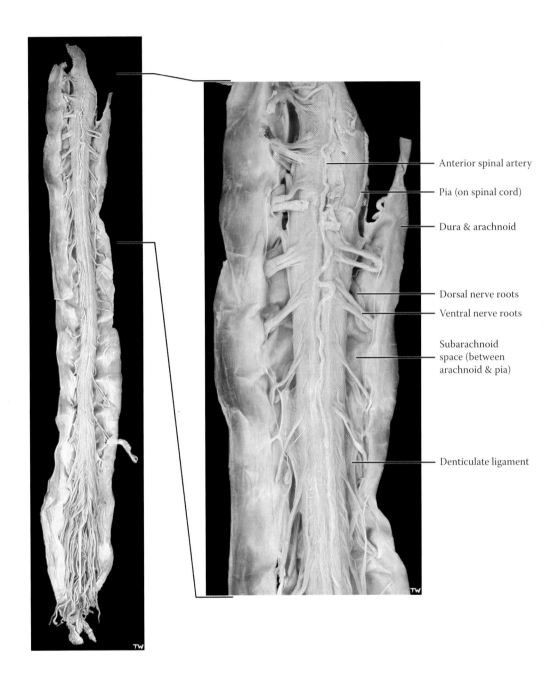

Anterior spinal artery

Pia (on spinal cord)

Dura & arachnoid

Dorsal nerve roots

Ventral nerve roots

Subarachnoid
space (between
arachnoid & pia)

Denticulate ligament

FIGURE 2B: Spinal Cord 3 — Cervical Region (photograph)

FIGURE 2C
SPINAL CORD 4

SPINAL CORD: CAUDA EQUINA (PHOTOGRAPH)

This is a higher magnification photographic image of the lowermost region of the spinal cord, the sacral region. The tapered end of the spinal cord is called the **conus medullaris**, and this lower portion of the cord corresponds approximately to the sacral segments.

The collection of dorsal and ventral nerve roots, below the level of the termination of the cord, is collectively called the **cauda equina**. These roots, which belong to the lumbar and sacral segments of the spinal cord, fill the expanded subarachnoid space in this region, known as the **lumbar cistern** (see Figure 3). The roots are traveling from the spinal cord levels to exit at their appropriate (embryological) intervertebral level (see Figure 1). The roots are floating in the CSF of the lumbar cistern.

The pia mater of the cord gathers at the tip of the conus medullaris into a ligament-like structure, the **filum terminale**, which attaches to the dura-arachnoid at the termination of the vertebral canal, at the level of (vertebral) S2. The three meningeal layers then continue and attach to the coccyx as the coccygeal ligament.

Sampling of CSF for the diagnosis of meningitis, an inflammation of the meninges, or for other neurological diseases, is done in the lumbar cistern. This procedure is called a **lumbar puncture** and must be performed using sterile technique. A trochar (which is a large needle with a smaller needle inside) is inserted *below* the termination of the spinal cord at L2, in the space between the vertebra, usually between the vertebra L4–L5 (see Figure 1). The trochar must pierce the very tough ligamentum flavum (shown in the next illustration), then the dura-arachnoid, and then "suddenly" enters into the lumbar cistern; the (inner) needle is withdrawn and CSF drips out to be collected in sterile vials. This is not a pleasant procedure for a patient and is especially unpleasant, if not frightening, when performed on children.

The nerve roots exit the spinal cord at the appropriate intervertebral level. The roots to the lower extremity, those exiting between L4–L5 and L5–S1, are the ones most commonly involved in the everyday back injuries that affect many adults. The student should be familiar with the signs and symptoms that accompany degenerative disc disease in the lumbar region (see also Figure 1).

Occasionally, neurologic deficits are seen in a pediatric patient, which indicates that the filum terminale is pulling on the spinal cord. If this is suspected clinically, further imaging studies are done, and in some cases the filum terminale must be surgically cut to relieve the tension on the spinal cord.

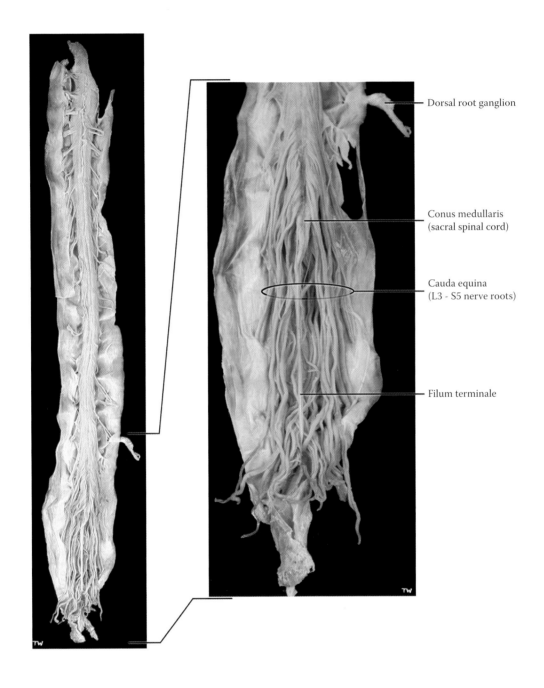

Dorsal root ganglion

Conus medullaris
(sacral spinal cord)

Cauda equina
(L3 - S5 nerve roots)

Filum terminale

FIGURE 2C: Spinal Cord 4 — Cauda Equina (photograph)

FIGURE 3
SPINAL CORD 5

SPINAL CORD MRI – T1: LONGITUDINAL VIEW (RADIOGRAPH)

This is a **magnetic resonance image** (MRI) of the vertebral column and spinal cord, viewed in a midsagittal plane. This is called a **TI**-weighted image, in which the cerebrospinal fluid (CSF) is dark. (The various radiological techniques used to image the nervous system are discussed below.) This image is from an adult, in which no pathology was found in the spinal cord radiological examination.

Because of the length of the spinal cord, it is being shown in two parts — upper and lower. The vertebral bodies, the intervertebral discs and the spinous processes posteriorly have been labeled, as well as the ligamentum flavum (discussed with the previous illustration). The vertebral bodies have been numbered at various levels — C2, T1, L1, and S1.

The UPPER portion shows the spinal cord to be a continuation of the medulla of the brainstem, at the lowermost border of the skull, the foramen magnum. The pons, medulla, and cerebellum are seen above foramen magnum occupying the posterior cranial fossa.

The spinal cord tissue is located in the middle of the vertebral column, surrounded by the meninges (which can dimly be visualized), with the dura-arachnoid separating the subarachnoid space containing CSF from the space outside the meninges, the epidural space, between the meninges and vertebra (see Figure 1). The epidural space in the lower thoracic region and in the lumbar and sacral regions often contains fat (epidural fat), which is seen as bright on this image.

The LOWER portion of the spinal cord shows the spinal cord itself, tapering as the conus medullaris and terminating around the level of vertebra L1–L2. Below that level is the enlarged subarachnoid space — called a cistern, the **lumbar cistern** — within which are the nerve roots, dorsal and ventral, for the lower extremity (shown in the previous illustration).

ADDITIONAL DETAIL

The sphenoid air sinus of the skull has been identified, as well as the air-containing (dark) nasal portion of the pharynx (the nasopharynx). The aorta (dark) is also labeled.

RADIOLOGICAL IMAGING

Ordinary x-rays show the skull and its bony structures but not the brain. A remarkable revolution occurred in clinical neurology and our understanding of the brain when imaging techniques were developed that allowed for visualization of the brain. This now includes:

- **Computed tomography (CT)** (often pronounced as a "CAT" scan, meaning computer assisted tomography see Figure 28A). This is done using x-rays, and there is a computer reconstruction of the brain after a series of views are taken from a large number of perspectives. In this view the bones of the skull are bright and the CSF is dark, with the brain tissue "gray" but not clear. This image can be obtained in several seconds, even with a very sick patient.
- **Magnetic resonance imaging (MRI)** does not use x-rays; the image is created by capturing the energy of the hydrogen ions of water. An extremely strong magnet is used for MRI, and capturing the images requires more time. Again, there is a computer reconstruction of the images. The brain itself looks "anatomic." This view can be *weighted* during the acquisition of the image so as to produce a **TI** image, in which the CSF is dark (this illustration), or a **T2** image, in which the CSF is bright (see Figure 28B). With MRI, the bones of the skull are dark, while fatty tissue (including the bone marrow) is bright. Other settings are now available to visualize the brain, such as FLAIR.

As imaging and technology improve, we are able to visualize the brain during functional activity — **functional MRIs** are becoming more widely available; this allows us to "see" which areas of the brain are particularly active during a certain task, based upon the increased blood supply to that area during the active period.

Other techniques are also used to visualize the living brain and its activity, such as positron emission tomography (**PET scan**); this technique utilizes a very short-acting radioactive compound, which is injected into the venous system. Its use is usually restricted to specialized neurological centers involved in research on the human brain.

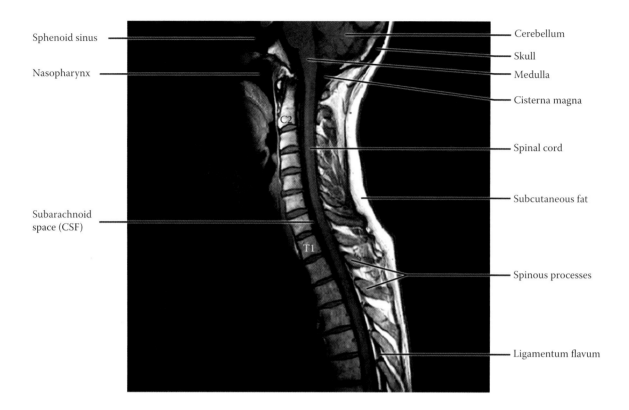

Sphenoid sinus — Cerebellum

Nasopharynx — Skull

Medulla

Cisterna magna

Spinal cord

Subcutaneous fat

Subarachnoid
space (CSF)

Spinous processes

Ligamentum flavum

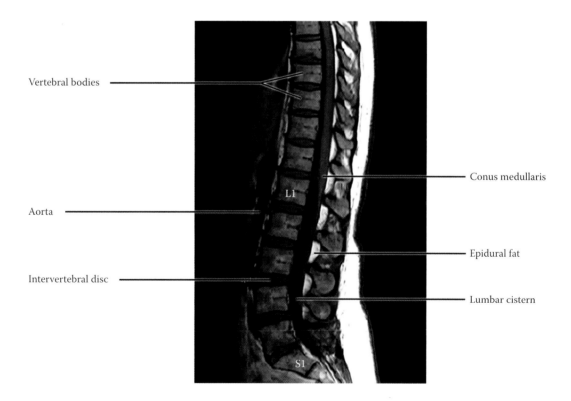

Vertebral bodies

Conus medullaris

Aorta

Epidural fat

Intervertebral disc

Lumbar cistern

FIGURE 3: **Spinal Cord 5** — MRI: Longitudinal View (radiograph)

FIGURE 4
SPINAL CORD 6

SPINAL CORD: CROSS-SECTIONAL VIEWS

UPPER DIAGRAM

The upper diagram is a cross-section through the spinal cord at the C8 level, the eighth cervical segmental level of the spinal cord (not the vertebral level, see Figure 1). The gray matter is said to be arranged in the shape of a butterfly (or somewhat like the letter H). The gray matter of the spinal cord contains a variety of cell groups (i.e. nuclei), which subserve different functions. Although it is rather difficult to visualize, these groups are continuous longitudinally throughout the length of the spinal cord.

The dorsal region of the gray matter, called the **dorsal or posterior horn**, is associated with the incoming (*afferent*) dorsal root, and is thus related to sensory functions. The cell body of these sensory fibers is located in the **dorsal root ganglion** (see Figure 1). The dorsal horn is quite prominent in this region because of the very large sensory input to this segment of the cord from the upper limb, particularly from the hand. The situation is similar in the lumbar region (as shown in the middle of the three lower illustrations).

The ventral gray matter, called the **ventral or anterior horn**, is the motor portion of the gray matter. The ventral horn has the large motor neurons, the anterior horn cells, which are *efferent* to the muscles (see Figure 44). These neurons, because of their location in the spinal cord, which is "below" the brain, are also known as **lower motor neurons**. (We will learn that the neurons in the cerebral cortex, at the "higher" level, are called upper motor neurons — discussed with Figure 45.) The ventral horn is again prominent at this level because of the large number of motor neurons supplying the small muscles of the hand. The situation is similar in the lumbar region, with the motor neurons supplying the large muscles of the thigh (as shown in the illustration below).

The area in between is usually called the **intermediate gray** and has a variety of cell groups with some association-type functions (see Figure 32 and Figure 44).

The **autonomic** nervous system to the organs of the chest, abdomen, and pelvis is controlled by neurons located in the spinal cord.

- Preganglionic **sympathetic** neurons form a distinctive protrusion of the gray matter, called the **lateral horn**, which extends throughout the thoracic region, from spinal cord level T1 to L2 (as shown in the first of the three lower illustrations). The post-ganglionic nerves supply the organs of the thorax, abdomen, and pelvis.
- **Parasympathetic** preganglionic neurons are located in the sacral area and do not form a separate horn (as shown in the illustration). This region of the spinal cord in the area of the conus medullaris (the last of the three lower illustrations) controls bowel and bladder function, subject to commands from higher centers, including the cerebral cortex.

The parasympathetic control of the organs of the thorax and abdomen comes from the vagus nerve, CN X, a cranial nerve (see Figure 6 and Figure 8A).

The central canal of the spinal cord (see Figure 20A, Figure 20B, and Figure 21) is located in the center of the commissural gray matter. This represents the remnant of the neural tube and is filled with CSF. In adults, the central canal of the spinal cord is probably not patent throughout the whole length of the spinal cord. A histological view of these levels of the spinal cord is shown in Figure 69 in Section C.

Note to the Learner: The white matter, which contains the ascending sensory and descending motor pathways, will be described with the pathways in Section B; a summary diagram with all the tracts is shown in Section C (see Figure 68).

ADDITIONAL DETAIL

The parasympathetic supply to the salivary glands travels with cranial nerves (CN) VII and IX (see Figure 8A).

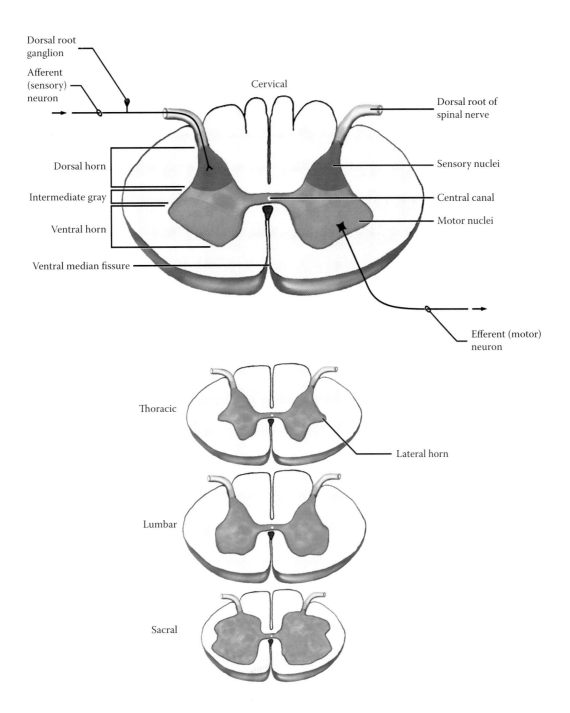

FIGURE 4: Spinal Cord 6 — Cross-Sectional Views

FIGURE 5
SPINAL CORD 7

SPINAL CORD MRI – T2: AXIAL VIEWS (RADIOGRAPH)

MRI views of the spinal cord are shown in the axial plane at the C4 (fourth cervical vertebral) level; the orientation should be noted with anterior (ventral) at the top. The CSF is bright in these **T2**-weighted images. The position of the spinal cord can be easily visualized within the vertebral canal, with the surrounding CSF space. The vertebral bodies and lamina are dark; the muscles of the neck can be visualized.

In both images it is possible to see the "butterfly" shape of the gray matter of the spinal cord (see Figure 1 and Figure 4). The orientation of the cord should be noted. In the upper image, the dorsal root and ventral root can be seen, as they head for the intervertebral foramen to form the spinal nerve (see Figure 1); neuroradiologists often call this the neural foramen. In the lower image, taken just a few millimeters below, the spinal nerve can be seen in the intervertebral (neural) foramen.

Note to the Learner: In viewing these radiographs, the left side of the image is in fact the right side of the patient and likewise on the other side — this is the convention. The veins, internal jugular and external jugular, appear white with MRI imaging; the common carotid artery appears dark because of the rapid flow of blood in the arteries; note the presence of the vertebral artery (dark) in the foramen in the transverse process.

CLINICAL ASPECT

Any abnormal protrusion of a vertebra or disc could be visualized, as well as tumors within the vertebral canal or of the cord itself (see also Figure 3). An enlargement of the central canal, called syringomyelia, is an unusual though not rare disease of the upper cord (discussed with Figure 32). A small arterio-venous (A-V) malformation may also be visualized with MRI within the spinal cord.

As discussed previously, the spinal cord may be transected following traumatic injuries. The immediate effect of an acute complete spinal cord transection in the human is a complete shutdown of all spinal cord activity. This is referred to as **spinal shock**. Neurologically, there is a loss of all muscle tone and an absence of all deep tendon reflexes, and no plantar response (i.e., no Babinski sign; discussed in Section B, Part III, Introduction). After a few weeks, intrinsic spinal reflexes appear, now no longer modified from higher control centers. (The details of the pathways involved will be discussed in Section B of this atlas.) The end result is a dramatic increase in muscle tone (spasticity) and hyperactive deep tendon reflexes (discussed with Figure 49B and also with Figure 68). Thereafter, there occur a number of abnormal or excessive reflex responses. Such patients require exceptional care by the nursing staff.

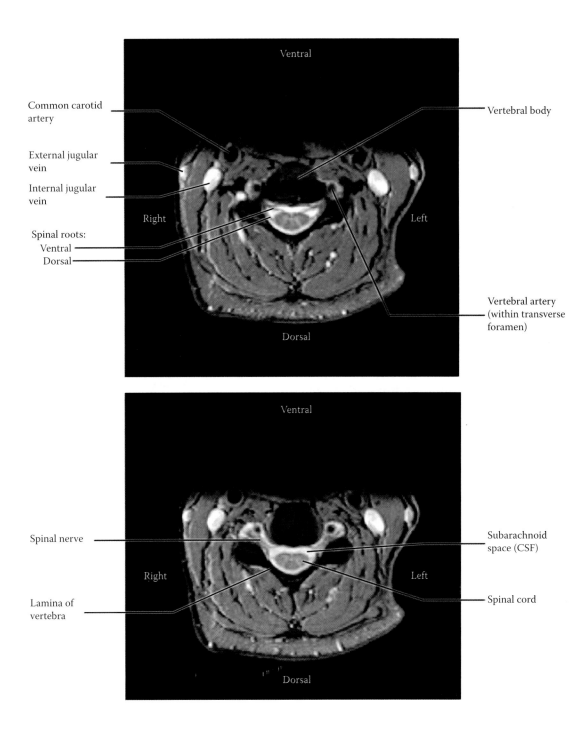

FIGURE 5: Spinal Cord 7 — MRI: Axial View (radiograph)

FIGURE 6
BRAINSTEM 1

BRAINSTEM AND DIENCEPHALON: VENTRAL VIEW

The brainstem is the lowermost part of the brain and is located above the spinal cord. It can be seen by viewing the brain from below (see Figure 15A; also Figure OA and Figure OL). This specimen has been obtained by dissecting out the brainstem, and cerebellum, along with the diencephalon; a photographic view of this specimen is shown in the next illustration (Figure 7). The diencephalon will be described subsequently (see Figure 11 and Figure 12).

In the human brain, the brainstem is a relatively small mass of brain tissue compared to the large hemispheres, but it is packed with various nuclei and tracts. Among these nuclei are those of 10 of the **cranial nerves** (CN III to CN XII). Many basic brain activities are located in the brainstem, including key vital functions (control of blood pressure, pulse, and respiration). Some motor functions are found at various brainstem levels, some as part of the reticular formation; the reticular formation is also part of a system that is responsible for consciousness. Most important, the ascending sensory and descending motor tracts/pathways that connect the spinal cord with "higher" areas of the brain pass through the brainstem (described in Section B). In addition, many of the connections to the cerebellum, including pathways and nuclei, are found in the brainstem. Finally, each part of the brainstem has a part of the ventricular system.

The brainstem is divided anatomically into three parts — the narrow midbrain, which is located under the diencephalon; the pons, with its ventral bulge; and the medulla, which connects with the spinal cord. Each of the parts has distinctive features that allow for the identification of the parts, both on the gross brain specimen or a microscopic cross-section.

- The **midbrain** region (mesencephalon) has two large "pillars" anteriorly called the cerebral peduncles, which consist of millions of axons descending from the cerebral cortex to various levels of the brainstem and spinal cord.
- The **pons** portion is distinguished by its bulge anteriorly, the pons proper, an area that is composed of nuclei (the pontine nuclei) that connect to the cerebellum.
- The **medulla** has two distinct elevations on either side of the midline, known as the pyramids; the direct voluntary motor pathway from the cortex to the spinal cord, the cortico-spinal tract, is located within the pyramid. Behind each is a prominent bulge, called the olive, the inferior olivary nucleus, which connects with the cerebellum.

CRANIAL NERVES AND THEIR ATTACHMENT

The cranial nerves of the brainstem will be presented in numerical order, starting at the midbrain level.

Midbrain Level

- CN III, the **oculomotor nerve**, emerges ventrally between the cerebral peduncles (in the interpeduncular fossa).
- CN IV, the **trochlear nerve**, which exits posteriorly, is a thin nerve that wraps around the lowermost border of the cerebral peduncle.

Pontine Level

- CN V, the **trigeminal nerve**, is a massive nerve attached along the middle cerebellar peduncle.
- CN VI, the **abducens nerve**, is seen exiting anteriorly at the junction between the pons and medulla.
- CN VII, the **facial nerve**, and CN VIII (the **vestibulocochlear nerve**), are both attached to the brainstem at the ponto-cerebellar angle.

Medullary Level

- CN IX, the **glossopharyngeal**, and CN X, the **vagus**, are attached to the lateral margin of the medulla, behind the inferior olive.
- CN XI, the **spinal accessory nerve**, from the uppermost region of the spinal cord, enters the skull and then exits from the skull as if it were a cranial nerve; by convention it is included as a cranial nerve.
- CN XII, the **hypoglossal nerve**, emerges by a series of rootlets between the inferior olive and the pyramid.

Information concerning the function of the cranial nerves will be discussed with Figure 8A and Figure 8B. The nuclei of the brainstem, including the cranial nerve nuclei, will be studied in cross-sections of the brainstem in Section C of this atlas (see Figure 64–Figure 67).

ADDITIONAL DETAIL

Structures labeled, such as the flocculus of the cerebellum, the pituitary stalk, and the mammillary bodies (nuclei), will be considered at the appropriate time.

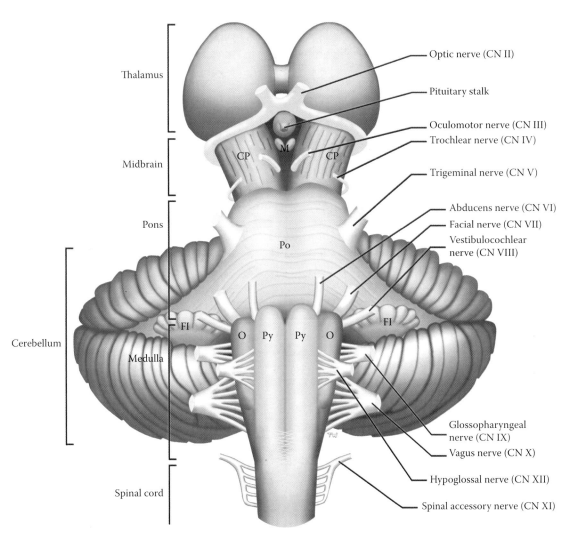

Thalamus

Optic nerve (CN II)

Pituitary stalk

Oculomotor nerve (CN III)

Trochlear nerve (CN IV)

Midbrain

CP   M   CP

Trigeminal nerve (CN V)

Pons

Abducens nerve (CN VI)

Facial nerve (CN VII)

Po

Vestibulocochlear
nerve (CN VIII)

Cerebellum

FI   FI

O   Py   Py   O

Medulla

Glossopharyngeal
nerve (CN IX)

Vagus nerve (CN X)

Hypoglossal nerve (CN XII)

Spinal cord

Spinal accessory nerve (CN XI)

M = Mammillary bodies
CP = Cerebral peduncle
Po = Pons
Py = Pyramid
O = Olive
FI = Flocculus

FIGURE 6: Brainstem 1 — Ventral View with Cranial Nerves

FIGURE 7
BRAINSTEM 2

BRAINSTEM AND DIENCEPHALON: VENTRAL (PHOTOGRAPHIC) VIEW

This specimen has been obtained by isolating the brainstem (and cerebellum) along with the diencephalon from the remainder of the brain. It is the same specimen as in the previous diagrammatic illustration (see Figure 6). The three parts of the brainstem can be differentiated on this ventral view (from above downward):

- The midbrain region has the two large "pillars" anteriorly called the **cerebral peduncles**. These contain fibers descending from the cerebral cortex to the spinal cord (cortico-spinal tract, see Figure 45), to the brainstem (cortico-bulbar tract, see Figure 46), and to the pontine nuclei (cortico-pontine fibers, see Figure 55).
- The pontine portion is distinguished by its bulge anteriorly, the **pons proper**, an area that is composed of the pontine nuclei; these relay to the cerebellum (see Figure 55).
- The medulla is distinguished by the **pyramids**, two distinct elevations on either side of the midline. The direct voluntary motor pathway from the cortex to the spinal cord, the cortico-spinal tract, actually forms these pyramids (see Figure 45). Behind each pyramid is the **olive**, a protrusion of the inferior olivary nucleus (discussed with Figure 55).

It should be noted that the cortico-spinal tract, from cortex to spinal cord, travels through the whole brainstem (see Figure 45), including the cerebral peduncles (see Figure 65A), within the pons proper (see Figure 66B), and then forms the pyramids in the medulla (see Figure 67C). This tract crosses the midline as the **pyramidal decussation**, demarcating the end of the medulla and the beginning of the spinal cord.

CRANIAL NERVE FUNCTIONS

Knowledge of the attachment of each cranial nerve (**CN**) to the brainstem is a marker of the location of the cranial nerve nucleus within the brainstem (see Figure 8A and Figure 8B), in almost all cases. In addition, it is necessary to know the function of each of the nerves.

Midbrain Level

- CN III, the oculomotor nerve, supplies several of the extraocular muscles, which move the eyeball. A separate part, called the Edinger-West-

phal nucleus, provides parasympathetic fibers to the pupil.
- CN IV, the trochlear nerve, supplies one extraocular muscle.

Pontine Level

- CN V, the trigeminal nerve — its major nucleus subserves a massive sensory function for structures of the face and head. A smaller nucleus supplies motor fibers to jaw muscles.
- CN VI, the abducens nerve, supplies one extraocular muscle.
- CN VII, the facial nerve — of its several nuclei, one supplies the muscles of the face and another nucleus is parasympathetic to salivary glands; a third nucleus subserves the sense of taste.
- CN VIII, the vestibulocochlear nerve — for the special senses of balance and hearing.

Medullary Level

- CN IX, the glossopharyngeal, and CN X, the vagus nerve — of its several nuclei, one supplies the muscles of the pharynx and larynx; the vagus nerve is primarily a parasympathetic nerve to the organs of the thorax and abdomen.
- CN XI, the spinal accessory nerve, innervates some of the muscles of the neck.
- CN XII, the hypoglossal nerve, supplies motor fibers to the muscles of the tongue.

More details concerning the innervation of each of the cranial nerves is given with Figure 8A for the motor cranial nerve nuclei, and with Figure 8B for the sensory cranial nerve nuclei.

CLINICAL ASPECT

Knowing the attachment of the cranial nerves to each part of the brainstem is fundamental to diagnosing lesions of the brainstem. For almost all of the cranial nerves, this attachment coincides with the location of the nucleus/nuclei of the cranial nerve within the brainstem. Not only does this assist in understanding the neuroanatomy of this region, but this knowledge is critical in determining the localization of a lesion of the brainstem region (discussed further in Section C of this atlas).

A lesion of the brainstem is likely to interrupt either one or more sensory or motor pathways as they pass through the brainstem. Because of the close relationship with the cerebellum, there may be cerebellar signs as well.

ADDITIONAL DETAIL

Structures belonging to the cerebellum are explained in Figure 54–Figure 57.

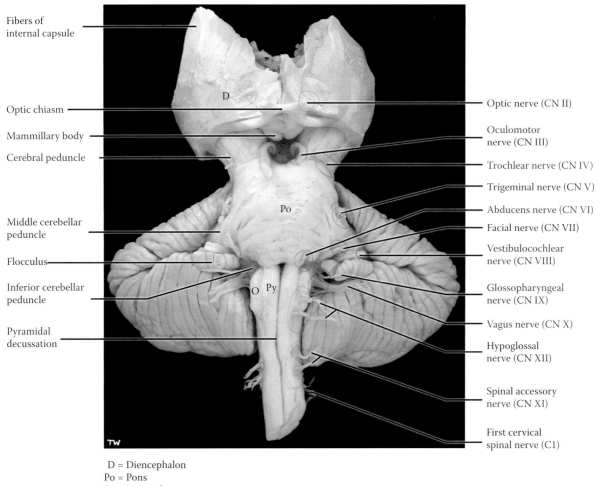

Fibers of
internal capsule

Optic chiasm

Mammillary body

Cerebral peduncle

Middle cerebellar
peduncle

Flocculus

Inferior cerebellar
peduncle

Pyramidal
decussation

D

Po

O   Py

TW

Optic nerve (CN II)

Oculomotor
nerve (CN III)

Trochlear nerve (CN IV)

Trigeminal nerve (CN V)

Abducens nerve (CN VI)

Facial nerve (CN VII)

Vestibulocochlear
nerve (CN VIII)

Glossopharyngeal
nerve (CN IX)

Vagus nerve (CN X)

Hypoglossal
nerve (CN XII)

Spinal accessory
nerve (CN XI)

First cervical
spinal nerve (C1)

D = Diencephalon
Po = Pons
Py = Pyramid
O = Inferior olive

FIGURE 7: Brainstem 2 — Ventral View (photograph)

FIGURE 8A
BRAINSTEM 3

CRANIAL NERVE NUCLEI: MOTOR

The cranial nerves are peripheral nerves that supply the head region, except for the olfactory (CN I) and optic (CN II) nerves. Each cranial nerve is unique and may have one or more functional components, either sensory, motor, or both, and some also have an autonomic (parasympathetic) component.

There are two kinds of motor functions:

1. The **motor** supply to the muscles derived from **somites**, including CN III, IV, VI, and XII, and to the muscles derived from the branchial arches, called **branchiomotor**, including CN V, VII, IX, and X (no distinction will be made between these muscle types in this atlas).
2. The **parasympathetic** supply to smooth muscles and glands of the head, a part of CN III, VII, and IX, and the innervation of the viscera in the thorax and abdomen with CN X.

This diagram shows the location of the motor nuclei of the cranial nerves, superimposed upon the ventral view of the brainstem. These nuclei are also shown in Figure 40, in which the brainstem is presented from a dorsal perspective. The details of the location of the cranial nerve nuclei within the brainstem will be described in Section C of this atlas (Neurological Neuroanatomy) with Figure 64–Figure 67.

MIDBRAIN LEVEL

- CN III, the oculomotor nerve, has both motor and autonomic fibers. The motor nucleus, which supplies most of the muscles of the eye, is found at the upper midbrain level. The parasympathetic nucleus, known as the **Edinger-Westphal nucleus**, supplies the pupillary constrictor muscle and the muscle that controls the curvature of the lens; both are part of the accommodation reflex (discussed with Figure 41C).
- CN IV, the trochlear nerve, is a motor nerve to one eye muscle, the superior oblique muscle. The trochlear nucleus is found at the lower midbrain level (see Figure 65B).

PONTINE LEVEL

- CN V, the trigeminal nerve, has a motor component to the muscles of mastication. The

nucleus is located at the midpontine level; the small motor nerve is attached to the brainstem at this level, along the middle cerebellar peduncle, with the much larger sensory root.
- CN VI, the abducens nerve, is a motor nerve that supplies one extraocular muscle, the lateral rectus muscle. The nucleus is located in the lower pontine region.
- CN VII, the facial nerve, is a mixed cranial nerve. The motor nucleus, which supplies the muscles of facial expression, is found at the lower pontine level. The parasympathetic fibers, to salivary and lacrimal glands, are part of CN VII (see Additional Details below).

MEDULLARY LEVEL

- CN IX, the glossopharyngeal nerve, and CN X, the vagus nerve, are also mixed cranial nerves. These supply the muscles of the pharynx (IX) and larynx (X), originating from the **nucleus ambiguus**. In addition, the parasympathetic component of CN X, coming from the **dorsal motor nucleus** of the vagus, supplies the organs of the thorax and abdomen. Both nuclei are found throughout the mid and lower portions of the medulla.
- Cranial nerve XI, the spinal accessory nerve, originates from a cell group in the upper 4–5 segments of the cervical spinal cord. This nerve supplies the large muscles of the neck (the sternomastoid and trapezius). As mentioned previously, CN XI enters the skull and exits again, as if it were a true cranial nerve.
- CN XII, the hypoglossal nerve, innervates all the muscles of the tongue. It has an extended nucleus in the medulla situated alongside the midline.

Note to the Learner: In this diagram, it appears that the nucleus ambiguus is the origin for CN XII. This is not the case but is a visualization problem. A clearer view can be found in Figure 48 and in the cross-sectional views (see Figure 67B and Figure 67C).

ADDITIONAL DETAIL

Two small parasympathetic nuclei are also shown but are rarely identified in brain sections — the superior and inferior salivatory nuclei. The superior nucleus supplies secretomotor fibers for cranial nerve VII (to the submandibular and sublingual salivary glands, as well as nasal and lacrimal glands). The inferior nucleus supplies the same fibers for cranial nerve IX (to the parotid salivary gland).

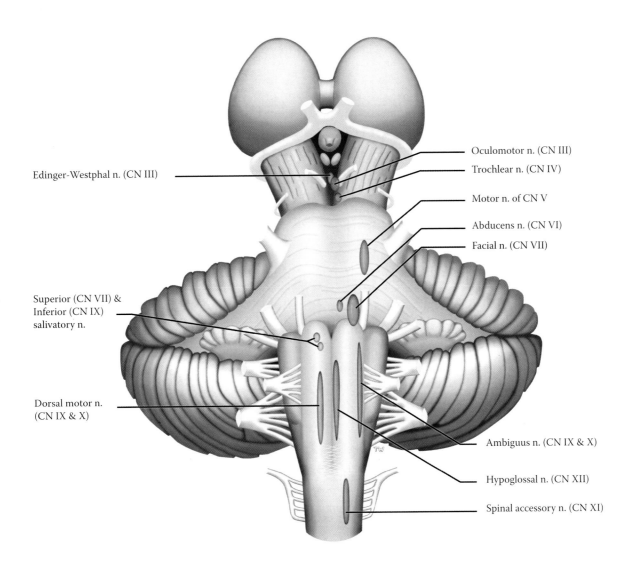

Edinger-Westphal n. (CN III)

Oculomotor n. (CN III)

Trochlear n. (CN IV)

Motor n. of CN V

Abducens n. (CN VI)

Facial n. (CN VII)

Superior (CN VII) &
Inferior (CN IX)
salivatory n.

Dorsal motor n.
(CN IX & X)

Ambiguus n. (CN IX & X)

Hypoglossal n. (CN XII)

Spinal accessory n. (CN XI)

FIGURE 8A: Brainstem 3 — Cranial Nerves Nuclei — Motor

FIGURE 8B
BRAINSTEM 4

CRANIAL NERVE NUCLEI: SENSORY

The cranial nerve nuclei with sensory functions are discussed in this diagram. It should be noted that the olfactory nerve (CN I) and the optic nerve (CN II) are not attached to the brainstem and not considered at this stage. Sensory information from the region of the head and neck includes the following:

- **Somatic afferents**: general sensations, consisting of touch (both discriminative and crude touch), pain and temperature; these come from the skin and the mucous membranes, via branches of the trigeminal nerve, CN V.
- **Visceral afferents**: sensory input from the pharynx and other homeostatic receptors of the neck (e.g., for blood pressure), and from the organs of the thorax and abdomen; this afferent input is carried mainly by the vagus, CN X, but also by the glossopharyngeal nerve, CN IX.
- **Special senses**: auditory (hearing) and vestibular (balance) afferents with the vestibulochoclear nerve, CN VIII, as well as the special sense of taste with CN VII and IX.

This diagram shows the location of the sensory nuclei of the cranial nerves, superimposed upon the ventral view of the brainstem. It is important to note that the location of the sensory nucleus of the cranial nerves inside the brainstem does not correspond exactly to the level of attachment of the nerve to the brainstem as seen externally, particularly in the case of CN V. (These nuclei are also shown in Figure 40, in which the brainstem is presented from a dorsal perspective.) The details of the location of the cranial nerve nuclei within the brainstem will be described in Section C of this atlas (Neurological Neuroanatomy) with Figure 64–Figure 67.

CN V, TRIGEMINAL NERVE

The major sensory nerve of the head region is the trigeminal nerve, CN V, through its three divisions peripherally (ophthalmic, maxillary, and mandibular). The sensory ganglion for this nerve, the trigeminal ganglion, is located inside the skull. The nerve supplies the skin of the scalp and face, the conjunctiva of the eye and the eyeball, the teeth, and the mucous membranes inside the head, including the surface of the tongue (but not taste — see below).

The sensory components of the trigeminal nerve are found at several levels of the brainstem. (See trigeminal pathways, Figure 35 and Figure 36):

- The **principal nucleus**, which is responsible for the discriminative aspects of touch, is located at the midpontine level, adjacent to the motor nucleus of CN V.
- A long column of cells that relays pain and temperature information, known as the **spinal nucleus of V** or the **descending trigeminal nucleus**, descends through the medulla and reaches the upper cervical levels of the spinal cord.
- Another group of cells extends into the midbrain region, the **mesencephalic** nucleus of V. These cells appear to be similar to neurons of the dorsal root ganglia and are thought to be the sensory proprioceptive neurons for the muscles of mastication.

CN VIII, VESTIBULOCOCHLEAR NERVE

Cochlear nuclei: The auditory fibers from the spiral ganglion in the cochlea are carried to the CNS in CN VIII, and form their first synapses in the cochlear nuclei, as it enters the brainstem at the uppermost level of the medulla (see Figure 6). The auditory pathway is presented in Section B (see Figure 37 and Figure 38).

Vestibular nuclei: Vestibular afferents enter the CNS as part of CN VIII. There are four nuclei: the medial and inferior, located in the medulla; the lateral, located at the ponto-medullary junction; and the small superior nucleus, located in the lower pontine region. The vestibular afferents terminate in these nuclei. The vestibular nuclei will be further discussed in Section B with the motor systems (see Figure 51A and Figure 51B).

VISCERAL AFFERENTS AND TASTE: SOLITARY NUCLEUS

The special sense of taste from the surface of the tongue is carried in CN VII and CN IX, and these terminate in the solitary nucleus in the medulla (see Figure 67A).

CLINICAL ASPECT

Trigeminal neuralgia is discussed with Figure 10.

ADDITIONAL DETAIL

The visceral afferents with CN IX and X from the pharynx, larynx, and internal organs are also received in the solitary nucleus (see Figure 67B and Figure 67C).

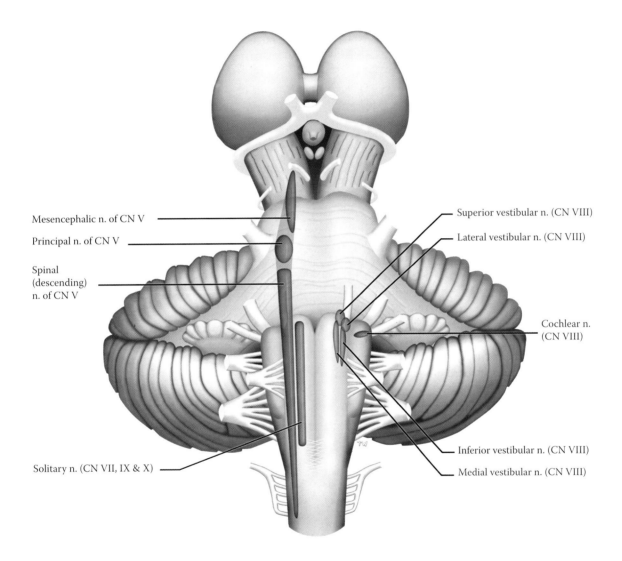

Mesencephalic n. of CN V

Principal n. of CN V

Spinal (descending) n. of CN V

Solitary n. (CN VII, IX & X)

Superior vestibular n. (CN VIII)

Lateral vestibular n. (CN VIII)

Cochlear n. (CN VIII)

Inferior vestibular n. (CN VIII)

Medial vestibular n. (CN VIII)

FIGURE 8B: Brainstem 4 — Cranial Nerves Nuclei — Sensory

FIGURE 9A
BRAINSTEM 5

BRAINSTEM AND CEREBELLUM: DORSAL (PHOTOGRAPHIC) VIEW

This specimen of the brainstem and diencephalon, with the cerebellum attached, is being viewed from the dorsal or posterior perspective. The third ventricle, the ventricle of the diencephalon, separates the thalamus of one side from that of the other (see Figure OA and Figure 20A; also Figure 17 and Figure 21, where the brain is separated down the midline in the midsagittal plane). The diencephalon is to be discussed with Figure 11.

Additional structures of the brainstem are seen from this perspective:

- The dorsal part of the midbrain is seen to have four elevations, named the superior and inferior colliculi (see also Figure 10). The upper ones are the **superior colliculi**, and they are functionally part of the visual system, a center for visual reflexes (see Figure 41C and Figure 51B). The lower ones are the **inferior colliculi**, and these are relay nuclei in the auditory pathway (see Figure 38). These colliculi form the "**tectum**," a term often used; a less frequently used term for these colliculi is the quadrigeminal plate. The **pineal**, a glandular structure, hangs down from the back of the diencephalon and sits between the colliculi.
- Although not quite in view in this illustration, the trochlear nerves (CN IV) emerge posteriorly at the lower level of the midbrain, below the inferior colliculi (see Figure 10).

This view also shows the back edge of the cerebral peduncle, the most anterior structure of the midbrain (see Figure 6 and Figure 7).

The posterior aspect of the pons and the medulla are hidden by the cerebellum — some of these structures will be seen in the next illustration (a photographic view, Figure 9B), and some are seen in a diagram with the cerebellum removed (Figure 10).

THE CEREBELLUM

The cerebellum, sometimes called the "little brain," is easily recognizable by its surface, which is composed of narrow ridges of cortex, called **folia** (singular **folium**). The cerebellum is located beneath a thick sheath of the meninges, the tentorium cerebelli, inferior to the occipital lobe of the hemispheres (see Figure 17 and Figure 30), in the posterior cranial fossa of the skull.

The cerebellum is involved with motor control and is part of the motor system, influencing posture, gait, and voluntary movements (discussed in more detail in Section B). Its function is to facilitate the performance of movements by coordinating the action of the various participating muscle groups. This is often spoken of simply as "smoothing out" motor acts. Although it is rather difficult to explain in words what the cerebellum does in motor control, damage to the cerebellum leads to quite dramatic alterations in ordinary movements (discussed with Figure 57). Lesions of the cerebellum result in the decomposition of the activity, or fractionation of movement, so that the action is no longer smooth and coordinated. Certain cerebellar lesions also produce a tremor, which is seen when performing voluntary acts, better known as an intention tremor.

Anatomically, the cerebellum can be described by looking at its appearance in a number of ways. The human cerebellum *in situ* has an upper or superior surface, as seen in this photograph, and a lower or inferior surface (shown in the next illustration). The central portion is known as the **vermis**. The lateral portions are called the **cerebellar hemispheres**.

Sulci separate the folia, and some of the deeper sulci are termed fissures. The **primary fissure** is located on the superior surface of the cerebellum, which is the view seen in this photograph. The **horizontal fissure** is located at the margin between the superior and inferior surfaces. Using these sulci and fissures, the cerebellar cortex has traditionally been divided into a number of different lobes, but many (most) of these do not have a distinctive functional or clinical importance, so only a few will be mentioned when the cerebellum is discussed (see Figure 54–Figure 57).

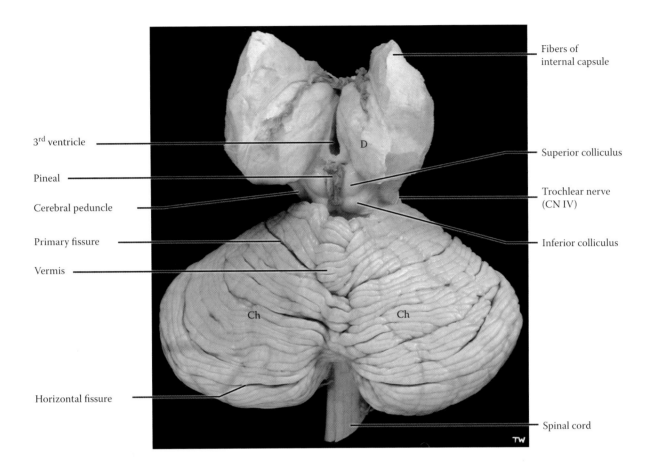

Fibers of
internal capsule

3rd ventricle

Pineal

Cerebral peduncle

Primary fissure

Vermis

Superior colliculus

Trochlear nerve
(CN IV)

Inferior colliculus

D

Ch   Ch

Horizontal fissure

Spinal cord

D = Diencephalon
Ch = Cerebellar hemisphere

FIGURE 9A: **Brainstem 5** — Dorsal View with Cerebellum (photograph)

FIGURE 9B
BRAINSTEM 6

BRAINSTEM AND CEREBELLUM: DORSAL INFERIOR (PHOTOGRAPHIC) VIEW

This is a photograph of the same specimen as Figure 9A, but the specimen is tilted to reveal the inferior aspect of the cerebellum and the posterior aspect of the medulla. The posterior aspect of the pons is still covered by the cerebellum (see Figure 10). The posterior aspect of the midbrain can no longer be seen. The upper end of the thalamus is still in view.

The horizontal fissure of the cerebellum is now clearly seen; it is used as an approximate divider between the superior and inferior surfaces of the cerebellum (see Figure 54). The vermis of the cerebellum is clearly seen between the hemispheres. Just below the vermis is an opening into a space — the space is the fourth ventricle (which will be described with Figure 20A, Figure 20B, and Figure 21) The opening is between the ventricle and the subarachnoid space outside the brain (discussed with Figure 21); the name of the opening is the **Foramen of Magendie**.

The part of the brainstem immediately below the foramen is the medulla, its posterior or dorsal aspect. The most significant structure seen here is a small elevation, repre-senting an important sensory relay nucleus, the **nucleus gracilis**. The pathway for discriminative touch sensation, called the gracilis tract (or fasciculus) continues up the posterior aspect of the spinal cord and synapses in the nucleus of the same name; the pathway then continues up to the cerebral cortex. (The details of this pathway will be discussed with Figure 33 and Figure 40). Beside it is another nucleus for a similar pathway with the same function, the nucleus cuneatus (see Figure 10). These nuclei will be discussed with the brainstem cross-sections in Section C (see Figure 67C). The medulla ends and the spinal cord begins where the C1 nerve roots emerge.

The cerebellar lobules adjacent to the medulla are known as the **tonsils** of the cerebellum (see ventral view of the cerebellum, Figure 7). The tonsils are found just inside the foramen magnum of the skull.

CLINICAL ASPECT

Should there be an increase in the mass of tissue occupying the posterior cranial fossa (e.g., a tumor, hemorrhage), the cerebellum would be pushed downward. This would force the cerebellar tonsils into the foramen magnum, thereby compressing the medulla. The compression, if severe, may lead to a compromising of function of the vital centers located in the medulla (discussed with Figure 6). The complete syndrome is known as **tonsillar herniation**, or coning. This is a life-threatening situation that may cause cardiac or respiratory arrest.

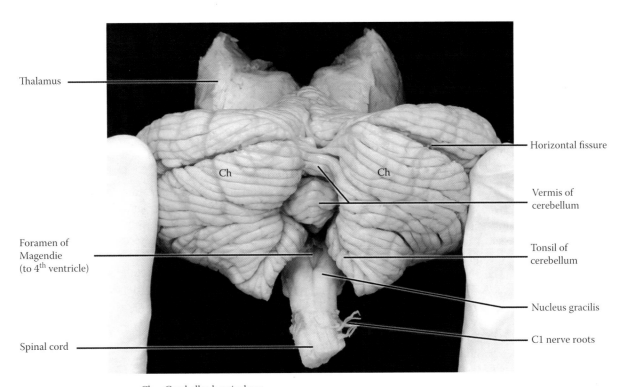

Thalamus

Horizontal fissure

Ch   Ch

Vermis of
cerebellum

Foramen of
Magendie
(to 4ᵗʰ ventricle)

Tonsil of
cerebellum

Nucleus gracilis

C1 nerve roots

Spinal cord

Ch = Cerebellar hemisphere

FIGURE 9B: Brainstem 6 — Dorsal Inferior View with Cerebellum (photograph)

FIGURE 10
BRAINSTEM 7

BRAINSTEM: DORSAL VIEW — CEREBELLUM REMOVED

This diagram shows the brainstem from the dorsal perspective, with the cerebellum removed. A similar view of the brainstem is used for some of the later diagrams (see Figure 40 and Figure 48). This dorsal perspective is useful for presenting the combined visualization of many of the cranial nerve nuclei and the various pathways of the brainstem.

MIDBRAIN LEVEL

The posterior aspect of the midbrain has the superior and inferior colliculi, as previously seen, as well as the emerging fibers of CN IV, the trochlear nerve. The posterior aspect of the cerebral peduncle is clearly seen.

PONTINE LEVEL

Now that the cerebellum has been removed, the dorsal aspect of the pons is seen. The space separating the pons from the cerebellum is the fourth ventricle — the ventricle has been "unroofed." (The ventricles of the brain will be discussed with Figure 20A, Figure 20B, and Figure 21.) The roof of the upper portion of the fourth ventricle is a sheet of nervous tissue and bears the name **superior medullary velum**; more relevant, it contains an important connection of the cerebellum, the superior cerebellar peduncles (discussed with Figure 57). The lower half of the roof of the fourth ventricle has choroid plexus (see Figure 21).

As seen from this perspective, the fourth ventricle has a "floor"; noteworthy are two large bumps, called the facial colliculus, where the facial nerve, CN VII, makes an internal loop (to be discussed with Figure 48 and also with the pons in Section C of this atlas, see Figure 66C).

As the cerebellum has been removed, the cut surfaces of the middle and inferior cerebellar peduncles are seen. The **cerebellar peduncles** are the connections between the brainstem and the cerebellum, and there are three pairs of them. The **inferior** cerebellar peduncle connects the medulla and the cerebellum, and the prominent **middle**

cerebellar peduncle brings fibers from the pons to the cerebellum. Both can be seen in the ventral view of the brainstem (see Figure 7). Details of the information carried in these pathways will be outlined when the functional aspects of the cerebellum are studied with the motor systems (see Figure 55). The **superior** cerebellar peduncles convey fibers from the cerebellum to the thalamus, passing through the roof of the fourth ventricle and the midbrain (see Figure 57). This peduncle can only be visualized from this perspective.

CN V emerges through the middle cerebellar peduncle (see also Figure 6 and Figure 7).

MEDULLARY LEVEL

The lower part of the fourth ventricle separates the medulla from the cerebellum (see Figure 21). The special structures below the fourth ventricle are two large protuberances on either side of the midline — the **gracilis and cuneatus nuclei**, relay nuclei which belong to the ascending somatosensory pathway (discussed with Figure 9B, Figure 33, and Figure 40).

The cranial nerves seen from this view include the entering nerve CN VIII. More anteriorly, from this oblique view, are the fibers of the glossopharyngeal (CN IX) and vagus (CN X) nerves, as these emerge from the lateral aspect of the medulla, behind the inferior olive.

A representative cross-section of the spinal cord is also shown, from this dorsal perspective.

ADDITIONAL DETAIL

The acoustic stria (not labeled) shown in the floor of the fourth ventricle are fibers of CN VIII, the auditory portion, which take an alternative route to relay in the lower pons, before ascending to the inferior colliculi of the midbrain.

Two additional structures are shown in the midbrain — the red nucleus (described with Figure 47 and Figure 65A), and the brachium of the inferior colliculus, a connecting pathway between the inferior colliculus and the medial geniculate body, all part of the auditory system (fully described with Figure 37 and Figure 38).

The medial and lateral geniculate nuclei belong with the thalamus (see Figure 11 and Figure 12). The lateral geniculate body (nucleus) is part of the visual system (see Figure 41A and Figure 41C).

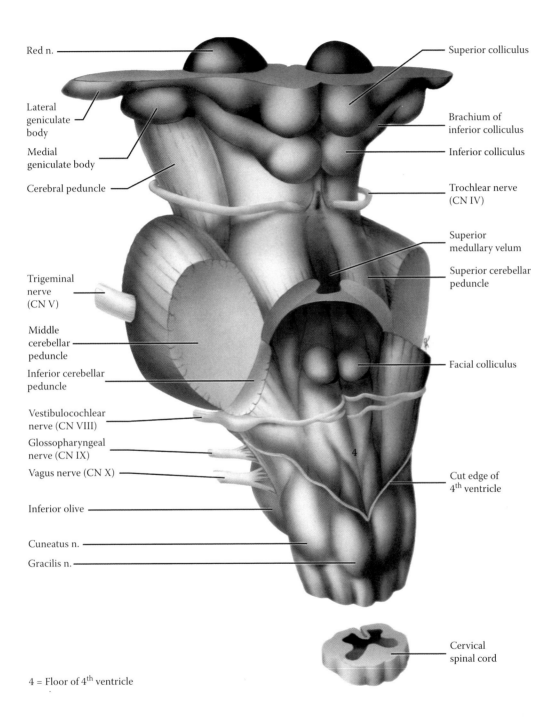

Red n.

Lateral
geniculate
body

Medial
geniculate body

Cerebral peduncle

Trigeminal
nerve
(CN V)

Middle
cerebellar
peduncle

Inferior cerebellar
peduncle

Vestibulocochlear
nerve (CN VIII)

Glossopharyngeal
nerve (CN IX)

Vagus nerve (CN X)

Inferior olive

Cuneatus n.

Gracilis n.

Superior colliculus

Brachium of
inferior colliculus

Inferior colliculus

Trochlear nerve
(CN IV)

Superior
medullary velum

Superior cerebellar
peduncle

Facial colliculus

Cut edge of
4th ventricle

Cervical
spinal cord

4 = Floor of 4th ventricle

FIGURE 10: Brainstem 7 — Dorsal View — Cerebellum Removed

FIGURE 11
THE DIENCEPHALON:
THALAMUS 1

THALAMUS: ORIENTATION

The diencephalon, which translates as "between brain," is the next region of the brain to consider. The diencephalon, including both thalamus and hypothalamus and some other subparts, is situated between the brainstem and the cerebral hemispheres, deep within the brain.

As shown diagrammatically (see Figure 6) and photographically (see Figure 7 and Figure 9A), the diencephalon sits "atop" the brainstem. The enormous growth of the cerebral hemispheres in the human brain has virtually hidden or "buried" the diencephalon (somewhat like a weeping willow tree) so that it can no longer be visualized from the outside except from the inferior view (see pituitary stalk and mammillary bodies in Figure 15A and Figure 15B, which are part of the hypothalamus).

In this section of the atlas, we will consider the **thalamus**, which makes up the bulk of the diencephalon. It is important to note that there are two thalami, one for each hemisphere of the brain, and these are often connected across the midline by nervous tissue, the massa intermedia (as seen in Figure 6). As has been noted, the third ventricle is situated between the two thalami (see Figure 9 and Figure 20B).

The thalamus is usually described as the gateway to the cerebral cortex (see Figure 63). This description leaves out an important principle of thalamic function, namely that most thalamic nuclei that project to the cerebral cortex also receive input from that area — these are called reciprocal connections. This principle does not apply, however, to all of the nuclei (see below).

The major function of the thalamic nuclei is to process information before sending it on to the select area of the cerebral cortex. This is particularly so for all the sensory systems, except the olfactory sense. It is possible that crude forms of sensation, including pain, are "appreciated" in the thalamus, but localization of the sensation to a particular spot on the skin surface requires the involvement of the cortex. Likewise, two subsystems of the motor systems, the basal ganglia and the cerebellum, relay in the thalamus before sending their information to the motor

areas of the cortex. In addition, the limbic system has circuits that involve the thalamus.

Other thalamic nuclei are related to areas of the cerebral cortex, which are called association areas, vast areas of the cortex that are not specifically related either to sensory or motor functions. Parts of the thalamus play an important role in the maintenance and regulation of the state of consciousness, and also possibly attention, as part of the ascending reticular activating system (ARAS, see Figure 42A).

Other parts of the Diencephalon:

- The **hypothalamus**, one in each hemisphere, is composed of a number of nuclei that regulate homeostatic functions of the body, including water balance. It will be discussed with the limbic system in Section D of this atlas (see Figure 78A).
- The **pineal** (visible in Figure 9A) is sometimes considered a part of the diencephalon. This gland is thought to be involved with the regulation of our circadian rhythm. Many people now take melatonin, which is produced by the pineal, to regulate their sleep cycle and to overcome jetlag.
- The **subthalamic nucleus** is described with the basal ganglia (see Figure 24).

ADDITIONAL DETAIL

As shown in the diagram, the diencephalon is situated within the brain below the level of the body of the lateral ventricles (see also Figure 17, Figure 18, and Figure 19A). In fact, the thalamus forms the "floor" of this part of the ventricle (see Figure 29). In a horizontal section of the hemispheres, the two thalami are located at the same level as the lentiform nucleus of the basal ganglia (see Figure OA and Figure OL; also Figure 26 and Figure 27). This important point will be discussed with the internal capsule (see Figure 26 and Figure 27).

Note to the Learner: The location of the thalamus within the substance of the brain is important for the understanding of the anatomical organization of the brain. This topographic information will make more sense after studying the hemispheres (see Figure 13–Figure 19) and basal ganglia (see Figure 22–Figure 30). The suggestion is made to review this material at that time.

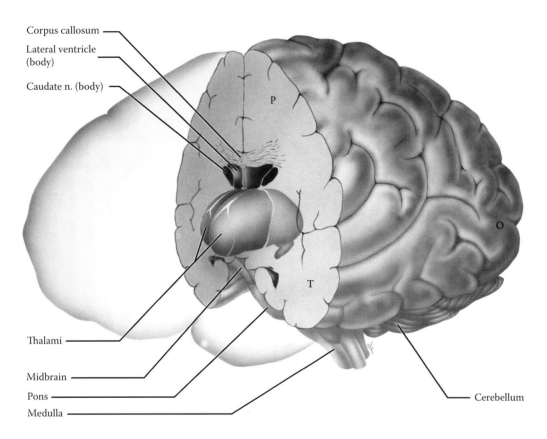

Corpus callosum

Lateral ventricle
(body)

Caudate n. (body)

P

O

T

Thalami

Midbrain

Pons

Medulla

Cerebellum

P = Parietal lobe
T = Temporal lobe
O = Occipital lobe

FIGURE 11: Thalamus 1 — Orientation

FIGURE 12
THALAMUS 2

THALAMUS: NUCLEI

In order to lay the groundwork for understanding the functional organization of the sensory and motor pathways (in Section B), it is necessary to have a familiarity with the nuclei of the thalamus, their organization, and names.

There are two ways of dividing up the nuclei of the thalamus, namely, topographically and functionally.

A. Topographically, the thalamus is subdivided by bands of white matter into a number of component parts. The main white matter band that runs within the thalamus is called the **internal medullary lamina** and it is shaped like the letter Y (see also the previous illustration). It divides the thalamus into a lateral mass, a medial mass, and an anterior group of nuclei.

B. Functionally, the thalamus has three different types of nuclei:
   - **Specific relay nuclei.** These nuclei relay sensory and motor information to specific sensory and motor areas of the cerebral cortex. Included with these are the medial and lateral geniculate bodies, relay nuclei for the auditory and visual systems. In addition, motor regulatory information from the basal ganglia and cerebellum is also relayed in the thalamus as part of this set of nuclei. These nuclei are located in the lateral nuclear mass.
   - **Association nuclei.** These are connected to broad areas of the cerebral cortex known as the association areas. One of the most important nuclei of this group is the dorsomedial nucleus, located in the medial mass of the thalamus.
   - **Nonspecific nuclei.** These scattered nuclei have other or multiple connections. Some of these nuclei are located within the internal medullary lamina and are often referred to as the **intralaminar** nuclei. This functional group of nuclei does not have the strong reciprocal connections with the cortex like the other nuclei. Some of these nuclei form part of the ascending reticular activating system, which is involved in the regulation of our state of consciousness and arousal (discussed with Figure 42A). The reticular nucleus, which lies on the outside of the thalamus is also part of this functional system.

The following detailed classification system is given at this point but will only be understood as the functional systems of the CNS are described (see Note to the Learner below).

Specific Relay Nuclei (and Function)

Their cortical connections are given at this point for information (<---> symbolizes a connection in both directions).

> **VA** — ventral anterior (motor) <---> premotor area and supplementary motor area
> **VL** — ventral lateral (motor) <---> precentral gyrus and premotor area
> **VPL** — ventral posterolateral (somatosensory) <---> postcentral gyrus
> **VPM** — ventral posteromedial (trigeminal) <---> postcentral gyrus
> **MGB** — medial geniculate (body) nucleus (auditory) <---> temporal cortex
> **LGB** — lateral geniculate (body) nucleus (vision) <---> occipital cortex

Association Nuclei (and Association Cortex)

These nuclei are reciprocally connected to association areas of the cerebral cortex.

> **DM** — dorsomedial nucleus <---> prefrontal cortex
> **AN** — anterior nucleus <---> limbic lobe
> **Pul** — pulvinar <---> visual cortex
> **LP** — lateral posterior <---> parietal lobe
> **LD** — lateral dorsal <---> parietal lobe

Nonspecific Nuclei (to Widespread Areas of the Cerebral Cortex)

> **IL** — intralaminar
> **CM** — centromedian
> **Ret** — reticular

ADDITIONAL DETAIL

For schematic purposes, this presentation of the thalamic nuclei, which is similar to that shown in a number of textbooks, is quite usable. Histological sections through the thalamus are challenging and beyond the scope of an introductory course.

Note to the Learner: The thalamus is being introduced at this point because it is involved throughout the study of the brain. The learner should learn the names and understand the general organization of the various nuclei at this point. It is advised to consult this diagram, as the cerebral cortex is described in the following illustrations. Each of the specific relay nuclei involved in one of the pathways will be introduced again with the functional systems (in Section B) and, at that point, the student should return to this illustration. A summary diagram showing the thalamus and the cortex with the detailed connections will be presented in Section C (see Figure 63). Various nuclei are also involved with the limbic system (see Section D).

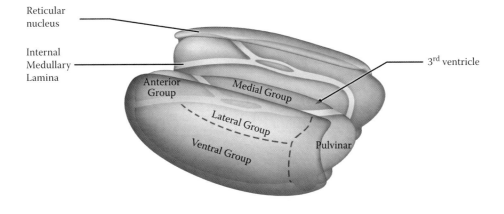

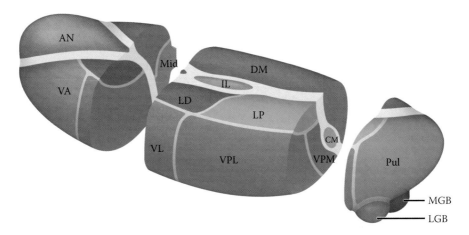

AN = Anterior nuclei

LD = Lateral dorsal
LP = Lateral posterior
Pul = Pulvinar

DM = Dorsomedial
Mid = Midline

LGB = Lateral geniculate body
MGB = Medial geniculate body

VA = Ventral anterior
VL = Ventral lateral
VPL = Ventral posterolateral
VPM = Ventral posteromedial

IL = Intralaminar
CM = Centromedian

FIGURE 12: Thalamus 2 — Nuclei

FIGURE 13
CEREBRAL HEMISPHERES 1

CEREBRAL CORTEX: DORSAL (PHOTOGRAPHIC) VIEW

When people talk about "the brain," they are generally referring to the cerebral hemispheres, also called the cerebrum. The brain of higher apes and humans is dominated by the cerebral hemispheres. The outer layer, the **cerebral cortex**, with its billions of neurons and its vast interconnections, is responsible for sensory perception, movement, language, thinking, memory, consciousness, and certain aspects of emotion. In short, we need the intact cerebral hemispheres to adapt to our ever-changing external environment.

The neurons of the cerebral cortex are organized in layers and generally there are six layers; this highly evolved cortex is called **neocortex**. Neurons in each of the layers differ in their functional contribution to cortical "processing." In formalin-fixed material, the cortex (which includes neurons, dendrites, and synapses) takes on a grayish appearance and is often referred to as the gray matter (see Figure 27 and Figure 29).

The cerebral hemispheres occupy the interior of the skull, the cranial cavity. The brain in this photograph is seen from above and from the side — one hemisphere has the meninges removed and the other is still covered with dura, the thick outer meningeal layer. The dural layer has additional folds within the skull that subdivide the cranial cavity and likely serve to keep the brain in place. The two major dural sheaths are the falx cerebri (between the hemispheres in the sagittal plane, see Figure 16) and the tentorium cerebelli (in the transverse plane between the occipital lobe and the cerebellum, see Figure 17 and Figure 30). Inside the dural layer are large channels, called venous sinuses, which convey blood from the surface of the hemispheres and return the blood to the heart via the internal jugular vein. The **superior sagittal sinus**, which is located at the upper edge of the interhemispheric fissure, is one of the major venous sinuses (see Figure 21). The **subarachnoid space**, between the arachnoid and pia, is filled with CSF (see Figure 21). Therefore, the brain is actually "floating" inside the skull.

The surface of the hemispheres in humans and some other species is thrown into irregular folds. These ridges are called **gyri** (singular **gyrus**), and the intervening crevices are called **sulci** (singular **sulcus**). This arrangement allows for a greater surface area to be accommodated within the same space (i.e., inside the skull). A very deep sulcus is called a **fissure**; two of these are indicated, the central fissure and the parieto-occipital fissure. These tend to be constant in all human brains.

Different parts of the cortex have different functions. Some parts have a predominantly motor function, whereas other parts are receiving areas for one of the major sensory systems. Most of the cerebral cortex in humans has an "**association** function," a term that can perhaps be explained functionally as interrelating the various activities in the different parts of the brain.

The basic division of each of the hemispheres is into **four lobes**: frontal, parietal, temporal, and occipital. Two prominent fissures allow this subdivision to be made — the central fissure and the lateral fissure. The **central fissure** divides the area anteriorly, the frontal lobe, from the area posteriorly, the parietal lobe. The parietal lobe extends posteriorly to the parieto-occipital fissure (see Figure 17). The brain area behind that fissure is the occipital lobe. The temporal lobe and the lateral fissure cannot be seen on this view of the brain (see next illustration).

The surface of the cerebral hemispheres can be visualized from a number of other directions — from the side (the dorsolateral view, see Figure 14), and from below (inferior view, see Figure 15A and Figure 15B); in addition, after dividing the two hemispheres along the interhemispheric fissure (in the midline), the hemispheres are seen to have a medial surface as well (see Figure 17).

CLINICAL ASPECT

Intracranial bleeds can occur between the skull and the dura (called epidural, usually arterial), between the dura and arachnoid (called subdural, usually venous), into the CSF space (called subarachnoid, usually arterial), or into the substance of the brain (brain hemorrhage). Since the brain is enclosed is a rigid box, the skull, any abnormal bleeding inside the head may lead to an increase in intracranial pressure (discussed with the Introduction to Section C).

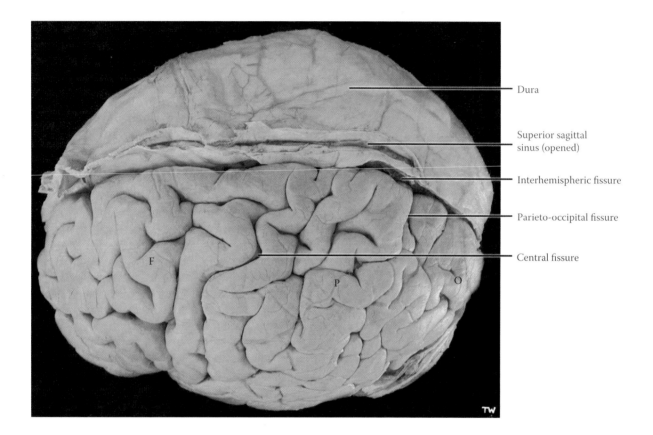

Dura

Superior sagittal
sinus (opened)

Interhemispheric fissure

Parieto-occipital fissure

Central fissure

F = Frontal lobe
P = Parietal lobe
O = Occipital lobe

FIGURE 13: Cerebral Hemispheres 1 — Dorsal View (photograph)

FIGURE 14A
CEREBRAL HEMISPHERES 2

CEREBRAL CORTEX: DORSOLATERAL (PHOTOGRAPHIC) VIEW

This is a photographic image of the same brain as shown in the previous illustration, tilted slightly, to show the dorsolateral aspect of the hemispheres. The edge of the other hemisphere (with meninges) is still in view. It is now possible to identify the sulci and fissures with more certainty. The **central fissure** (often called the fissure of Rolando) is seen more completely, dividing the frontal lobe anteriorly from the parietal lobe posteriorly. The deep **lateral fissure** is clearly visible (see below).

Some cortical areas are functionally directly connected with either a sensory or motor system; these are known as the **primary areas**. The gyrus in front of the central fissure is called the **precentral** gyrus, also called area 4, and it is the primary **motor** area, specialized for the control of voluntary movements (see Figure 53 and Figure 60). The area in front of this gyrus is called the **(lateral) premotor area**, also called area 6, which is likewise involved with voluntary motor actions (see also Figure 53 and Figure 60). An area in the frontal lobe (outlined) has a motor function in regards to eye movements; this is called the **frontal eye field** (area 8). The gyrus behind the central fissure is the **postcentral gyrus**, including areas 1, 2 and 3 (see Figure 36 and Figure 60), and it has a **somatosensory** function for information from the skin (and joints). (Other sensory primary areas will be identified at the appropriate time.)

The remaining cortical areas that are not directly linked to either a sensory or motor function are called **association** cortex. The most anterior parts of the frontal lobe are the newest in evolution and are known as the **prefrontal cortex** (in front of the frontal eye fields previously mentioned). This broad cortical area seems to be the chief "executive" part of the brain. The **parietal areas** are connected to sensory inputs and have a major role in integrating sensory information from the various modalities. In the parietal lobe, there are two special gyri, the **supramarginal** and **angular** gyri; these areas, particularly on the nondominant side, seem to be involved in visuospatial activities.

Some cortical functions are not equally divided between the two hemispheres. One hemisphere is therefore said to be dominant for that function. This is the case for **language** ability, which, in most people, is located in the left hemisphere. This photograph of the left hemisphere shows the two language areas: **Broca's** area for the motor aspects of speech and **Wernicke's** area for the comprehension of written and spoken language (near the auditory area).

The lateral fissure (also known as the fissure of Sylvius) divides the temporal lobe below from the frontal and parietal lobes above. Extending the line of the lateral fissure posteriorly continues the demarcation between the temporal and parietal lobes. The **temporal lobe** seen on this view is a large area of association cortex whose function is still being defined, other than the portions involved with the auditory system (see Figure 38 and Figure 39) and language (on the dominant side). Other portions of the temporal lobe include the inferior parts (to be discussed with the following illustrations) and the medial portion, which is part of the limbic system (see Section D).

The location of the parieto-occipital fissure is indicated on this photograph (see also previous illustration). This fissure, which separates the parietal lobe from the occipital lobe, is best seen when the medial aspect of the brain is visualized after dividing the hemispheres (see Figure 17). The occipital lobe is concerned with the processing of visual information.

The cerebellum lies below the occipital lobe, with the large dural sheath, the tentorium cerebelli (not labeled, see Figure 17) separating these parts of the brain.

CLINICAL ASPECTS

It is most important to delineate anatomically the functional areas of the cortex. This forms the basis for understanding the clinical implications of damage (called lesions) to the various parts of the brain. Clinicians are now being assisted in their tasks by modern imaging techniques, including CT (see Figure 28A) and MRI (see Figure 28B).

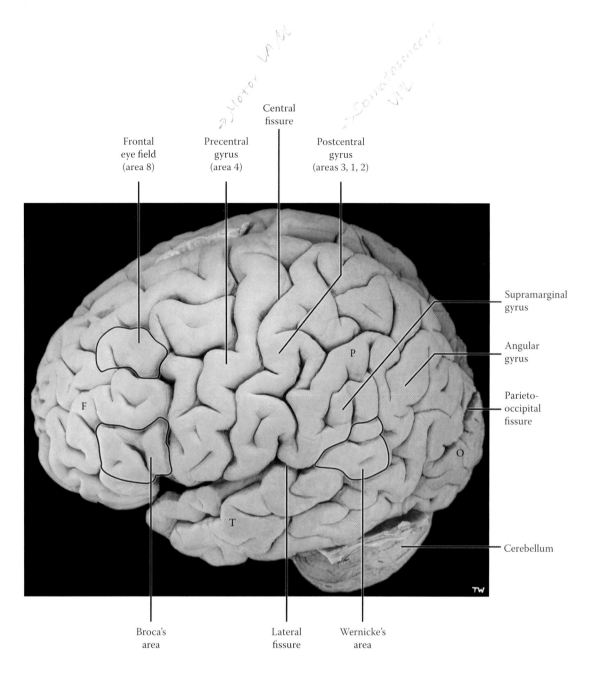

F = Frontal lobe
P = Parietal lobe
T = Temporal lobe
O = Occipital lobe (areas 18, 19)

FIGURE 14A: Cerebral Hemispheres 2 — Dorsolateral View (photograph)

FIGURE 14B
CEREBRAL HEMISPHERES 3

THE INSULA

The lateral fissure has been "opened" to reveal some buried cortical tissue; this area is called the **insula**. The function of this cortical area has been somewhat in doubt over the years. It seems that this is the area responsible for receiving taste sensations, relayed from the brainstem (see Figure 8B and Figure 67A). Sensations from our internal organs may reach the cortical level in this area.

The specialized cortical gyri for hearing (audition) are also to be found within the lateral fissure, but they are part of the upper surface of the superior temporal gyrus (as shown in Figure 38 and Figure 39).

It should be noted that the lateral fissure has within it a large number of blood vessels, which have been removed —branches of the middle cerebral artery (discussed with Figure 58). Branches to the interior of the brain, the striate arteries, are given off in the lateral fissure (see Figure 62). The insular cortex can be recognized on a horizontal section of the brain (see Figure 27) and also on coronal views of the brain (see Figure 29), as well as with brain imaging (CT and MRI).

CLINICAL ASPECT

A closed head injury that affects the brain is one of the most serious forms of accidents. The general term for this is a **concussion**, a bruising of the brain. There are various degrees of concussion depending upon the severity of the trauma. The effects vary from mild headache to unconsciousness and may include some memory loss, usually temporary. Everything possible should be done to avoid a brain injury, particularly when participating in sport activities. Proper headgear in the form of a helmet should be worn by children and adults while cycling, skiing, snowboarding, and skating (winter and inline). Closed head injuries occur most frequently with motor vehicle accidents, and the use of seatbelts and proper seats for children reduces the risk.

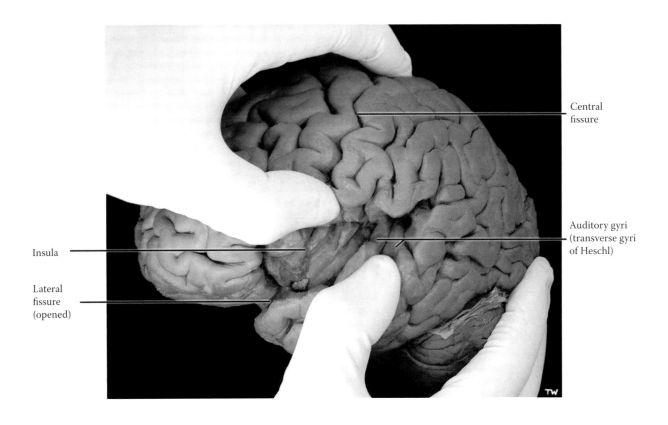

Central
fissure

Auditory gyri
(transverse gyri
of Heschl)

Insula

Lateral
fissure
(opened)

FIGURE 14B: Cerebral Hemispheres 3 — The Insula (photograph)

FIGURE 15A
CEREBRAL HEMISPHERES 4

CEREBRAL CORTEX: INFERIOR (PHOTOGRAPHIC) VIEW WITH BRAINSTEM

This is a photographic view of the same brain seen from below, the inferior view, a view that includes the brainstem and the cerebellum. The medulla and pons, parts of the brainstem can be identified (see Figure 6 and Figure 7), but the midbrain is mostly hidden from view. The cranial nerves are still attached to the brainstem, and some of the arteries to the brain are also present.

The frontal lobe occupies the anterior cranial fossa of the skull. The inferior surface of the frontal lobe extends from the frontal pole to the anterior tip of the temporal lobe (and the beginning of the lateral fissure). These gyri rest on the roof of the orbit and are sometimes referred to as the **orbital gyri**. This is association cortex and these gyri have strong connections with the limbic system (discussed in Section D).

The next area is the inferior surface of the **temporal lobe**. This lobe occupies the middle cranial fossa of the skull. The temporal lobe extends medially toward the midbrain and ends in a blunt knob of tissue known as the **uncus**. Moving laterally from the uncus, the first sulcus visible is the collateral sulcus/fissure (seen clearly on the left side of this photograph). The **parahippocampal gyrus** is the gyrus medial to this sulcus; it is an extremely important gyrus of the limbic system (discussed with Figure 74). It should be noted that the uncus is the most medial protrusion of this gyrus. (The clinical significance of the uncus and uncal herniation will be discussed with the next illustration.)

The olfactory tract and optic nerve (and chiasm) are seen on this view. Both are, in fact, CNS pathways and are not peripheral cranial nerves, even though they are routinely called CN I and CN II. The olfactory bulb is the site of termination of the olfactory nerve filaments from the nose; these filaments are, in fact, the peripheral nerve

CN I (see Figure 79). Olfactory information is then carried in the olfactory tract to various cortical and subcortical areas of the temporal lobe (discussed with Figure 79). The optic nerves (CN II) exit from the orbit and continue to the optic chiasm, where there is a partial crossing of visual fibers, which then continue as the optic tract (see Figure 41A). Posterior to the chiasm is the area of the hypothalamus, part of the diencephalon, including the pituitary stalk and the mammillary bodies, which will be seen more clearly in the next illustration.

The brainstem and cerebellum occupy the posterior part of this brain from this inferior perspective. These structures occupy the posterior cranial fossa of the skull. In fact, the cerebellum obscures the visualization of the occipital lobe (which is shown in the next photograph, after removing most of the brainstem and cerebellum). Various cranial nerves can be identified as seen previously (see Figure 7). The oculomotor nerve, CN III, should be noted as it exits from the midbrain; the slender trochlear nerve (CN IV) can also be seen.

Part of the arterial system is also seen in this brain specimen (the arterial supply is discussed with Figure 58–Figure 62). The initial part, vertebral arteries and the formation of the basilar artery, are missing, as are the three pairs of cerebellar arteries. The basilar artery, which is situated in front of the pons, ends by dividing into the posterior cerebral arteries to supply the occipital regions of the brain. The cut end of the internal carotid artery is seen, but the remainder of the arterial circle of Willis is not dissected on this specimen (see Figure 58); the arterial supply to the cerebral hemispheres will be fully described in Section C (see Figure 60 and Figure 61).

Note to the Learner: The specimen of the brainstem and diencephalon shown in Figure 7 was created by dissecting these parts of the brain free of the hemispheres. This has been done by cutting the fibers going to and from the thalamus, as well as all the fibers ascending to and descending from the cerebral cortex (called projection fibers, discussed with Figure 16). The diagrams of such a specimen are shown in Figure 6, Figure 8A, and Figure 8B.

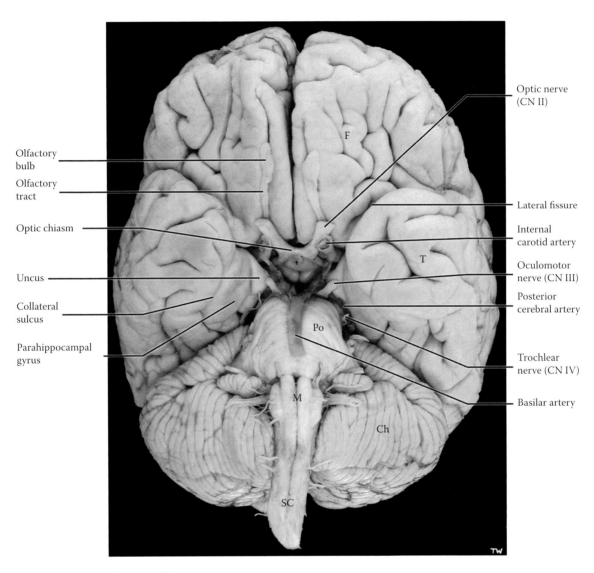

Olfactory bulb

Olfactory tract

Optic chiasm

Uncus

Collateral sulcus

Parahippocampal gyrus

Optic nerve (CN II)

Lateral fissure

Internal carotid artery

Oculomotor nerve (CN III)

Posterior cerebral artery

Trochlear nerve (CN IV)

Basilar artery

F = Frontal lobe
T = Temporal lobe
Po = Pons
M = Medulla
SC = Spinal cord
Ch = Cerebellar hemisphere

FIGURE 15A: **Cerebral Hemispheres 4** — Inferior View with Brainstem (photograph)

FIGURE 15B
CEREBRAL HEMISPHERES 5

INFERIOR SURFACE: INFERIOR (PHOTOGRAPHIC) VIEW WITH MIDBRAIN

This is another brain specimen showing the inferior surface of the brain, in which the brainstem has been sectioned through at the level of the midbrain, removing most of the brainstem and the attached cerebellum. The cut surface of the midbrain is exposed, showing a linear area of brain tissue, which is black in coloration; this elongated cluster of cells is the nucleus of the midbrain called the **substantia nigra**, and consists of neurons with pigment inside the cells (discussed with Figure 65). The functional role of the substantia nigra is discussed with the basal ganglia (see Figure 24 and Figure 52).

This dissection reveals the inferior surface of both the temporal and the occipital lobes. It is not possible to define the boundary between these two lobes on this view. Some of these inferior gyri are involved with the processing of visual information, including color, as well as facial recognition. The parahippocampal gyri should be noted on both sides, with the collateral sulcus demarcating the lateral border of this gyrus (seen in the previous illustration; discussed with Figure 72A and Figure 72B).

The optic nerves (cut) lead to the optic chiasm, and the regrouped visual pathway continues, now called the optic tract (see Figure 41A and Figure 41C). Behind the optic chiasm are the median eminence and then the mammillary (nuclei) bodies, both of which belong to the hypothalamus. The **median eminence** (not labeled) is an elevation of tissue that contains some hypothalamic nuclei. The **pituitary stalk**, identified on the previous illustration, is attached to the median eminence, and this stalk connects the hypothalamus to the pituitary gland. Behind this are the paired **mammillary bodies**, two nuclei of the hypothalamus (which will be discussed with the limbic system, see Figure 78A).

Also visible on this specimen is the posterior thickened end of the corpus callosum (discussed with the next illustration) called the splenium (see Figure 17 and Figure 19A).

A thick sheath of dura separates the occipital lobe from the cerebellum below — the **tentorium cerebelli** (as it covers over the cerebellum). The cut edge of the tentorium can be seen in Figure 17, and its location is seen in Figure 18, above the cerebellum. The tentorium divides the cranial cavity into an area above it, the supratentorial space, a term that is used often by clinicians to indicate a problem in any of the lobes of the brain. The area below the tentorium, the infratentorial space, corresponds to the posterior cranial fossa. The tentorial sheath of dura, the tentorium cerebelli, splits around the brainstem at the level of the midbrain; this split in the tentorium is called the **tentorial notch** (hiatus).

CLINICAL ASPECT

The uncus has been clearly identified in the specimens, with its blunted tip pointed medially. The uncus is in fact positioned just above the free edge of the tentorium cerebelli. Should the volume of brain tissue increase above the tentorium, due to brain swelling, hemorrhage, or a tumor, accompanied by an increase in intracranial pressure (ICP), the hemispheres would be forced out of their supratentorial space. The only avenue to be displaced is in a downward direction, through the tentorial notch, and the uncus becomes the leading edge of this pathological event. The whole process is clinically referred to as "**uncal herniation**."

Since the edges of the tentorium cerebelli are very rigid, the extra tissue in this small area causes a compression of the brain matter, leading to compression of the brainstem; this is followed by a progressive loss of consciousness. CN III is usually compressed as well, damaging it, and causing a fixed and dilated pupil on that side, an ominous sign in any lesion of the brain. This is a medical emergency! Continued herniation will lead to further compression of the brainstem and a loss of vital functions, followed by rapid death.

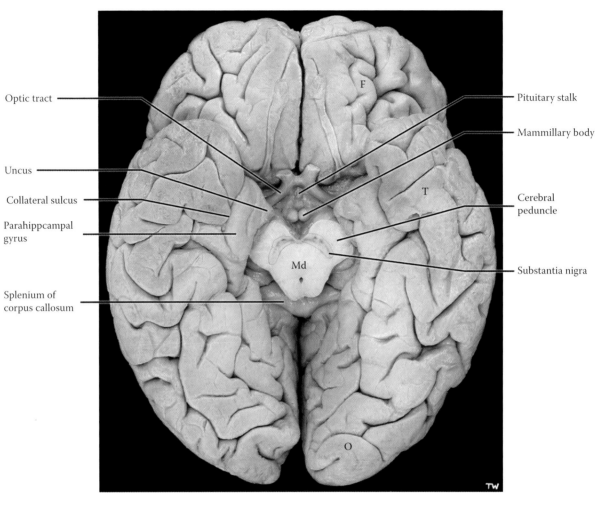

Optic tract

Uncus

Collateral sulcus

Parahippcampal
gyrus

Splenium of
corpus callosum

Pituitary stalk

Mammillary body

Cerebral
peduncle

Substantia nigra

F = Frontal lobe
T = Temporal lobe
O = Occipital lobe
Md = Midbrain (cut)

FIGURE 15B: Cerebral Hemispheres 5 — Inferior View with Midbrain (photograph)

FIGURE 16
CEREBRAL HEMISPHERES 6

CORPUS CALLOSUM: SUPERIOR
(PHOTOGRAPHIC) VIEW

In this photograph, the brain is again being viewed from directly above (see Figure 13), with the interhemispheric fissure opened. The dural fold between the hemispheres, the falx cerebri, has been removed from the interhemispheric fissure. This thick sheath of dura keeps the two halves of the hemispheres in place within the cranial cavity. A whitish structure is seen in the depths of the fissure — the **corpus callosum**.

One of the other major features of the cerebral cortex is the vast number of neurons that are devoted to communicating with other neurons of the cortex. These interneurons are essential for the processing and elaboration of information, whether generated in the external world or internally by our "thoughts." This intercommunicating network is reflected in the enormous number of interconnections between cortical areas. These interconnecting axons are located within the depths of the hemispheres. They have a white coloration after fixation in formalin, and these regions are called the **white matter** (see Figure 27 and Figure 29).

The white matter bundles within the hemispheres are of three kinds:

- **Commissural** bundles — connecting cortical areas across the midline
- **Association** bundles — interconnecting the cortical areas on the same side
- **Projection** fibers — connecting the cerebral cortex with subcortical structures, including the basal ganglia, thalamus, brainstem, and spinal cord

All such connections are bidirectional, including the projection fibers.

The corpus callosum is the largest of the commissural bundles, as well as the latest in evolution. This is the anatomic structure required for each hemisphere to be kept informed of the activity of the other hemisphere. The axons connect to and from the lower layers of the cerebral cortex, and in most cases the connections are between homologous areas and are reciprocal. If the brain is sectioned in the sagittal plane along the interhemispheric fissure, the medial aspect of the brain will be revealed (see next illustration). The corpus callosum will be divided in the process. The fibers of the corpus callosum can be followed from the midline to the cortex (see Figure 19A).

It is difficult on this view to appreciate the depth of the corpus callosum within the interhemispheric fissure. In fact, there is a considerable amount of cortical tissue on the medial surface of the hemispheres, as represented by the frontal, parietal, and occipital lobes (see the next illustration).

In this specimen, the blood vessels supplying the medial aspect of the hemispheres are present. These vessels are the pericallosal arteries, a continuation of the anterior cerebral arteries (to be fully described with Figure 58 and Figure 61; see also Figure 70B). It should also be noted that the cerebral ventricles are located below (i.e., inferior to) the corpus callosum (see Figure 17 and Figure 19A).

The other white matter bundles, the association and projection fibers, will be discussed with other photographic views of the brain (see Figure 19A and Figure 19B). The anterior commissure is an older and smaller commissure connecting the anterior portions of the temporal lobe and limbic structures (see Figure 70A).

The clinical aspect of the corpus callosum is discussed with Figure 19A.

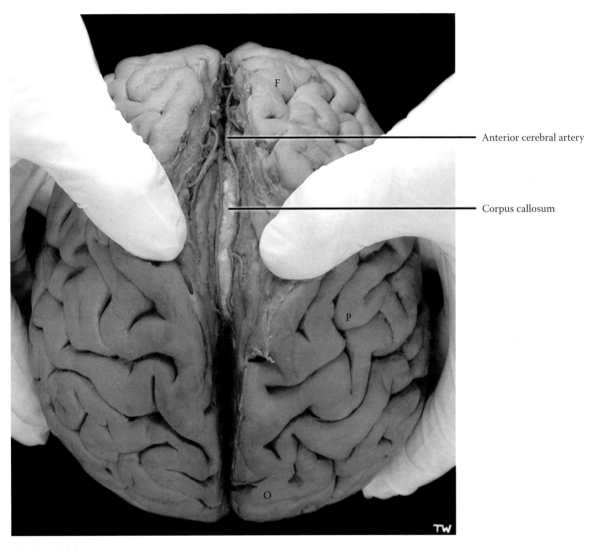

Anterior cerebral artery

Corpus callosum

F = Frontal lobe
P = Parietal lobe
O = Occipital lobe

FIGURE 16: Cerebral Hemispheres 6 — Superior View (photograph)

FIGURE 17
CEREBRAL HEMISPHERES 7

CEREBRAL HEMISPHERES: MEDIAL (PHOTOGRAPHIC) VIEW

This view of the brain sectioned in the midline (mid-sagittal plane) is probably the most important view for understanding the gross anatomy of the hemispheres, the diencephalon, the brainstem, and the ventricles. The section has divided the corpus callosum, gone in between the thalamus of each hemisphere (through the third ventricle), and passed through all parts of the brainstem.

The medial aspects of the lobes of the brain are now in view. The central fissure does extend onto this part of the brain (although not as deep as on the dorsolateral surface). The medial surface of the frontal lobe is situated anterior to the fissure; the inferior gyri of the frontal lobe sit on the bone that separates the anterior cranial fossa from the orbits (see Figure 15A and Figure 15B). The parietal lobe lies between the central fissure and the deep **parieto-occipital fissure**. The **occipital lobe** is now visible, posterior to this fissure. The main fissure that divides this lobe is the calcarine fissure (see Figure 41B); the primary visual area, commonly called **area 17** is situated along its banks (see Figure 41A and Figure 41B).

The corpus callosum in this specimen has the expected "white matter" appearance. Inside each cerebral hemisphere is a space filled with CSF, the lateral ventricle (see Figure 20A and Figure 20B). The **septum pellucidum**, a membranous septum that divides the anterior portions of the lateral ventricles of one hemisphere from that of the other side, has been torn during dissection, revealing the lateral ventricle of one hemisphere behind it (see Figure OL and Figure 28A). The fornix, a fiber tract of the limbic system, is located in the free lower edge of the septum. Above the corpus callosum is the **cingulate gyrus,** an important gyrus of the limbic system (see Figure 70A).

The sagittal section goes through the midline third ventricle (see Figure OA, Figure 9A, Figure 20A, and Figure 20B), thereby revealing the diencephalic region. (This region is shown at a higher magnification in Figure 41B). On this medial view, the thalamic portion of the diencephalon is separated from the hypothalamic part by a groove, the **hypothalamic sulcus**. This sulcus starts at the foramen of Monro (the interventricular foramen, discussed with the ventricles, see Figure 20A and Figure 20B) and ends at the aqueduct of the midbrain. The optic chiasm is found at the anterior aspect of the hypothalamus, and behind it is the mammillary body (see Figure 15B).

The three parts of the brainstem can be distinguished on this view — the midbrain, the pons with its bulge anteriorly, and the medulla (refer to the ventral views shown in Figure 6 and Figure 7). Through the midbrain is a narrow channel for CSF, the aqueduct of the midbrain (see Figure 20A and Figure 20B). The midbrain (behind the aqueduct) includes the superior and inferior colliculi, referred to as the tectum (see Figure 9A, Figure 10, and Figure 18).

The aqueduct connects the third ventricle with the fourth ventricle, a space with CSF that separates the pons and medulla from the cerebellum (see Figure 20A and Figure 20B). CSF escapes from the ventricular system at the bottom of the fourth ventricle through the foramen of Magendie (see Figure 21), and the ventricular system continues as the narrow central canal of the spinal cord (see Figure 4).

The cerebellum lies behind (or above) the fourth ventricle. It has been sectioned through its midline portion, the **vermis** (see Figure 54). Although it is not necessary to name all of its various parts, it is useful to know two of them — the lingula and the nodulus. (The reason for this will become evident when describing the cerebellum, see Figure 54). The tonsil of the cerebellum can also be seen in this view (not labeled, see Figure 9B and Figure 56).

The cut edge of the tentorium cerebelli, the other main fold of the dura, is seen separating the cerebellum from the occipital lobe. One of the dural venous sinuses, the straight sinus, runs in the midline of the tentorium (see next illustration). This view clarifies the separation of the supratentorial space, namely the cerebral hemispheres, from the infratentorial space, the brainstem, and the cerebellum in the posterior cranial fossa.

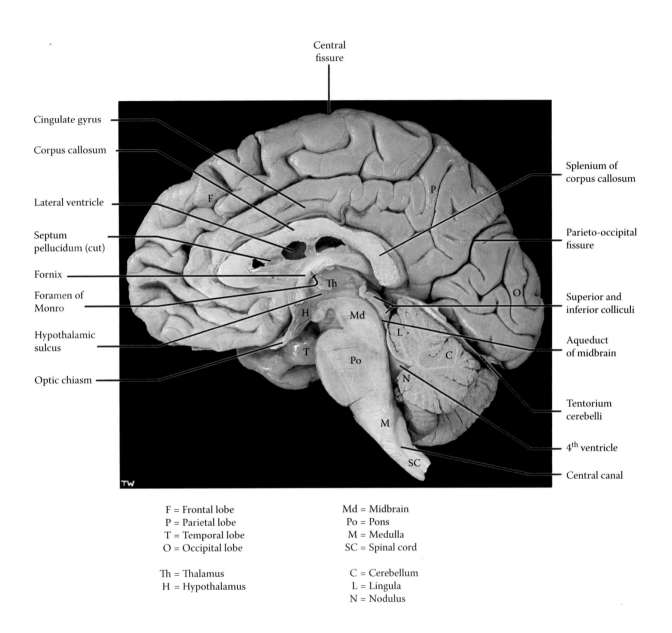

Central
fissure

Cingulate gyrus

Corpus callosum

Lateral ventricle

Septum
pellucidum (cut)

Fornix

Foramen of
Monro

Hypothalamic
sulcus

Optic chiasm

Splenium of
corpus callosum

Parieto-occipital
fissure

Superior and
inferior colliculi

Aqueduct
of midbrain

Tentorium
cerebelli

4th ventricle

Central canal

F = Frontal lobe
P = Parietal lobe
T = Temporal lobe
O = Occipital lobe

Th = Thalamus
H = Hypothalamus

Md = Midbrain
Po = Pons
M = Medulla
SC = Spinal cord

C = Cerebellum
L = Lingula
N = Nodulus

FIGURE 17: Cerebral Hemispheres 7 — Medial View (photograph)

FIGURE 18
CEREBRAL HEMISPHERES 8

MRI: T1 SAGITTAL VIEW (RADIOGRAPH)

This radiological image, obtained by magnetic resonance imaging (**MRI**), shows the brain as clearly as the actual brain itself (review the NOTE on radiologic imaging with Figure 3). This is the way the brain will be seen in the clinical setting. The view presented is called a **T1-weighted** image. Note that the CSF is dark in this image, including the ventricles, the subarachnoid space, and cisterns (see Figure 21). The bones (tables) of the skull are visible as a dark space, while the bone marrow, including its replacement by fatty tissue, and layers of soft tissue (and fatty tissue) of the scalp are well demarcated (white). The superior sagittal sinus can also be seen (see Figure 13 and Figure 21).

The various structures of the brain can easily be identified by comparing this view with the photographic view of the brain shown in the previous illustration, including the lobes of the brain. The corpus callosum can be easily identified, with the cingulate gyrus just above it and the lateral ventricle just below it (see also Figure 30). Various fissures (e.g., parieto-occipital, calcarine) can also be identified along with some cortical gyri (e.g., area 17, see Figure 41B). The space below the occipital lobe is occupied by the tentorium cerebelli (discussed with Figure 15B); the straight sinus, one of the dural venous sinuses, runs in the midline of the tentorium (see Figure 21).

The thalamus (the diencephalon) is seen below the lateral ventricle, and the tract immediately above it is the fornix (see Figure 70A). The structure labeled septum pellucidum separates the lateral ventricles of the hemispheres from each other (shown clearly in Figure 28A and Figure 30). The pineal is visible on this radiograph at the posterior end of the thalamus (just below the splenium of the corpus callosum); the pineal gland is cystic in this case, making it easy to identify. The pituitary gland is situated within the pituitary bony fossa, the sella turcica (see Figure 21).

Below the thalamus is the brainstem — its three parts, midbrain, pons, and medulla, can be identified. The tectum (with its four colliculi) is seen behind the aqueduct of the midbrain (see Figure 21). Posterior to the tectum is a CSF cistern (see Figure 28A, the guadrigeminal cistern). The fourth ventricle separates the cerebellum from the pons and medulla. The medulla ends at the foramen magnum and becomes the spinal cord.

The cerebellar folia are quite distinct on this image. The location of the cerebellar tonsil(s) should be noted, adjacent to the medulla and immediately above the foramen magnum, the "opening" at the base of the skull (see discussion on tonsillar herniation with Figure 9). The location of the cerebello-medullary cistern, the **cisterna magna**, behind the medulla and just above the foramen magnum is easily seen (see Figure 3 and Figure 21).

The remaining structures are those of the nose and mouth, which are not within our subject matter in this atlas.

CLINICAL ASPECT

This is a most important view for viewing the brain in the clinical setting. Abnormalities of structures, particularly in the posterior cranial fossa, can be easily visualized. Displacement of the brainstem into the foramen magnum because of a developmental disorder, known as an **Arnold-Chiari malformation**, will cause symptoms related to compression of the medulla at that level; in addition, there may be blockage of the CSF flow causing hydrocephalus (see Figure 21).

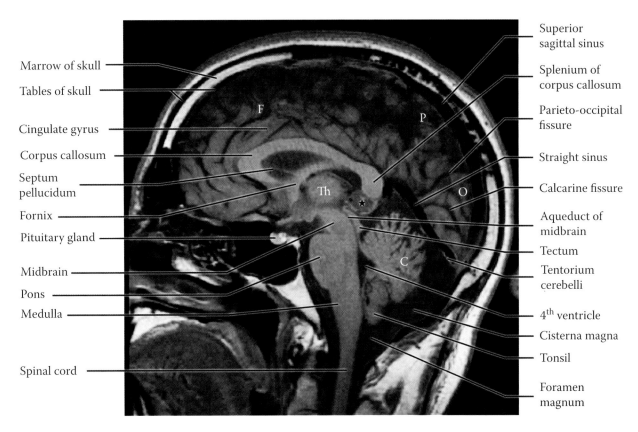

Marrow of skull

Tables of skull

Cingulate gyrus

Corpus callosum

Septum
pellucidum

Fornix

Pituitary gland

Midbrain

Pons

Medulla

Spinal cord

Superior
sagittal sinus

Splenium of
corpus callosum

Parieto-occipital
fissure

Straight sinus

Calcarine fissure

Aqueduct of
midbrain

Tectum

Tentorium
cerebelli

4th ventricle

Cisterna magna

Tonsil

Foramen
magnum

F = Frontal lobe
P = Parietal lobe
O = Occipital lobe

Th = Thalamus

C = Cerebellum

★ = Pineal cyst

FIGURE 18: Cerebral Hemispheres 8 — MRI: Sagittal View (radiograph)

FIGURE 19A
CEREBRAL HEMISPHERES 9

WHITE MATTER: MEDIAL DISSECTED VIEW — CORPUS CALLOSUM (PHOTOGRAPH)

The structures that are found within the depths of the cerebral hemispheres include the white matter, the cerebral ventricles, and the basal ganglia (see Figure OA and Figure OL). The white matter consists of the myelinated axonal fibers connecting brain regions. In the spinal cord these were called tracts; in the hemispheres these bundles are classified in the following way (also discussed with Figure 16) — association bundles, projections fibers, and commissural connections.

The dissection of this specimen needs some explanation. The brain is again seen from the medial view. (Its anterior aspect is on the left side of this photograph.) Cortical tissue has been removed from a brain (such as the one shown in Figure 17), using blunt dissection techniques. If done successfully, the fibers of the corpus callosum can be followed, as well as other white matter bundles (see Figure 19B). These fibers intermingle with other fiber bundles that make up the mass of white matter in the depth of the hemisphere.

The corpus callosum is the massive commissure of the forebrain, connecting homologous regions of the two hemispheres of the cortex across the midline (see also Figure 16). This dissection shows the white matter of the corpus callosum, followed to the cortex. In the midline, the thickened anterior aspect of the corpus callosum is called the genu, and the thickened posterior portion is the splenium (neither has been labeled).

If one looks closely, looping U-shaped bundles of fibers can be seen connecting adjacent gyri; these are part of the local association fibers.

The lateral ventricle is situated under the corpus callosum, while the diencephalon (the thalamus) is below the ventricle. Inside the anterior horn of the ventricle (see Figure 20A) there is a bulge that is formed by the head of the caudate nucleus; the caudate bulge is also seen on horizontal views of the brain (see Figure 27 and Figure 28A).

CLINICAL ASPECT

Although the connections of the corpus callosum are well described, its function under normal conditions is hard to discern. In rare cases, persons are born without a corpus callosum, a condition called agenesis of the corpus callosum, and these individuals as children and adults usually cannot be distinguished from normal individuals, unless specific testing is done.

The corpus callosum has been sectioned surgically in certain individuals with intractable epilepsy, that is, epilepsy which has not been controllable using anti-convulsant medication. The idea behind this surgery is to stop the spread of the abnormal discharges from one hemisphere to the other. Generally, the surgery has been helpful in well-selected cases, and there is apparently no noticeable change in the person, nor in his or her level of brain function.

Studies done in these individuals have helped to clarify the role of the corpus callosum in normal brain function. Under laboratory conditions, it has been possible to demonstrate in these individuals how the two hemispheres of the brain function independently, after the sectioning of the corpus callosum. These studies show how each hemisphere responds differently to various stimuli, and the consequences in behavior of the fact that information is not getting transferred from one hemisphere to the other hemisphere.

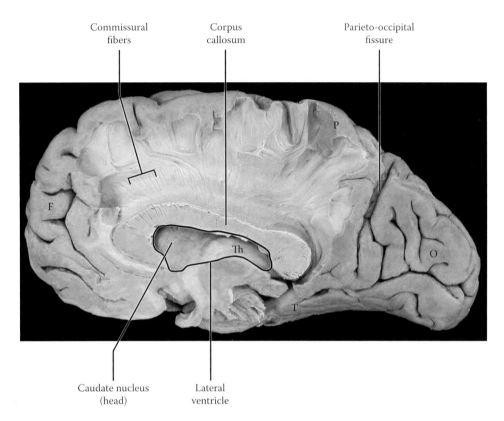

Commissural   Corpus   Parieto-occipital
fibers   callosum   fissure

F = Frontal lobe
P = Parietal lobe
T = Temporal lobe
O = Occipital lobe

Th = Thalamus (cut)

Caudate nucleus   Lateral
(head)   ventricle

FIGURE 19A: Cerebral Hemispheres 9 — Medial Dissected View: Corpus Callosum (photograph)

FIGURE 19B
CEREBRAL HEMISPHERES 10

WHITE MATTER: LATERAL DISSECTED VIEW
— ASSOCIATION BUNDLES
(PHOTOGRAPH)

The dorsolateral aspect of the brain is being viewed in this photograph (see Figure 14A). The lateral fissure has been opened, with the temporal lobe below; deep within the lateral fissure is the insula (see Figure 14B and Figure 39).

Under the cerebral cortex is the white matter of the brain. It is possible to dissect various fiber bundles (not easily) using a blunt instrument (e.g., a wooden tongue depressor). Some of these, functionally, are the association bundles, fibers that interconnect different parts of the cerebral cortex on the same side (classified with Figure 16).

This specimen has been dissected to show two of the association bundles within the hemispheres. The **superior longitudinal fasciculus** (fasciculus is another term for a bundle of axons) interconnects the posterior parts of the hemisphere (e.g., the parietal lobe) with the frontal lobe. There are other association bundles present in the hemispheres connecting the various portions of the cerebral cortex. The various names of these association bundles usually are not of much importance in a general introduction to the CNS and only will be mentioned if need be.

Shorter association fibers are found between adjacent gyri (see previous illustration).

These association bundles are extremely important in informing different brain regions of ongoing neuronal processing, allowing for integration of our activities (for example sensory with motor and limbic). One of the major functions of these association bundles in the human brain seems to be bringing information to the frontal lobes, especially to the prefrontal cortex, which acts as the "executive director" of brain activity (see Figure 14A).

One of the most important association bundles, the **arcuate bundle**, connects the two language areas. It connects Broca's area anteriorly with Wernicke's area in the superior aspect of the temporal lobe, in the dominant (left) language hemisphere (see Figure 14A).

CLINICAL ASPECT

Damage to the arcuate bundle due to a lesion, such as an infarct or tumor, in that region leads to a specific disruption of language, called conduction aphasia. **Aphasia** is a general term for a disruption or disorder of language. In conduction aphasia, the person has normal comprehension (intact Wernicke's area) and fluent speech (intact Broca's area). The only language deficit seems to be an inability to repeat what has been heard. This is usually tested by asking the patient to repeat single words or phrases whose meaning cannot be readily understood (e.g., the phrase "no ifs, ands, or buts"). There is some uncertainty whether this is in fact the only deficit, since isolated lesions of the arcuate bundle have not yet been described.

Superior longitudinal
fasciculus

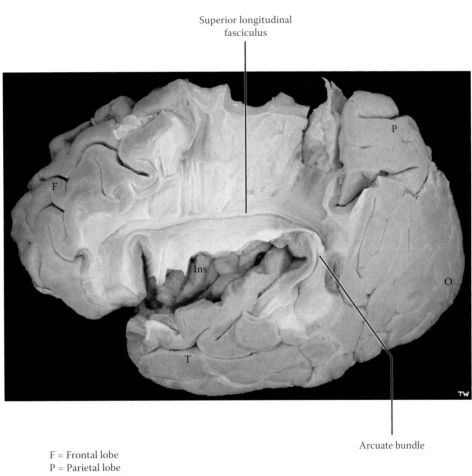

Arcuate bundle

F = Frontal lobe
P = Parietal lobe
T = Temporal lobe
O = Occipital lobe

Ins = Insula

FIGURE 19B: Cerebral Hemispheres 10 — Lateral Dissected View: Association Bundles (photograph)

FIGURE 20A
VENTRICLES 1

VENTRICLES: LATERAL VIEW

The ventricles are cavities within the brain filled with **CSF**. The formation, circulation, and locations of the CSF will be explained with Figure 21.

The ventricles of the brain are the spaces within the brain that remain from the original neural tube, the tube that was present during development. The cells of the nervous system, both neurons and glia, originated from a germinal matrix that was located adjacent to the lining of this tube. The cells multiply and migrate away from the walls of the neural tube, forming the nuclei and cerebral cortex. As the nervous system develops, the mass of tissue grows and the size of the tube diminishes, leaving various spaces in different parts of the nervous system (see Figure OA and Figure OL).

The parts of the tube that remain in the hemispheres are called the cerebral ventricles, also called the **lateral ventricles**. The lateral ventricle of the hemispheres, shown here from the lateral perspective, is shaped like the letter C (in reverse); it curves posteriorly and then enters into the temporal lobe. Its various parts are: the **anterior horn**, which lies deep to the frontal lobes; the central portion, or **body**, which lies deep to the parietal lobes; the **atrium** or **trigone**, where it widens and curves and then enters into the temporal lobe as the **inferior horn**. In addition, there may be an extension into the occipital lobes, the **occipital** or posterior horn, and its size varies. These lateral ventricles are also called ventricles **I and II** (assigned arbitrarily).

Each lateral ventricle is connected to the midline third ventricle by an opening, the **foramen of Monro** (interventricular foramen — seen in the medial view of the brain, Figure 17 and Figure 41B; also Figure 20B and Figure 21). The **third ventricle** is a narrow slit-like ventricle between the thalamus on either side and could also be called the ventricle of the diencephalon (see Figure

9B). Sectioning through the brain in the midline (as in Figure 17) passes through the third ventricle. Note that the "hole" in the middle of the third ventricle represents the interthalamic adhesion, linking the two thalami across the midline (see Figure 6; discussed with Figure 11; see also Figure 41B).

The ventricular system then narrows considerably as it goes through the midbrain and is now called the **aqueduct of the midbrain**, the cerebral aqueduct, or the aqueduct of Sylvius (see Figure 17, Figure 18, and Figure 20B; also Figure 41B and Figure 65). In the hindbrain region, the area consisting of pons, medulla, and cerebellum, the ventricle widens again to form the **fourth ventricle** (see Figure 17, Figure 20B, and Figure 66). The channel continues within the CNS and becomes the very narrow **central canal** of the spinal cord (see Figure 17, Figure 20B, Figure 21, and Figure 69).

Specialized tissue, the **choroid plexus,** the tissue responsible for the formation of the CSF, is located within the ventricles. It is made up of the lining cells of the ventricles, the ependyma, and pia with blood vessels (discussed with Figure 21). This diagram shows the choroid plexus in the body and inferior horn of the lateral ventricle; the tissue forms large invaginations into the ventricles in each of these locations (see Figure 27 and Figure 74 for a photographic view of the choroid plexus). The blood vessel supplying this choroid plexus comes from the middle cerebral artery (shown here schematically; see Figure 58). Choroid plexus is also found in the roof of the third ventricle and in the lower half of the roof of the fourth ventricle (see Figure 21).

CSF flows through the ventricular system, from the lateral ventricles, through the interventricular foramina into the third ventricle, then through the narrow aqueduct and into the fourth ventricle (see Figure 21). At the bottom of the fourth ventricle, CSF flows out of the ventricular system via the major exit, the foramen of Magendie, in the midline, and enters the subarachnoid space. There are two additional exits of the CSF laterally from the fourth ventricle — the foramina of Luschka, which will be seen in another perspective (in the next illustration).

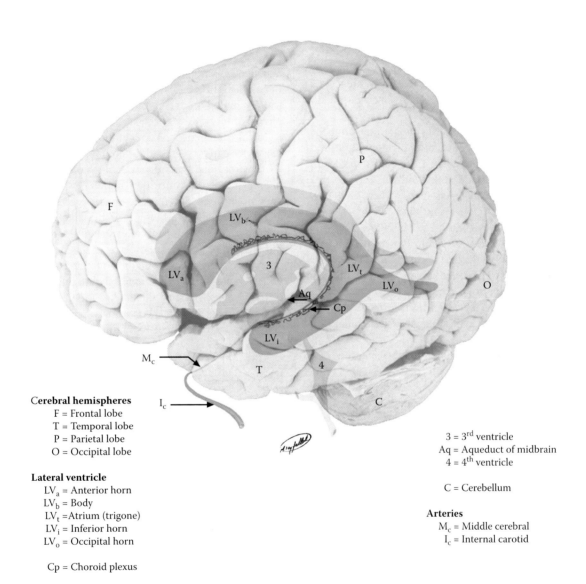

Cerebral hemispheres
  F = Frontal lobe
  T = Temporal lobe
  P = Parietal lobe
  O = Occipital lobe

Lateral ventricle
  LV_a = Anterior horn
  LV_b = Body
  LV_t = Atrium (trigone)
  LV_i = Inferior horn
  LV_o = Occipital horn

  Cp = Choroid plexus

  3 = 3rd ventricle
  Aq = Aqueduct of midbrain
  4 = 4th ventricle

  C = Cerebellum

Arteries
  M_c = Middle cerebral
  I_c = Internal carotid

FIGURE 20A: Ventricles 1 — Lateral View

FIGURE 20B
VENTRICLES 2

VENTRICLES: ANTERIOR VIEW

The ventricular system is viewed from the anterior perspective in this illustration. One can now see both lateral ventricles and the short interventricular foramen (of Monro) on both sides, connecting each lateral ventricle with the midline third ventricle (see Figure 28B and Figure 29). It is important to note that the thalamus (diencephalon) is found on either side of the third ventricle (see also Figure 9A).

CSF flows from the third ventricle into the aqueduct of the midbrain. This ventricular channel continues through the midbrain, and then CSF enters the fourth ventricle, which also straddles the midline. The ventricle widens into a diamond-shaped space, when seen from the anterior perspective. This ventricle separates the pons and medulla anteriorly from the cerebellum posteriorly. The lateral recesses carry CSF into the cisterna magna, the CSF cistern outside the brain (see Figure 21), through the **foramina of Luschka**, the lateral apertures, one on each side. The space then narrows again, becoming a narrow channel at the level of the lowermost medulla, which continues as the central canal of the spinal cord (see Figure 4).

Sections of the brain in the coronal (frontal) axis, if done at the appropriate plane, will reveal the spaces of the lateral ventricles within the hemispheres (see Figure 29 and Figure 74). Likewise, sections of the brain in the horizontal axis, if done at the appropriate level, will show the ventricular spaces of the lateral and third ventricles (see Figure 27). These can also be visualized with radiographic imaging (CT and MRI, see Figure 28A, Figure 28B, and Figure 30).

CLINICAL ASPECT

It is quite apparent that the flow of CSF can be interrupted or blocked at various key points within the ventricular system. The most common site is the aqueduct of the midbrain, the cerebral aqueduct (of Sylvius). Most of the CSF is formed upstream, in the lateral (and third) ventricles. A blockage at the narrowest point, at the level of the aqueduct of the midbrain, will create a damming effect. In essence, this causes a marked enlargement of the ventricles, called **hydrocephalus**. The CSF flow can be blocked for a variety of reasons, such as developmentally, following meningitis, or by a tumor in the region. Enlarged ventricles can be seen with brain imaging (e.g., CT scan).

Hydrocephalus in infancy occurs not uncommonly, for unknown reasons. Since the sutures of the infant's skull are not yet fused, this leads to an enlargement of the head and may include the bulging of the anterior fontanelle. Clinical assessment of all infants should include measuring the size of the head and charting this in the same way one charts height and weight. Untreated hydrocephalus will eventually lead to a compression of the nervous tissue of the hemispheres and damage to the developing brain. Clinical treatment of this condition, after evaluation of the causative factor, includes shunting the CSF out of the ventricles into one of the body cavities.

In adults, hydrocephalus caused by a blockage of the CSF flow leads to an increase in intracranial pressure (discussed in the introduction to Section C). Since the sutures are fused, skull expansion is not possible. The cause in adults is usually a tumor, and in addition to the specific symptoms, the patient will most commonly complain of headache, often in the early morning.

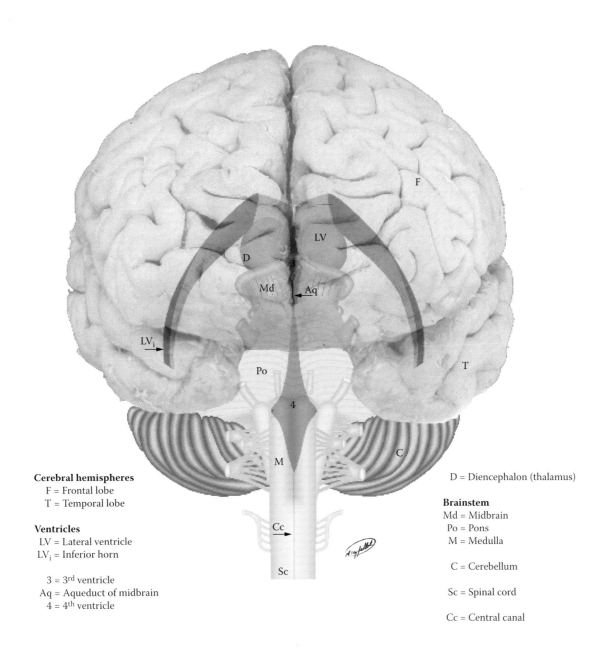

Cerebral hemispheres
  F = Frontal lobe
  T = Temporal lobe

Ventricles
  LV = Lateral ventricle
  LV$_i$ = Inferior horn

  3 = 3rd ventricle
  Aq = Aqueduct of midbrain
  4 = 4th ventricle

D = Diencephalon (thalamus)

Brainstem
Md = Midbrain
Po = Pons
M = Medulla

C = Cerebellum

Sc = Spinal cord

Cc = Central canal

FIGURE 20B: Ventricles 2 — Anterior View

FIGURE 21
VENTRICLES 3

CSF CIRCULATION

This is a representation of the production, circulation, and reabsorption of CSF, the ventricles of the brain, and the subarachnoid spaces around the brain, enlargements of which are called cisterns.

The ventricles of the brain are lined with a layer of cells known as the ependyma. In certain loci within each of the ventricles, the ependymal cells and the pia meet, thus forming the **choroid plexus**, which invaginates into the ventricle. Functionally, the choroid plexus has a vascular layer, i.e., the pia, on the inside, and the ependymal layer on the ventricular side. CSF is actively secreted by the choroid plexus. The blood vessels of the choroid plexus are freely permeable, but there is a cellular barrier between the interior of the choroid plexus and the ventricular space — the **blood-CSF barrier** (B-CSF-B). The barrier consists of tight junctions between the ependymal cells that line the choroid plexus. CSF is actively secreted by the choroid plexus, and an enzyme is involved. The ionic and protein composition of CSF is different from that of serum.

Choroid plexus is found in the lateral ventricles (see Figure 20A), the roof of the third ventricle, and the lower half of the roof of the fourth ventricle. CSF produced in the lateral ventricles flows via the foramen of Monro (from each lateral ventricle) into the third ventricle, and then through the aqueduct of the midbrain into the fourth ventricle. CSF leaves the ventricular system from the fourth ventricle, as indicated schematically in the diagram. In the intact brain, this occurs via the medially placed foramen of Magendie and the two laterally placed foramina of Luschka (as described in the previous illustrations) and enters the enlargement of the subarachnoid space under the cerebellum, the cerebello-medullary cistern, the **cisterna magna**. The cisterna magna is found inside the skull, just above the foramen magnum (see Figure 18).

CSF flows through the subarachnoid space, between the pia and arachnoid. The CSF fills the enlargements of the subarachnoid spaces around the brainstem — the various cisterns (each of which has a separate name). The CSF then flows upward around the hemispheres of the brain and is found in all the gyri and fissures. CSF also flows in the subarachnoid space downward around the spinal cord to fill the lumbar cistern (see Figure 1, Figure 2C, and Figure 3).

This slow circulation is completed by the return of CSF to the venous system. The return is through the **arachnoid villi**, protrusions of arachnoid into the venous sinuses of the brain, particularly along the superior sagittal sinus (see Figure 18). These can sometimes be seen on the specimens as collections of villi, called **arachnoid granulations**, on the surface of the brain lateral to the interhemispheric fissure.

There is no real barrier between the intercellular tissue of the brain and the CSF through the ependyma lining the ventricles (at all sites other than the choroid plexus). Therefore, substances found in detectable amounts in the intercellular spaces of the brain may be found in the CSF.

On the other hand, there is a real barrier, both structural and functional, between the blood vessels and the brain tissue. This is called the **blood-brain barrier** (**BBB**), and it is situated at the level of the brain capillaries where there are tight junctions between the endothelial cells. Only oxygen, carbon dioxide, glucose, and other (select) small molecules are normally able to cross the BBB.

CLINICAL ASPECT

The CSF flows down around the spinal cord and into the lumbar cistern. Sampling of CSF for clinical disease, including inflammation of the meninges (meningitis), is performed in the lumbar cistern (see Figure 1, Figure 2C, and Figure 3). The CSF is then analyzed, for cells, proteins, and other constituents to assist or confirm a diagnosis.

The major arteries of the circle of Willis travel through the subarachnoid space (see Figure 58). An aneurysm of these arteries that "bursts" (discussed with Figure 59A) will do so within the CSF space; this is called a subarachnoid hemorrhage.

Hydrocephalus has been discussed with the previous illustration.

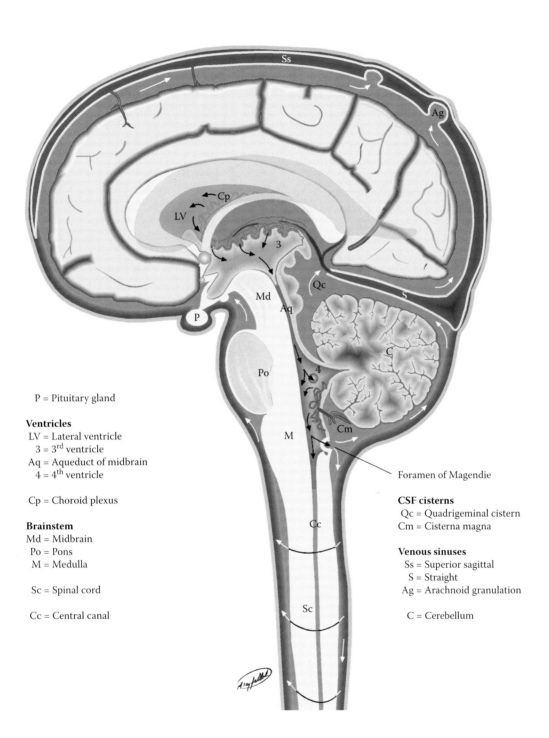

P = Pituitary gland

Ventricles
LV = Lateral ventricle
 3 = 3rd ventricle
Aq = Aqueduct of midbrain
 4 = 4th ventricle

Cp = Choroid plexus

Brainstem
Md = Midbrain
Po = Pons
M = Medulla

Sc = Spinal cord

Cc = Central canal

Foramen of Magendie

CSF cisterns
 Qc = Quadrigeminal cistern
Cm = Cisterna magna

Venous sinuses
 Ss = Superior sagittal
 S = Straight
Ag = Arachnoid granulation

 C = Cerebellum

FIGURE 21: Ventricles 3 — CSF Circulation

FIGURE 22
BASAL GANGLIA 1

BASAL GANGLIA: ORIENTATION

There are large collections of gray matter within the hemispheres, belonging to the forebrain, in addition to the white matter and the ventricles already described. These neuronal groups are collectively called the **basal ganglia**. Oftentimes the term **striatum** is used for the basal ganglia, but this term is not always used with neuroanatomical precision. Our understanding of the functional role of the basal ganglia is derived largely from disease states affecting these neurons. In general, humans with lesions in the basal ganglia have some form of motor dysfunction, a **dyskinesia**, that is, a movement disorder. But, as will be discussed, these neurons have connections with both neocortical and limbic areas, and are definitely involved in other brain functions.

The description of the basal ganglia will be done in a series of illustrations. This diagram is for orientation and terminology; the following diagrams will discuss more anatomical details and the functional aspects. The details of the connections and the circuitry involving the basal ganglia will be described in Section C (see Figure 52 and Figure 53).

From the strictly anatomical point of view, the basal ganglia are collections of neurons located within the hemispheres. Traditionally, this would include the **caudate** nucleus, the **putamen**, the **globus pallidus**, and the amygdala (see Figure OA and Figure OL). The caudate and putamen are also called the **neostriatum**; histologically these are the same neurons but in the human brain they are partially separated from each other by projection fibers (see Figure 26). The putamen and globus pallidus are anatomically grouped together in the human brain and are called the **lentiform** or **lenticular nucleus** because of the lens-like configuration of the two nuclei, yet these are functionally distinct.

The development of the human brain includes the evolution of a temporal lobe and many structures "migrate" into this lobe, including the lateral ventricle. The caudate nucleus organization follows the curvature of the lateral ventricle into the temporal lobe (see Figure OL and Figure 25).

These basal ganglia are involved in the control of complex patterns of motor activity, such as skilled movements (e.g., writing). There are two aspects to this involvement. The first concerns the initiation of the movement. The second concerns the quality of the performance of the motor task. It seems that different parts of the basal ganglia are concerned with how rapidly a movement is to be performed and the magnitude of the movement. In addition, some of the structures that make up the basal ganglia are thought to influence cognitive aspects of motor control, helping to plan the sequence of tasks needed for purposeful activity. This is sometimes referred to as the selection of motor strategies.

Functionally, the basal ganglia system acts as a sub-loop of the motor system by altering cortical activity (to be fully discussed with Figure 52 and Figure 53). In general terms, the basal ganglia receive much of their input from the cortex, from the motor areas, and from wide areas of association cortex, as well as from other nuclei of the basal ganglia system. There are intricate connections between the various parts of the system (see Figure 52), involving different neurotransmitters; the output is directed via the thalamus mainly to premotor, supplementary motor, and frontal cortical areas (see Figure 53).

The amygdala, also called the amygdaloid nucleus, is classically one of the basal ganglia, because it is a subcortical collection of neurons (in the temporal lobe, anteriorly, see Figure OL and Figure 25). All the connections of the amygdala are with limbic structures, and so the discussion of this nucleus will be done in Section D (see Figure 75A and Figure 75B).

There are now known to be other subcortical nuclei in the forebrain, particularly in the basal forebrain region. These have not been grouped with the basal ganglia and will be described with the limbic system (in Section D).

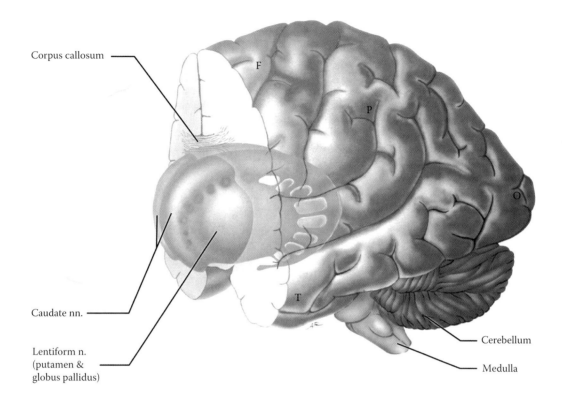

Corpus callosum

Caudate nn.

Lentiform n.
(putamen &
globus pallidus)

Cerebellum

Medulla

F = Frontal lobe
P = Parietal lobe
T = Temporal lobe
O = Occipital lobe

FIGURE 22: Basal Ganglia 1 — Orientation

FIGURE 23
BASAL GANGLIA 2

BASAL GANGLIA: NUCLEI — LATERAL VIEW

The basal ganglia, from the point of view of strict neuroanatomy, consist of three major nuclei in each of the hemispheres. (The reader is reminded that this illustration has been enlarged from the previous figure, and that these structures are located within the forebrain.)

- The **caudate**
- The **putamen**
- The **globus pallidus**

- The caudate nucleus is anatomically associated with the lateral ventricle and follows its curvature. It is described as having three portions (see Figure 25):
  - The head, located deep within the frontal lobe
  - The body, located deep in the parietal lobe
  - The tail, which goes in to the temporal lobe

The basal ganglia are shown in this illustration from the lateral perspective, as well as from above, allowing a view of the caudate nucleus of both sides. The various parts of the caudate nucleus are easily recognized — head, body, and tail. The head of the caudate nucleus is large and actually intrudes into the space of the anterior horn of the lateral ventricle (see Figure 27 and Figure 28A). The body of the caudate nucleus tapers and becomes considerably smaller and is found beside the body of the lateral ventricle (see Figure 29 and Figure 76). The tail follows the inferior horn of the lateral ventricle into the temporal lobe (see Figure 76). As the name implies, this is a slender extended group of neurons, even more difficult to identify in sections of the temporal lobe (see Figure 74).

The **lentiform** or lenticular nucleus, so named because it is lens-shaped, in fact is composed of two nuclei (see next illustration) — the **putamen** and the **globus pallidus**.

The lentiform nucleus is situated laterally and deep in the hemispheres, within the central white matter. Sections of the brain in the horizontal plane (see Figure 27) and in the coronal (frontal) plane (see Figure 29) show the location of the lentiform nucleus in the depths of the hemispheres, and this can be visualized with brain imaging (see Figure 28A and Figure 28B).

The lentiform (lenticular) nucleus is only a descriptive name, which means lens-shaped. The nucleus is in fact composed of two functionally distinct parts — the putamen laterally, and the globus pallidus medially (see Figure OA, Figure 27, and Figure 52). When viewing the basal ganglia from the lateral perspective, one sees only the putamen part (see Figure OL and Figure 73).

The caudate and the putamen contain the same types of neurons and have similar connections; often they are collectively called the **neostriatum**. Strands of neuronal tissue are often seen connecting the caudate nucleus with the putamen. A very distinct and important fiber bundle, the internal capsule, separates the head of the caudate nucleus from the lentiform nucleus (see next illustration). These fiber bundles "fill the spaces" in between the cellular strands.

ADDITIONAL DETAIL

The inferior or ventral portions of the putamen and globus pallidus are found at the level of the anterior commissure. Both have a limbic connection (discussed with Figure 80B). The amygdala, though part of the basal ganglia by definition, has its functional connections with the limbic system and will be discussed at that time (see Figure 75A and Figure 75B).

NOTE on terminology: Many of the names of structures in the neuroanatomical literature are based upon earlier understandings of the brain, with terminology that is often descriptive and borrowed from other languages. As we learn more about the connections and functions of brain areas, this terminology often seems awkward if not obsolete, yet it persists.

The term ganglia, in the strict use of the term, refers to a collection of neurons in the peripheral nervous system. Therefore, the anatomically correct name for the neurons in the forebrain should be the *basal nuclei*. Few texts use this term. Most clinicians would be hard-pressed to change the name from basal ganglia to something else, so the traditional name remains.

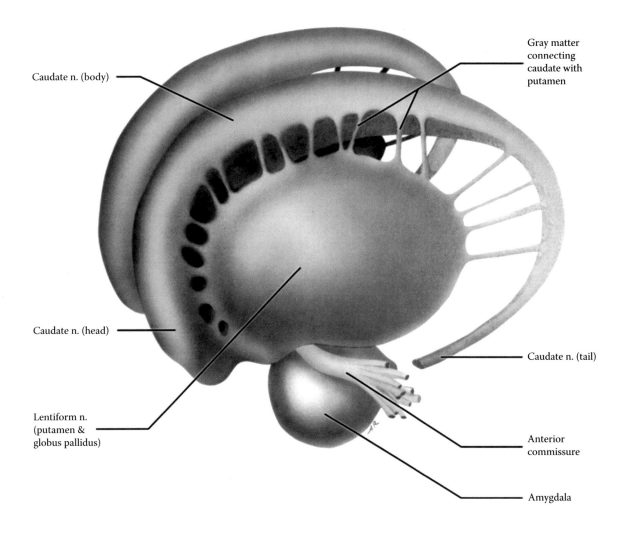

Caudate n. (body)

Gray matter connecting caudate with putamen

Caudate n. (head)

Caudate n. (tail)

Lentiform n. (putamen & globus pallidus)

Anterior commissure

Amygdala

FIGURE 23: Basal Ganglia 2 — Nuclei: Lateral View

FIGURE 24
BASAL GANGLIA 3

BASAL GANGLIA: NUCLEI — MEDIAL VIEW

This view has been obtained by removing all parts of the basal ganglia of one hemisphere, except the tail of the caudate and the amygdala. This exposes the caudate nucleus and the lentiform nucleus of the "distal" side; the lentiform nucleus is thus being visualized from a medial perspective.

The lentiform nucleus is now seen to be composed of its two portions, the putamen, laterally, and the globus pallidus, which is medially placed. In fact, the globus pallidus has two parts, an external (lateral) segment and an internal (medial) segment.

Functionally, the nuclei of the basal ganglia are organized in the following way. The input from the cerebral cortex and from other sources (thalamus, substantia nigra) is received by the caudate and putamen (see Figure 52). This information is relayed to the globus pallidus. It is composed of two segments, the medial and lateral segments, also known as internal and external segments, respectively. (These can also be seen in the horizontal section of the brain, see Figure 27). This subdivision of the globus pallidus is quite important functionally, as each of the segments has distinct connections. The globus pallidus, internal segment, is the major efferent nucleus of the basal ganglia (see Figure 53).

From the functional point of view, and based upon the complex pattern of interconnections, two other nuclei, which are not in the forebrain, should be included with the description of the basal ganglia — the **subthalamic nucleus** (part of the diencephalon), and the **substantia nigra** (located in the midbrain). The functional connections of these nuclei will be discussed as part of the motor systems (see Figure 52 and Figure 53).

A distinct collection of neurons is found in the ventral region of the basal ganglia — the **nucleus accumbens**. The nucleus accumbens is somewhat unique, in that it seems to consist of a mix of neurons from the basal ganglia and from the limbic structures in the region. This nucleus is involved with what is termed "reward" behavior and seems to be the part of the brain most implicated in drug addiction (discussed with the limbic system, see Figure 80B).

CLINICAL ASPECT

The functional role of this large collection of basal ganglia neurons is best illustrated by clinical conditions in which this system does not function properly. These disease entities manifest abnormal movements, such as chorea (jerky movements), athetosis (writhing movements), and tremors (rhythmic movements).

The most common condition that affects this functional system of neurons is **Parkinson's** disease. The person with this disease has difficulty initiating movements, the face takes on a mask-like appearance with loss of facial expressiveness, there is muscular rigidity, a slowing of movements (bradykinesia), and a tremor of the hands at rest, which goes away with purposeful movements (and in sleep). Some individuals with Parkinson's also develop cognitive and emotional problems, implicating these neurons in brain processes other than motor functions.

People with Parkinson's disease also develop **rigidity**. In rigidity, there is an increased resistance to passive movement of the limb, which involves both the flexors and extensors, and the response is not velocity dependent. There is no alteration of reflex responsiveness, nor is there clonus (discussed with Figure 49B). In this clinical state, the plantar response is normal (see Section B, Part III, Introduction).

The other major disease that affects the Basal Ganglia is **Huntington's Chorea**, an inherited degenerative condition. This disease, which starts in midlife, leads to severe motor dysfunction, as well as cognitive decline. The person whose name is most associated with this disease is Woody Guthrie, a legendary folk singer. There is now a genetic test for this disease that predicts whether the individual, with a family history of Huntington's, will develop the disease.

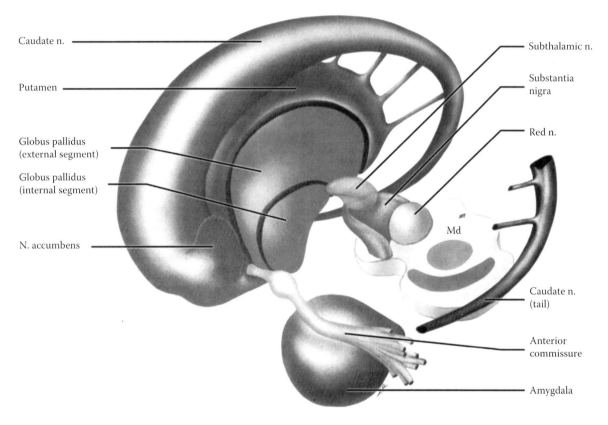

Caudate n.

Putamen

Globus pallidus
(external segment)

Globus pallidus
(internal segment)

N. accumbens

Subthalamic n.

Substantia
nigra

Red n.

Md

Caudate n.
(tail)

Anterior
commissure

Amygdala

Md = Midbrain

FIGURE 24: Basal Ganglia 3 — Nuclei: Medial View

FIGURE 25
BASAL GANGLIA 4

BASAL GANGLIA AND VENTRICLES

In humans, the three nuclei of the basal ganglia have a complex and finite arrangement in the hemispheres of the brain. Visualization of their location is made easier by understanding their relationship with the cerebral ventricles (see Figure OA and Figure OL).

The lateral ventricles of the hemispheres are shown in this view, from the lateral perspective (as in Figure OL and Figure 20A). The way in which all three parts of the caudate nucleus, the head, body, and tail, are situated adjacent to the lateral ventricle can be clearly seen, with the tail following the ventricle into the temporal lobe (see Figure 22 and also Figure 76).

The various parts of the basal ganglia include the caudate nucleus, the lentiform nucleus, and also the amygdala. The lentiform nucleus, including putamen and globus pallidus, is located deep within the hemispheres, not adjacent to the ventricle. This "nucleus" is found lateral to the thalamus, which locates the lentiform nucleus as lateral to third ventricle in a horizontal section of the brain (see Figure 27). The lentiform nucleus, actually the putamen, is seen in a dissection of the brain from the lateral perspective (see Figure 73).

In this diagram one can see that the caudate and the lentiform nuclei are connected anteriorly. In addition, there are connecting strands of tissue between the caudate and putamen. (These connecting strands have been shown in the previous diagrams.) As fiber systems develop, namely the projection fibers, these nuclei become separated from each other, specifically by the anterior limb of the internal capsule (see next illustration).

Again, it should be noted that basal ganglia occupy a limited area in the depths of the hemispheres. Sections taken more anteriorly or more posteriorly (see Figure 74), or above the ventricles, will not have any parts of these basal ganglia.

In summary, both the caudate and the lentiform nuclei are found below the plane of the corpus callosum. The head of the caudate nucleus and the lentiform nucleus are found at the same plane as the thalamus, as well as the anterior horns of the lateral ventricles (see Figure 27). As will be seen, this is also the plane of the lateral fissure and the insula. These are important aspects of neuroanatomy to bear in mind when the brain is seen neuroradiologically with CT and MRI (see Figure 28A and Figure 28B).

From this lateral perspective, the third ventricle, occupying the midline, is almost completely hidden from view by the thalamus, which lies adjacent to this ventricle and forms its lateral boundaries (see Figure 9, Figure OA, Figure OL, and Figure 20B).

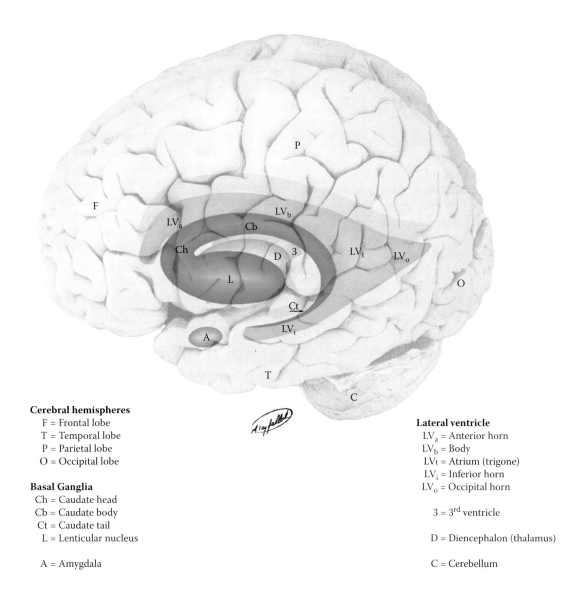

Cerebral hemispheres
 F = Frontal lobe
 T = Temporal lobe
 P = Parietal lobe
 O = Occipital lobe

Basal Ganglia
 Ch = Caudate head
 Cb = Caudate body
 Ct = Caudate tail
 L = Lenticular nucleus

 A = Amygdala

Lateral ventricle
 LV_a = Anterior horn
 LV_b = Body
 LV_t = Atrium (trigone)
 LV_i = Inferior horn
 LV_o = Occipital horn

 3 = 3rd ventricle

 D = Diencephalon (thalamus)

 C = Cerebellum

FIGURE 25: Basal Ganglia 4 — Nuclei and Ventricles

FIGURE 26
BASAL GANGLIA 5

INTERNAL CAPSULE: PROJECTION FIBERS

The white matter bundles that course between parts of the basal ganglia and the thalamus are collectively grouped together and called the **internal capsule**. These are projection fibers, axons going to and coming from the cerebral cortex. The internal capsule is defined as a group of fibers located at a specific plane within the cerebral hemispheres in a region that is situated between the head of the caudate, the lentiform, and the thalamus (see Figure OA, Figure OL, and Figure 25).

The internal capsule has three parts:

- **Anterior limb**. A group of fibers separates the two parts of the neostriatum from each other, the head of the caudate from the putamen. This fiber system carries axons that are coming down from the cortex, mostly to the pontine region, which are then relayed to the cerebellum (see Figure 55). Other fibers in the anterior limb relay from the thalamus to the cingulate gyrus (see Figure 77A) and to the prefrontal cortex (see Figure 77B).
- **Posterior limb**. The fiber system that runs between the thalamus (medially) and the lentiform nucleus (laterally) is the posterior limb of the internal capsule. The posterior limb carries three extremely important sets of fibers
  - Sensory information from thalamus to cortex, as well as the reciprocal connections from cortex to thalamus.
  - Most of the descending fibers to the brainstem (cortico-bulbar, see Figure 46) and spinal cord (cortico-spinal, see Figure 45).
  - In addition, there are fibers from other parts of the cortex that are destined for the cerebellum, after synapsing in the pontine nuclei (discussed with Figure 55).
- **The genu**. In a horizontal section, the internal capsule (of each side) is seen to be V-shaped (see Figure 27). Both the anterior limb and the posterior limb have been described — the bend of the "V" is called the genu, and it points medially (also seen with neuroradiological imaging, both CT, see Figure 28A, and MRI, see Figure 28B).

The internal capsule fibers are also seen from the medial perspective in a dissection in which the thalamus has been removed (see Figure 70B). The fibers of the internal capsule are also shown in a dissection of the brain from the lateral perspective, just medial to the lentiform nucleus (see Figure 73).

Below the level of the internal capsule is the midbrain. The descending fibers of the internal capsule continue into the midbrain and are next located in the structure called the cerebral peduncle of the midbrain (see Figure 6, Figure 7, Figure 45, and Figure 46; also seen in cross-sections of the brainstem in Figure 65).

In summary, at the level of the internal capsule, there are both the ascending fibers from thalamus to cortex, as well as descending fibers from widespread areas of the cerebral cortex to the thalamus, the brainstem and cerebellum, and the spinal cord. These ascending and descending fibers are all called projection fibers (discussed with Figure 16). This whole fiber system is sometimes likened to a funnel, with the top of the funnel being the cerebral cortex and the stem the cerebral peduncle. The base of the funnel, where the funnel narrows, would be the internal capsule. The main point is that the various fiber systems, both ascending and descending, are condensed together in the region of the internal capsule.

Note to the Learner: Many students have difficulty understanding the concept of the internal capsule, and where it is located. One way of thinking about it is to look at the projection fibers as a busy two-lane highway. The internal capsule represents one section of this pathway, where the roadway is narrowed.

CLINICAL ASPECT

The posterior limb of the internal capsule is a region that is apparently particularly vulnerable for small vascular bleeds. These small hemorrhages destroy the fibers in this region. Because of the high packing density of the axons in this region, a small lesion can cause extensive disruption of descending motor or ascending sensory pathways. This is one of the most common types of cerebrovascular accidents, commonly called a "stroke." (The details of the vascular supply to this region will be discussed with Figure 62.)

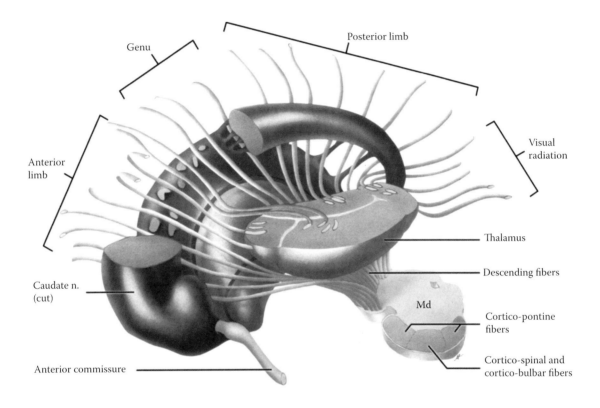

FIGURE 26: Basal Ganglia 5 — Internal Capsule and Nuclei

FIGURE 27
BASAL GANGLIA 6

HORIZONTAL SECTION OF HEMISPHERES
(PHOTOGRAPHIC VIEW)

In this photograph, the brain has been sectioned in the horizontal plane. From the dorsolateral view (the small figure on the upper left), the level of the section is just above the lateral fissure and at a slight angle downward from front to back. Using the medial view of the brain (the figure on the upper right), the plane of section goes through the anterior horn of the lateral ventricle, the thalamus and the occipital lobe.

This brain section exposes the white matter of the hemispheres, the basal ganglia, and parts of the ventricular system. Understanding this particular depiction of the brain is vital to the study of the forebrain. The structures seen in this view are also of immeasurable importance clinically, and this view is most commonly used in neuroimaging studies, both CT and MRI (shown in Figure 28A and Figure 28B).

The basal ganglia are present when the brain is sectioned at this level (see Figure 25). The head of the caudate nucleus protrudes into the anterior horn of the lateral ventricle (seen in the CT, Figure 28A). The lentiform nucleus, shaped somewhat like a lens, is demarcated by white matter. Since the putamen and caudate neurons are identical, therefore, the two nuclei have the same grayish coloration. The globus pallidus is functionally different and contains many more fibers, and therefore is lighter in color. Depending upon the level of the section, it is sometimes possible (in this case on both sides) to see the two subdivisions of the globus pallidus, the internal and external segments (see Figure 24).

The white matter medial to the lentiform nucleus is the internal capsule (see Figure 26 and Figure 73). It is divisible into an anterior limb and a posterior limb and genu. The **anterior limb** separates the lentiform nucleus from the head of the caudate nucleus. The **posterior limb** of the internal capsule separates the lentiform nucleus from the thalamus. Some strands of gray matter located within the internal capsule represent the strands of gray matter between the caudate and the putamen (as shown in Figure 23). The base of the "V" is called the **genu**.

The anterior horn of the lateral ventricle is cut through its lowermost part and is seen in this photograph as a small cavity (see Figure 20A). The plane of the section has passed through the connection between the lateral ventricles and the third ventricle, the foramina of Monro (see Figure 20B). The section has also passed through the lateral ventricle as it curves into the temporal lobe to become the inferior horn of the lateral ventricle, the area called the atrium or trigone (better seen on the left side of this photograph; see Figure 20A and Figure 25). The choroid plexus of the lateral ventricle, which follows the inner curvature of the ventricle, is present on both sides (not labeled; see Figure 20A).

The section is somewhat asymmetrical in that the posterior horn of the lateral ventricle is fully present in the occipital lobe on the left side and not on the right side of the photograph. On the right side, a group of fibers is seen streaming toward the posterior pole, and these represent the visual fibers, called the optic radiation (discussed with Figure 41A and Figure 41B). The small size of the tail of the caudate nucleus alongside the lateral ventricle can be appreciated (see Figure 23 and Figure 25).

The third ventricle is situated between the thalamus of both sides (see Figure 9). The pineal is seen attached to the back end of the ventricle. A bit of the cerebellar vermis is visible posteriorly, behind the thalamus and between the occipital lobes.

CLINICAL ASPECT

This is the plane of view that would be used to look for small bleeds, called lacunes, in the posterior limb of the internal capsule (discussed with Figure 62). The major ascending sensory tracts and the descending motor tracts from the cerebral cortex are found in the posterior limb.

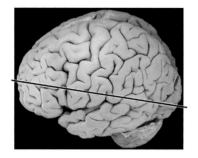
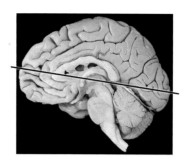

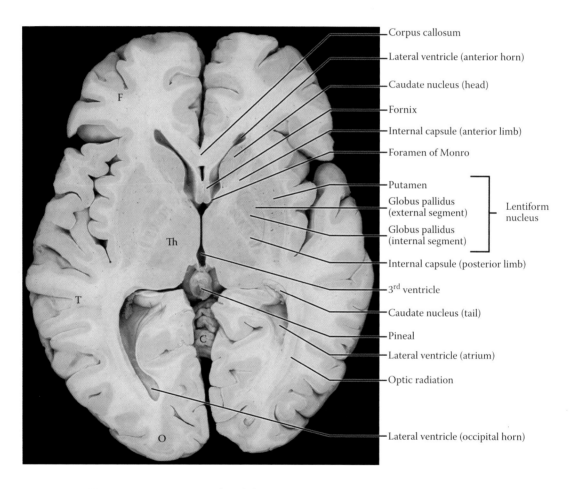

Corpus callosum

Lateral ventricle (anterior horn)

Caudate nucleus (head)

Fornix

Internal capsule (anterior limb)

Foramen of Monro

Putamen

Globus pallidus (external segment)   Lentiform nucleus

Globus pallidus (internal segment)

Internal capsule (posterior limb)

3rd ventricle

Caudate nucleus (tail)

Pineal

Lateral ventricle (atrium)

Optic radiation

Lateral ventricle (occipital horn)

F = Frontal lobe   Th = Thalamus
T = Temporal lobe
O = Occipital lobe   C = Cerebellum (vermis)

FIGURE 27: Basal Ganglia 6 — Horizontal Section (photograph)

FIGURE 28A
BASAL GANGLIA 7

HORIZONTAL VIEW: CT SCAN (RADIOGRAPH)

This radiological view of the brain is not in exactly the same horizontal plane as the anatomical specimen shown in the previous illustration. The radiological images of the brain are often done at a slight angle in order to minimize the exposure of the stuctures of the orbit, the retina and the lens, to the potential damaging effects of the x-rays used to generate a CT scan.

A CT image shows the skull bones (in white) and the relationship of the brain to the skull. A piece of the falx cerebri can also be seen. The outer cortical tissue is visible, with gyri and sulci, but not in as much detail as an MRI (shown in the next illustration). The structures seen in the interior of the brain include the white matter, and the ventricular spaces, the lateral ventricles with the septum pellucidum, and the third ventricle. Note that the CSF is dark (black). The cerebellum can be recognized, with its folia, but there is no sharp delineation between it and the cerebral hemispheres.

Although the basal ganglia and thalamus can be seen, there is little tissue definition. Note that the head of the caudate nucleus "protrudes" (bulges) into the anterior horn of the (lateral) ventricle (as in the previous brain section). The lentiform nucleus is identified and the internal capsule can be seen as well, with both the anterior and posterior limbs, and the genu.

The CSF cistern is seen behind the tectal plate (the colliculi; also known as the tectum or the quadrigeminal plate, see Figure 9A and Figure 10) — called the quadrigeminal plate cistern (seen also in the mid-sagittal views, Figure 17 and Figure 18, but not labeled); its "wings" are called the cisterna ambiens, a very important landmark for the neuroradiologist.

CLINICAL ASPECT

A regular CT can show an area of hemorrhage (blood has increased density), an area of decreased density (e.g., following an infarct), as well as changes in the size and shifting of the ventricles. This examination is invaluable in the assessment of a neurological patient in the acute stage of an illness or following a head injury and is most frequently used because the image can be captured in seconds. A CT can also be "enhanced" by injecting an iodinated compound into the blood circulation and noting whether it "escapes" into the brain tissue because of leakage in the BBB (discussed with Figure 21), for example, with tumors of the brain.

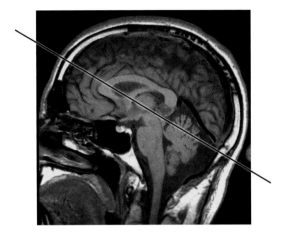

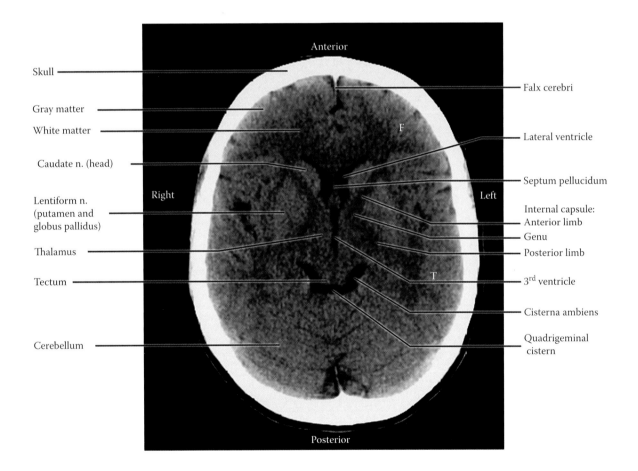

F = Frontal lobe
T = Temporal lobe

FIGURE 28A: Basal Ganglia 7 — CT: Horizontal View (radiograph)

FIGURE 28B
BASAL GANGLIA 8

HORIZONTAL VIEW: T2 MRI (RADIOGRAPH)

This radiograph is a view of the brain taken in the same plane, horizontally, closer to the plane of the brain section (see Figure 27), but a little higher than the previous radiograph. Parameters of the MRI have been adjusted to generate a **T2**-weighted image (see explanation with Figure 3). In this view, the CSF of the ventricles is white, while the bones of the skull are dark.

This MRI shows the brain as if it were an anatomical specimen (compare with the previous illustration) — there is a good differentiation between the gray matter and the white matter. There is a clear visualization of the basal ganglia and its subdivisions (head of the caudate, lentiform nucleus), as well as the thalamus. (**Note**: The line separating the putamen from the globus pallidus can "almost" be seen on the right side of the photograph.) In addition, the area of the internal capsule is also clearly seen.

The anterior horns of the lateral ventricle are present, and the section has passed through the formina of Monro (see Figure 20A, Figure 20B, and Figure 21). The lateral ventricle posteriorly is cut at the level of its widening, the atrium or trigone, as it curves into the temporal lobe (see Figure 20A). The third ventricle is in the midline, between the thalami (see Figure 6, Figure 7, and Figure 9A).

The linear marking in the white matter behind the atrium likely represents the optic radiation, from the lateral geniculate to the calcarine cortex (see Figure 41A and Figure 41C).

CLINICAL ASPECT

The MRI has proved to be invaluable in assessing lesions of the CNS — infarcts, tumors, plaques of multiple sclerosis, and numerous other lesions. An MRI can also be enhanced with intravenous gadolinium, which escapes with the blood when there is a breakdown of the BBB, and helps in the evaluation of pathology, such as tumors.

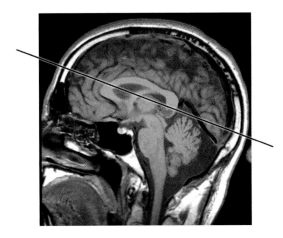

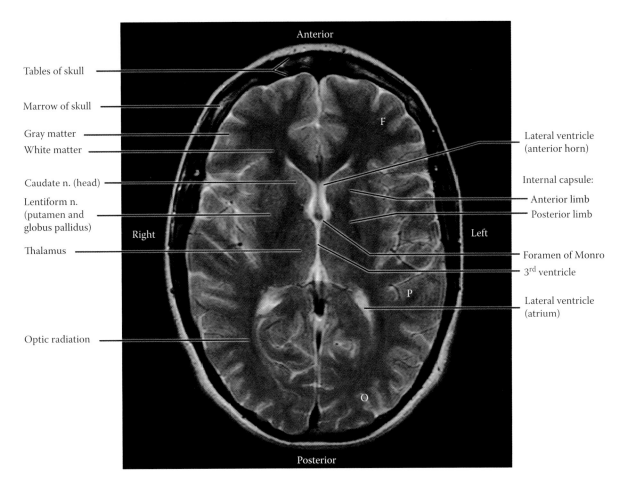

Anterior

Tables of skull

Marrow of skull

Gray matter

White matter

Caudate n. (head)

Lentiform n.
(putamen and
globus pallidus)

Thalamus

Right

Optic radiation

F

Left

P

O

Lateral ventricle
(anterior horn)

Internal capsule:

Anterior limb

Posterior limb

Foramen of Monro

3rd ventricle

Lateral ventricle
(atrium)

Posterior

F = Frontal lobe
P = Parietal lobe
O = Occipital lobe

FIGURE 28B: Basal Ganglia 8 — MRI: Horizontal View (radiograph)

FIGURE 29
BASAL GANGLIA 9

CORONAL SECTION OF HEMISPHERES
(PHOTOGRAPHIC VIEW)

This photographic view of the brain is sectioned in the coronal plane and shows the internal aspect of the hemispheres. On the dorsolateral view (small figure, upper left) the plane of section goes through both the frontal and the temporal lobes and would include the region of the basal ganglia. From the medial perspective (the figure on the upper right), the section includes the body of the lateral ventricles with the corpus callosum above, the anterior portion of the thalamus, and the third ventricle; the edge of the section also passes through the hypothalamus, the mammillary nucleus, and includes the optic tracts. The section passes in front of the anterior part of the midbrain, the cerebral peduncles, and the front tip of the pons.

The cerebral cortex, the gray matter, lies on the external aspect of the hemispheres and follows its outline into the sulci in between, wherever there is a surface. The deep interhemispheric fissure is seen between the two hemispheres, above the corpus callosum (not labeled, see Figure 16 and Figure 17). The lateral fissure is also present, well seen on the left side of the photograph (also not labeled), with the insula within the depths of this fissure (see Figure 14B and Figure 39).

The white matter is seen internally; it is not possible to separate out the various fiber systems of the white matter (see Figure 19A and Figure 19B). Below the corpus callosum are the two spaces, the cavities of the lateral ventricle, represented at this plane by the body of the ventricles (see Figure 20B, Figure 25, and Figure 76). The small gray matter on the side of the lateral ventricle is the body of the caudate nucleus (see Figure 23, Figure 25,

Figure 27, and Figure 76). Because the section was not cut symmetrically, the inferior horn of the lateral ventricle is found only on the right side of this photograph, in the temporal lobe.

The brain is sectioned in the coronal plane through the diencephalic region. The gray matter on either side of the third ventricle is the thalamus (see Figure 11). Lateral to this is a band of white matter, which by definition is part of the internal capsule, with the lentiform nucleus on its lateral side. In order to identify which part this is, the learner should refer to the section in the horizontal plane (see Figure 26 and Figure 27); the portion between the thalamus and lentiform nucleus is the posterior limb.

The parts of the lentifrom nucleus seen in this view include the putamen as well as the two portions of the globus pallidus, the external and internal segments. Since the brain has not been sectioned symmetrically, the two portions are more easily identified on the right side of the photograph. The claustrum has also been labeled (see below). The structures noted in this section should be compared with a similar (coronal) view of the brain taken more posteriorly (see Figure 74).

The gray matter within the temporal lobe, best seen on the left side of the photograph, is the amygdala (see Figure OL, Figure 25, and Figure 75A). It is easy to understand why this nucleus is considered one of the basal ganglia, by definition. Its function, as well as that of the fornix, will be explained with the limbic system section of this atlas (Section D).

ADDITIONAL DETAIL

Lateral to the lentiform nucleus is another thin strip of gray matter, the claustrum. The functional contribution of this small strip of tissue is not really known. The claustrum is also seen in the horizontal section (see Figure 27). Lateral to this is the cortex of the insula, inside the lateral fissure (see Figure 14B and Figure 39).

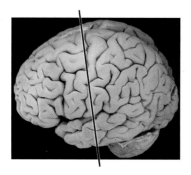
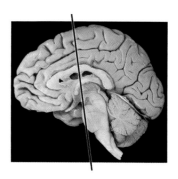

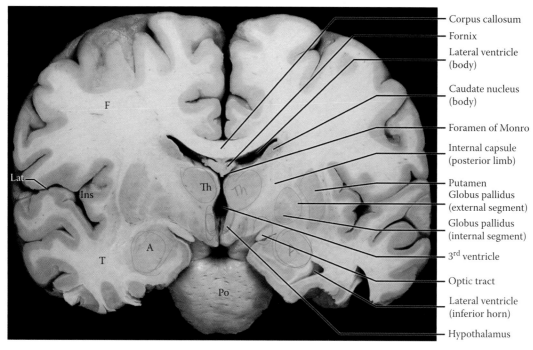

Corpus callosum

Fornix

Lateral ventricle
(body)

Caudate nucleus
(body)

Foramen of Monro

Internal capsule
(posterior limb)

Putamen
Globus pallidus
(external segment)

Globus pallidus
(internal segment)

3rd ventricle

Optic tract

Lateral ventricle
(inferior horn)

Hypothalamus

F = Frontal lobe   Th = Thalamus
T = Temporal lobe
                            A = Amygdala
Lat = Lateral fissure
Ins = Insula   Po = Pons

FIGURE 29: Basal Ganglia 9 — Coronal Section (photograph)

FIGURE 30
BASAL GANGLIA 10

CORONAL VIEW: MRI (RADIOGRAPH)

This is a view of the brain similar to the previous brain section, in the coronal plane. The T2 MRI has been adjusted on the viewing screen to invert the displayed image (sometimes called an inverted video view). The distinction between the gray matter and the white matter is enhanced with this view; the CSF is dark. Note that the tables of the skull are now white, and the bone marrow is dark. The superior sagittal sinus is seen in the midline, at the top of the falx cerebri (bright).

The cortex and white matter can be easily differentiated. The corpus callosum is seen crossing the midline. The caudate nucleus is diminishing in size, from the head (anteriorly) to the body (posteriorly — compare with another coronal section of the brain, see Figure 74). The

lentiform nucleus is still present and the thalamus can be seen adjacent to the third ventricle.

By definition, the section has passed through the posterior limb of the internal capsule (see Figure 26). Its fibers are seen as continuing to become the cerebral peduncle (see Figure 6 and Figure 7). The plane of section includes the lateral fissure, and the insula (see Figure 17B). The temporal lobe includes the hippocampal formation and the inferior horn of the lateral ventricle (see Figure 20A, Figure 20B, and Figure 74).

The lateral ventricle is seen, divided by the septum pellucidum into one for each hemisphere (see also Figure 62). Again, the plane of section has passed through the foramina of Monro, connecting to the third ventricle, which is situated between the thalamus on either side.

This view also includes the brainstem — the midbrain (the cerebral peduncles), the pons (the ventral portion), and the medulla. The trigeminal nerve has been identified at the midpontine level. The tentorium cerebelli can now be clearly seen (see Figure 17 and Figure 41B), with its opening (also called incisura) at the level of the midbrain (discussed with uncal herniation, see Figure 15B).

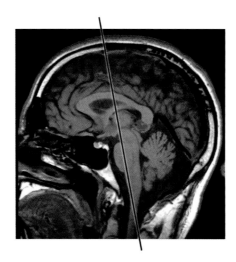

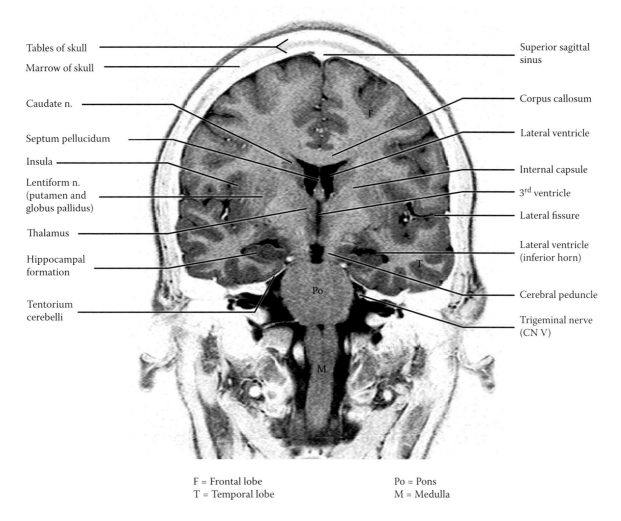

Tables of skull

Marrow of skull

Caudate n.

Septum pellucidum

Insula

Lentiform n.
(putamen and
globus pallidus)

Thalamus

Hippocampal
formation

Tentorium
cerebelli

Superior sagittal
sinus

Corpus callosum

Lateral ventricle

Internal capsule

3rd ventricle

Lateral fissure

Lateral ventricle
(inferior horn)

Cerebral peduncle

Trigeminal nerve
(CN V)

F = Frontal lobe   Po = Pons
T = Temporal lobe   M = Medulla

FIGURE 30: Basal Ganglia 10 — MRI: Coronal View (radiograph)

Section B

FUNCTIONAL SYSTEMS

INTRODUCTION

This section explains how the nervous system is organized to assess sensory input and execute motor actions. The functioning nervous system has a hierarchical organization to carry out its activities.

Incoming sensory fibers, called *afferents*, have their input into the spinal cord as well as the brainstem, except for the special senses of vision and olfaction (which will be discussed separately). This sensory input is processed by relay nuclei, including the thalamus, before the information is analyzed by the cortex. In the cortex, there are primary areas that receive the information, other cortical association areas that elaborate the sensory information, and still other areas that integrate the various sensory inputs.

On the motor side, the outgoing motor fibers, called *efferents*, originate from motor neurons in the brainstem and the spinal cord. These motor nuclei are under the control of motor centers in the brainstem and cerebral cortex. In turn, these motor areas are influenced by other cortical areas and by the basal ganglia, as well as by the cerebellum.

Simpler motor patterns are organized as reflexes. In all cases, except for the myotatic (muscle) reflex, called the stretch reflex (discussed with Figure 44), there is some processing that occurs in the CNS, involving interneurons in the spinal cord, brainstem, thalamus, or cortex.

The processing of both sensory and motor activities, beyond simple reflexes, therefore involves a series of neuronal connections, creating functional systems. These include nuclei of the CNS at the level of the spinal cord, brainstem, and thalamus. In almost all functional systems in humans, the cerebral cortex is also involved. The axonal connections between the nuclei in a functional system usually run together forming a distinct bundle of fibers, called a **tract or pathway**. These tracts are named according to the direction of the pathway, for example spinothalamic, means that the pathway is going from the spinal cord to the thalamus; cortico-spinal means the pathway is going from the cortex to the spinal cord. Along their way, these axons may distribute information to several other parts of the CNS by means of **axon collaterals**.

In Part I of this section, we will be concerned with the sensory tracts or pathways and their connections in the CNS. Part II introduces the reticular formation, which has both sensory, motor, and other "integrative" functions. In Part III we will discuss the pathways and brain regions concerned with motor control.

PART I: SENSORY SYSTEMS

Sensory systems, also called modalities (singular **modality**), share many features. All sensory systems begin with **receptors**, sometimes free nerve endings and others that are highly specialized, such as those in the skin for touch and vibration sense, and the hair cells in the cochlea for hearing, as well as the rods and cones in the retina. These receptors activate the peripheral sensory fibers appropriate for that sensory system. The peripheral nerves have their cell bodies in **sensory ganglia**, which belong to the peripheral nervous system (**PNS**). For the body (neck down), these are the dorsal root ganglia, located in the intervertebral spaces (see Figure 1). The trigeminal ganglion inside the skull serves the sensory fibers of the head. The central process of these peripheral neurons enters the CNS and synapses in the nucleus appropriate for that sensory system (this is hard-wired).

Generally speaking, the older systems both peripherally and centrally involve axons of small diameter that are thinly myelinated or unmyelinated, with a slow rate of conduction. In general, these pathways consist of fibers-synapses-fibers, with collaterals, creating a multisynaptic chain with many opportunities for spreading the information, but thereby making transmission slow and quite insecure. The newer pathways that have evolved have larger axons that are more thickly myelinated and therefore conduct more rapidly. These form rather direct connections with few, if any, collaterals. The latter type of pathway transfers information more securely and is more specialized functionally.

Because of the upright posture of humans, the sensory systems go upward or ascend to the cortex — the **ascending systems**. The sensory information is "processed" by various nuclei along the pathway. Three systems are concerned with sensory information from the skin, two from the body region and one (with subparts) from the head:

- The **dorsal column — medial lemniscus** pathway, a newer pathway for the somatosensory sensory modalities of disriminative touch, joint position, and "vibration." **Discriminative touch** is the ability to discriminate whether the skin is being touched by one or two points simultaneously; it is usually tested by asking the patient to identify objects (e.g., a coin) placed in the hand, with the eyes closed; in fact, this act requires interpretation by the cortex. **Joint position** is tested by moving a joint and asking the patient to report the direction of the movement (again with the eyes closed). **Vibration** is tested by placing a tuning fork that has been set into motion onto a bony prominence (e.g., the wrist, the ankle). These sensory receptors in the skin and the joint surfaces are quite specialized; the fibers carrying the afferents to the CNS are large in diameter and thickly myelinated, meaning that the information is carried quickly and with a high degree of fidelity.
- The **anterolateral system**, an older system that carries pain and temperature, and some less discriminative forms of touch sensations, was formerly called the lateral spino-thalamic and ventral (anterior) spino-thalamic tracts, respectively.
- The **trigeminal pathway**, carrying sensations from the face and head area (including discriminative touch, pain, and temperature), involves both newer and older types of sensation.

Some of the special senses will be studied in detail, namely the auditory and visual systems. Each has unique features that will be described. Other sensory pathways, such as vestibular (balance) and taste also will be reviewed. All these pathways, except for olfaction, relay in the thalamus before going on to the cerebral cortex (see Figure 63); the olfactory system (smell) will be considered with the limbic system (see Figure 79).

PART II: RETICULAR FORMATION

Interspersed with the consideration of the functional systems is the reticular formation, located in the core of the brainstem. This group of nuclei comprises a rather old system with multiple functions — some generalized and some involving the sensory or the motor systems. Some sensory pathways have collaterals to the reticular formation, some do not.

The reticular formation is partially responsible for setting the level of activity of motor neurons; in addition, some motor pathways originate in the reticular formation. The explanation of the reticular formation will be presented after the sensory pathways; the motor aspects will be discussed with the motor systems.

CLINICAL ASPECT

Destruction of the nuclei and pathways due to disease or injury leads to a neurological loss of function. How does the physician or neurologist diagnose what is wrong? He or she does so on the basis of a detailed knowledge of the pathways and their position within the central nervous system; this is a prerequisite for the part of the diagnosis that locates *where* the disease is occurring in the nervous system, i.e., **localization**. The disease that is causing the loss of function, the *etiological* diagnosis, can sometimes be recognized by experienced physicians on the basis of the pattern of the disease process; at other times, specialized investigations are needed to make the disease-specific diagnosis.

There is an additional caveat — almost all of the pathways cross the midline, each at a unique and different location; this is called a **decussation**. The important clinical correlate is that destruction of a pathway may affect the opposite side of the body, depending upon the location of the lesion in relation to the level of the decussation.

Note on Use of the CD-ROM: The pathways in this section are presented on the CD-ROM with flash animation demonstrating activation of the pathway. After studying the details of a pathway with the text and illustration, the learner should then view the same figure on the CD for a better understanding of the course of the tract, the synaptic relays, and the decussation of the fibers.

FIGURE 31
PATHWAYS AND X-SECTIONS

ORIENTATION TO DIAGRAMS

The illustrations of the sensory and motor pathways in this section of the atlas are all done in a standard manner:

- On the left side, the CNS is depicted, including spinal cord, brainstem, thalamus, and a coronal section through the hemispheres, with small diagrams of the hemisphere at the top showing the area of the cerebral cortex involved.
- On the right side, cross-sections (X-sections) of the brainstem and spinal cord, at standardized levels are depicted; the exact levels are indicated by arrows on the diagram on the left. In all, there are 10 cross-sections — 8 through the brainstem and 2 through the spinal cord. For each of the pathways, 5 of these will be used.

The diagram of the hemispheres is a coronal section, similar to the one already described in Section A, at the plane of the lenticular nucleus (see Figure 29). Note the basal ganglia, the thalamus, the internal capsule, and the ventricles; these labels will not be repeated in the following diagrams. This diagram will be used to convey the overall course of the tract and, particularly, at what level the fibers cross (i.e., decussate).

The X-sections (cross-sections) of the brainstem and the spinal cord include:

- Two levels through the midbrain — upper and lower
- Three levels through the pons — upper, mid, and lower
- Three levels through the medulla — upper, mid, and lower
- Two levels through the spinal cord — cervical and lumbar

The exact position of the tract under consideration is indicated in these cross-sections. It is important to note that only some of the levels are used in describing each of the pathways.

These brainstem and spinal cord cross-sections are the same as those shown in Section C of this atlas (see Figure 64–Figure 69). In that section, details of the histological anatomy of the spinal cord and brainstem are given. We have titled that section of the atlas Neurological Neuroanatomy because it allows precise location of the tracts, which is necessary for the localization of an injury or disease. The learner may wish to consult these detailed diagrams at this stage.

LEARNING PLAN

Studying pathways in the central nervous system necessitates visualizing the pathways, a challenging task for many. The pathways that are under study extend longitudinally through the CNS, going from spinal cord and brainstem to thalamus and cortex for sensory (ascending) pathways, and from cortex to brainstem and spinal cord for motor (descending) pathways. As is done in other texts and atlases, diagrams are used to facilitate this visualization exercise for the learner; color adds to the ability to visualize these pathways, as does the illustration on a CD-ROM.

CLINICAL ASPECT

This section is a foundation for the student in correlating the anatomy of the pathways with the clinical symptomatology.

Note to the Learner: In this presentation of the pathways, the learner is advised to return to the description of the thalamus and the various specific relay nuclei (see Figure 12 and Figure 63). Likewise, referring to the cortical illustrations (see Figure 13–Figure 17) will inform the learner which areas of the cerebral cortex are involved in the various sensory modalities. This will assist in integrating the anatomical information presented in the previous section.

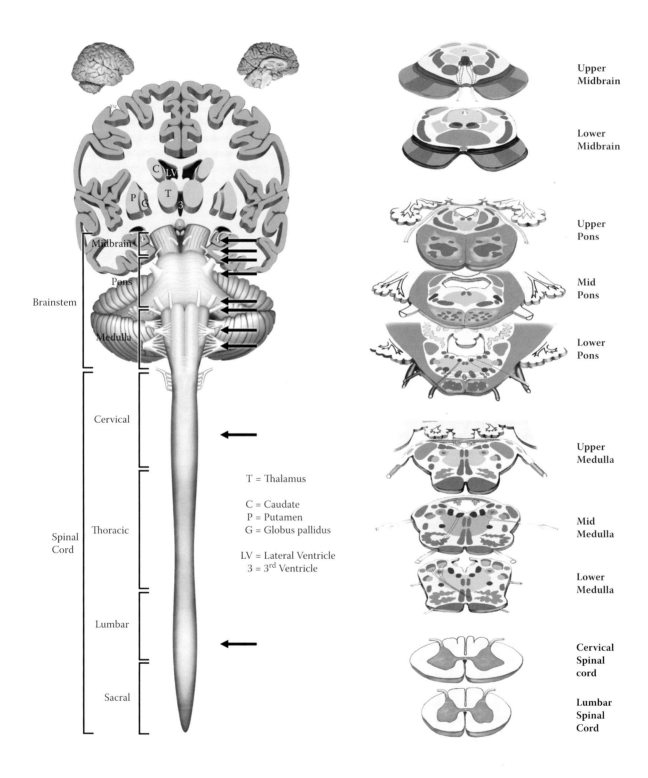

FIGURE 31: Pathways — Orientation to Diagrams

Upper
Midbrain

Lower
Midbrain

Upper
Pons

Mid
Pons

Lower
Pons

Upper
Medulla

Mid
Medulla

Lower
Medulla

Cervical
Spinal
cord

Lumbar
Spinal
Cord

Brainstem

Midbrain

Pons

Medulla

Spinal
Cord

Cervical

Thoracic

Lumbar

Sacral

C LV
P T
G 3

T = Thalamus

C = Caudate
P = Putamen
G = Globus pallidus

LV = Lateral Ventricle
3 = 3rd Ventricle

PART I: SENSORY SYSTEMS
FIGURE 32
SPINAL CORD X-SECTION

SENSORY: NUCLEI AND AFFERENTS

This is a representation of a spinal cord cross-section, at the cervical level (see Figure 4), with a focus on the sensory afferent side. All levels of the spinal cord have the same sensory organization, although the size of the nuclei will vary with the number of afferents.

UPPER FIGURE

The dorsal horn of the spinal cord has a number of nuclei related to sensory afferents, particularly pain and temperature, as well as crude touch. The first nucleus encountered is the posteromarginal, where some sensory afferents terminate. The next and most prominent nucleus is the **substantia gelatinosa**, composed of small cells, where many of the pain afferents terminate. Medial to this is the **proper sensory nucleus**, which is a relay site for these fibers; neurons in this nucleus project across the midline and give rise to a tract — the anterolateral tract (see below and Figure 34).

There is a small local tract that carries pain and temperature afferents up and down the spinal cord for a few segments, called the **dorsolateral fasciulus** (of Lissauer).

The other sensory-related nucleus is the **dorsal nucleus** (of Clarke). This is a relay nucleus for muscle afferents that project to the cerebellum. In the lower illustration, the fibers from this nucleus are seen to ascend, on the same side, as the **dorsal spino-cerebellar tract** (see Figure 55 and Figure 68).

LOWER FIGURE

This illustration shows the difference at the entry level between the two sensory pathways — the dorsal column tracts and the anterolateral system. The cell bodies for these peripheral nerves are located in the **dorsal root ganglion**, the **DRG** (see Figure 1).

On the left side, the afferent fibers carrying discriminative touch, position sense, and vibration enter the dorsal horn and immediately turn upward. The fibers may give off local collaterals (e.g., to the intermediate gray), but the information from these rapidly conducting, heavily myelinated fibers is carried upward in the two tracts that lie between the dorsal horns, called collectively the **dorsal columns**. The first synapse in this pathway occurs at the level of the lower medulla (see Figure 33).

On the right side, the afferents carrying the pathways for pain, temperature, and crude touch enter and synapse in the nuclei of the dorsal horn. The nerves conveying this sensory input into the spinal cord are thinly myelinated or unmyelinated, and conduct slowly. After several synapses, these fibers cross the midline in the white matter in front of the commissural gray matter (the gray matter joining the two sides), called the **ventral (anterior) white commissure** (see upper illustration). The fibers then ascend as the spino-thalamic tracts, called collectively the **anterolateral system** (see Figure 34).

CLINICAL ASPECT

The effect of a lesion of one side of the spinal cord will therefore affect the two sensory systems differently because of this arrangement. The sensory modalities of the dorsal column system will be disrupted on the same side. The pain and temperature pathway, having crossed, will lead to a loss of these modalities on the opposite side.

Any lesion that disrupts just the crossing pain and temperature fibers at the segmental level will lead to a loss of pain and temperature of just the levels affected. There is an uncommon disease called **syringomyelia** that involves a pathological cystic enlargement of the central canal. The cause for this is largely unknown but sometimes can be related to a previous traumatic injury. The enlargement of the central canal interrupts the pain and temperature fibers in their crossing anteriorly in the anterior white commissure. Usually this occurs in the cervical region and the patients complain of the loss of these modalities in the upper limbs and hand, in what is called a cape-like distribution. The enlargement can be visualized with MRI.

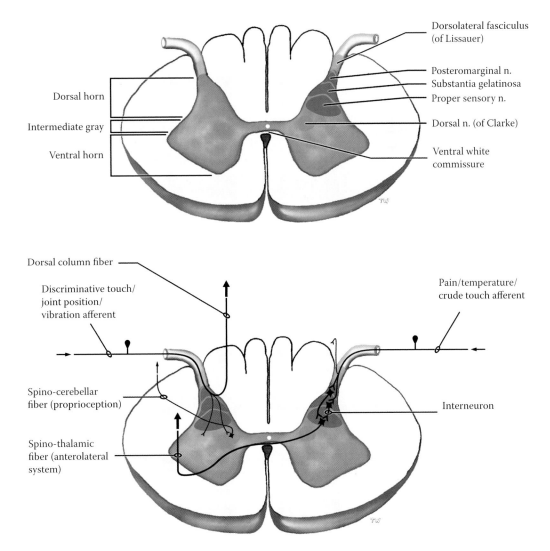

Dorsolateral fasciculus
(of Lissauer)

Posteromarginal n.
Substantia gelatinosa
Proper sensory n.

Dorsal horn

Intermediate gray

Ventral horn

Dorsal n. (of Clarke)

Ventral white
commissure

Dorsal column fiber

Discriminative touch/
joint position/
vibration afferent

Pain/temperature/
crude touch afferent

Spino-cerebellar
fiber (proprioception)

Interneuron

Spino-thalamic
fiber (anterolateral
system)

FIGURE 32: Spinal Cord Nuclei — Sensory

FIGURE 33
DORSAL COLUMN — MEDIAL LEMNISCUS PATHWAY

DISCRIMINATIVE TOUCH, JOINT POSITION, VIBRATION

This pathway carries the modalities **discriminative touch, joint position**, and the somewhat artificial "sense" of **vibration** from the body. Receptors for these modalities are generally specialized endings in the skin and joint capsule.

The axons enter the spinal cord and turn upward, with no synapse (see Figure 32). Those fibers entering below spinal cord level T6 (sixth thoracic spinal segmental level) form the **fasciculus gracilis**, the gracile tract; those entering above T6, particularly those from the upper limb, form the **fasciculus cuneatus**, the cuneate tract, which is situated more laterally. These tracts ascend the spinal cord between the two dorsal horns, forming the **dorsal column** (see Figure 32, Figure 68, and Figure 69).

The first synapse in this pathway is found in two nuclei located in the lowermost part of the medulla, in the **nuclei gracilis and cuneatus** (see Figure 9B, Figure 40, and Figure 67C). Topographical representation, also called **somatotopic** organization, is maintained in these nuclei, meaning that there are distinct populations of neurons that are activated by areas of the periphery that were stimulated.

After neurophysiological processing, axons emanate from these two nuclei, which will cross the midline. This stream of fibers, called the **internal arcuate fibers,** can be recognized in suitably stained sections of the lower medulla, (see Figure 40 and Figure 67C). The fibers then group together to form the **medial lemniscus,** which ascends through the brainstem. This pathway does not give off collaterals to the reticular formation in the brainstem. This pathway changes orientation and position as it ascends through the pons and midbrain (see Figure 40 and Figure 65–Figure 67).

The medial lemniscus terminates (i.e., synapses) in the **ventral posterolateral nucleus** of the thalamus, the **VPL** (see Figure 12 and Figure 63). The fibers then enter the internal capsule, its posterior limb, and travel to the somatosensory cortex, terminating along the **post-central gyrus, areas 1, 2, and 3** (see Figure 14A and Figure 63).

The representation of the body on this gyrus is not proportional to the size of the area being represented; for example, the fingers, particularly the thumb, are given a much larger area of cortical representation than the trunk; this is called the **sensory "homunculus."** The lower limb, represented on the medial aspect of the hemisphere (see Figure 17), has little cortical representation.

NEUROLOGICAL NEUROANATOMY

The cross-sectional levels for this pathway include the lumbar and cervical spinal cord levels, and the brainstem levels, lower medulla, mid-pons, and upper midbrain.

In the spinal cord, the pathways are found between the two dorsal horns, as a well myelinated bundle of fibers, called the dorsal column(s). The tracts have a topographical organization, with the lower body and lower limb represented in the medially placed gracile tract, and the upper body and upper limb in the laterally placed cuneate tract. After synapsing in their respective nuclei and the crossing of the fibers in the lower medulla (internal arcuate fibers), the medial lemniscus tract is formed. This heavily myelinated tract that is easily seen in myelin-stained sections of the brainstem (e.g., see Figure 67C), is located initially between the inferior olivary nuclei and is oriented in the dorsal-ventral position (see Figure 40 and Figure 67B). The tract moves more posteriorly, shifts laterally, and also changes orientation as it ascends (see Figure 40; also Figure 65A, Figure 66A, and Figure 67A). The fibers are topographically organized, with the leg represented laterally and the upper limb medially. The medial lemniscus is joined by the anterolateral system and trigeminal pathway in the upper pons (see Figure 36 and Figure 40).

CLINICAL ASPECT

Lesions involving this tract will result in the loss of the sensory modalities carried in this pathway. A lesion of the dorsal column in the spinal cord will cause a loss on the same side; after the crossing in the lower brainstem, any lesion of the medial lemniscus will result in the deficit occurring on the opposite side of the body. Lesions occurring in the midbrain and internal capsule will usually involve the fibers of the anterolateral pathway, as well as the modalities carried in the trigeminal pathway (to be discussed with Figure 36 and Figure 40). With cortical lesions, the part of the body affected will be determined by the area of the post-central gyrus involved.

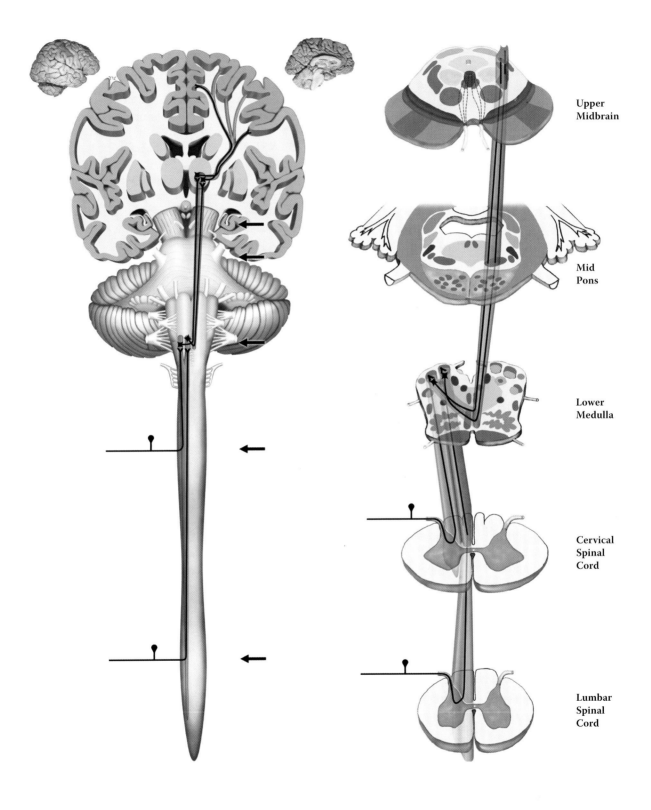

Upper
Midbrain

Mid
Pons

Lower
Medulla

Cervical
Spinal
Cord

Lumbar
Spinal
Cord

FIGURE 33: Dorsal Column — Medial Lemniscus — Discriminative Touch, Joint Position, and Vibration

FIGURE 34
ANTEROLATERAL SYSTEM

PAIN, TEMPERATURE, CRUDE TOUCH

This pathway carries the modalities of **pain and temperature** and a form of touch sensation called **crude or light touch**. The sensations of itch and tickle, and other forms of sensation (e.g., "sexual") are likely carried in this system. In the periphery the receptors are usually simply free nerve endings, without any specialization.

These incoming fibers (sometimes called the first order neuron) enter the spinal cord and synapse in the dorsal horn (see Figure 4 and Figure 32). There are many collaterals within the spinal cord that are the basis of several protective reflexes (see Figure 44). The number of synapses formed is variable, but eventually a neuron is reached that will project its axon up the spinal cord (sometimes referred to as the second order neuron). This axon will cross the midline, decussate, in the **ventral** (anterior) **white commissure**, usually within two to three segments above the level of entry of the peripheral fibers (see Figure 4 and Figure 32).

These axons now form the **anterolateral** tract, located in that portion of the white matter of the spinal cord. It was traditional to speak of two pathways — one for pain and temperature, the **lateral spino-thalamic tract**, and another for light (crude) touch, the **anterior (ventral) spino-thalamic tract**. Both are now considered together under one name.

The tract ascends in the same position through the spinal cord (see Figure 68 and Figure 69). As fibers are added from the upper regions of the body, they are positioned medially, pushing the fibers from the lower body more laterally. Thus, there is a topographic organization to this pathway in the spinal cord. The axons of this pathway are either unmyelinated or thinly myelinated. In the brainstem, collaterals are given off to the reticular formation, which are thought to be quite significant functionally. Some of the ascending fibers terminate in the **ventral posterolateral (VPL) nucleus** of the thalamus (sometimes referred to as the third order neuron in a sensory pathway), and some in the nonspecific intralaminar nuclei (see Figure 12 and Figure 63).

There is a general consensus that pain sensation has two functional components. The older (also called the **paleospinothalamic**) pathway involves the reported sensation of an ache, or diffuse pain that is poorly localized. The fibers underlying this pain system are likely unmyelinated both peripherally and centrally, and the central connections are probably very diffuse; most likely these fibers terminate in the nonspecific thalamic nuclei and influence the cortex widely. The newer pathway, sometimes called the **neospinothalamic** system, involves thinly myelinated fibers in the PNS and CNS, and likely ascends to the VPL nucleus of the thalamus and from there is relayed to the postcentral (sensory) gyrus. Therefore, the sensory information in this pathway can be well localized. The common example for these different pathways is a paper cut — immediately one knows exactly where the cut has occurred; this is followed several seconds later by a diffuse poorly localized aching sensation.

NEUROLOGICAL NEUROANATOMY

The cross-sectional levels for this pathway include the lumbar and cervical spinal cord levels, and the brainstem levels mid-medulla, mid-pons, and upper midbrain.

In the spinal cord, this pathway is found among the various pathways in the anterolateral region of the white matter (see Figure 32, Figure 68, and Figure 69), hence its name. Its two parts cannot be distinguished from each other or from the other pathways in that region. In the brainstem, the tract is small and cannot usually be seen as a distinct bundle of fibers. In the medulla, it is situated dorsal to the inferior olivary nucleus; in the uppermost pons and certainly in the midbrain, the fibers join the medial lemniscus (see Figure 40).

CLINICAL ASPECT

Lesions of the anterolateral pathway from the point of crossing in the spinal cord upward will result in a loss of the modalities of pain and temperature and crude touch on the opposite side of the body. The exact level of the lesion can be quite accurately ascertained, as the sensation of pain can be quite simply tested at the bedside by using the end of a pin. (The tester should be aware that this is quite uncomfortable or unpleasant for the patient being tested.)

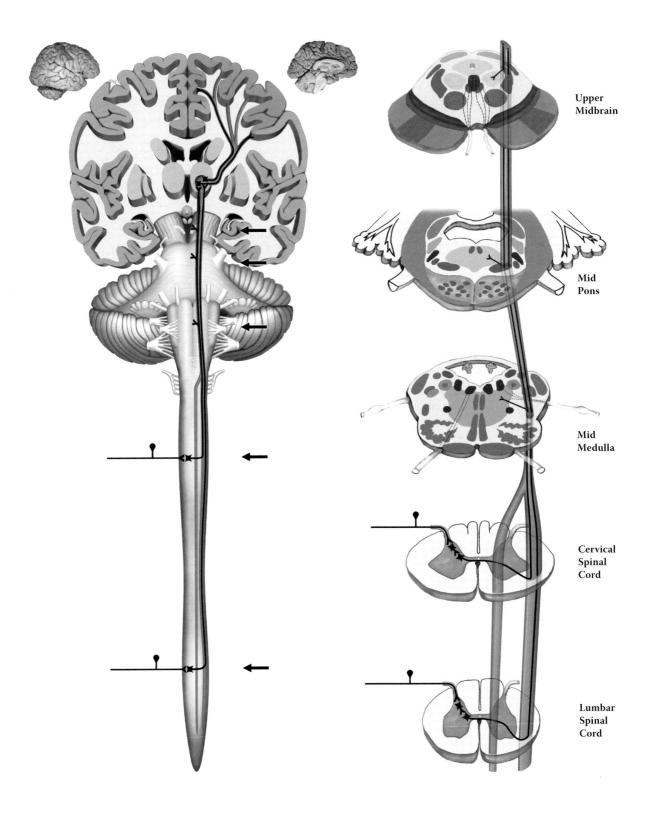

Upper
Midbrain

Mid
Pons

Mid
Medulla

Cervical
Spinal
Cord

Lumbar
Spinal
Cord

FIGURE 34: Anterolateral System — Pain, Temperature, and Crude Touch

FIGURE 35
TRIGEMINAL PATHWAYS

DISCRIMINATIVE TOUCH, PAIN, TEMPERATURE

The sensory fibers include the modalities discriminative touch as well as pain and temperature. The sensory input comes from the face, particularly from the lips, all the mucous membranes inside the mouth, the conjunctiva of the eye, and the teeth. The fiber sizes and degree of myelination are similar to the sensory inputs below the neck. The cell bodies of these fibers are found in the trigeminal ganglion inside the skull.

The fibers enter the brainstem along the middle cerebellar peduncle (see Figure 6 and Figure 7). Within the CNS there is a differential handling of the modalities, comparable to the previously described pathways in the spinal cord.

Those fibers carrying the sensations of discriminative touch will synapse in the **principal (main) nucleus** of CN V, in the mid-pons, at the level of entry of the nerve (see Figure 8B and Figure 66B). The fibers then cross the midline and join the medial lemniscus, terminating in the **ventral posteromedial (VPM) nucleus** of the thalamus (see Figure 12 and Figure 63). They are then relayed via the posterior limb of the internal capsule to the postcentral gyrus, where the face area is represented on the dorsolateral surface (see Figure 14A); the lips and tongue are very well represented on the sensory homunculus.

Those fibers carrying the modalities of pain and temperature descend within the brainstem. They form a tract that starts at the mid-pontine level, descends through the medulla, and reaches the upper level of the spinal cord (see Figure 8B) called the **descending or spinal tract of V**, also called the **spinal trigeminal tract**. Immediately medial to this tract is a nucleus with the same name. The fibers terminate in this nucleus and, after synapsing, cross to the other side and ascend (see Figure 40). Therefore, these fibers decussate over a wide region and do not form a compact bundle of crossing fibers; they also send collaterals to the reticular formation. These trigeminal fibers join with those carrying touch, forming the **trigeminal pathway** in the mid-pons. They terminate in the VPM and

other thalamic nuclei, similar to those of the anterolateral system (see Figure 34; also Figure 12 and Figure 63). The trigeminal pathway joins the medial lemniscus in the upper pons, as does the anterolateral pathway (see Figure 36 and Figure 40).

NEUROLOGICAL NEUROANATOMY

The cross-sectional levels for this pathway include the three medullary levels of the brainstem, the mid-pons, and the lower midbrain.

The principal nucleus of CN V is seen at the midpontine level (see also Figure 66B). The descending trigeminal tract is found in the lateral aspect of the medulla, with the nucleus situated immediately medially (see Figure 67A and Figure 67B). The crossing pain and temperature fibers join the medial lemniscus over a wide area and are thought to have completely crossed by the lower pontine region (see Figure 66A). The collaterals of these fibers to the reticular formation are shown.

CLINICAL ASPECT

Trigeminal neuralgia is an affliction of the trigeminal nerve of uncertain origin which causes severe "lightning" pain in one of the branches of CN V; often there is a trigger such as moving the jaw, or an area of skin. The shooting pains may occur in paroxysms lasting several minutes. An older name for this affliction is tic douloureux. Treatment of these cases, which cause enormous pain and suffering, is difficult, and used to involve the possibility of surgery involving the trigeminal ganglion inside the skull, an extremely difficult if not risky treatment; nowadays most cases can be managed with medical therapy.

A vascular lesion in the lateral medulla will disrupt the descending pain and temperature fibers and result in a loss of these sensations on the same side of the face, while leaving the fibers for discriminative touch sensation from the face intact. This lesion, known as the lateral medullary syndrome (of Wallenberg), includes other deficits (see Figure 40 and discussed with Figure 67B). A lesion of the medial lemniscus above the mid-pontine level will involve all trigeminal sensations on the opposite side. Internal capsule and cortical lesions cause a loss of trigeminal sensations from the opposite side, as well as involving other pathways.

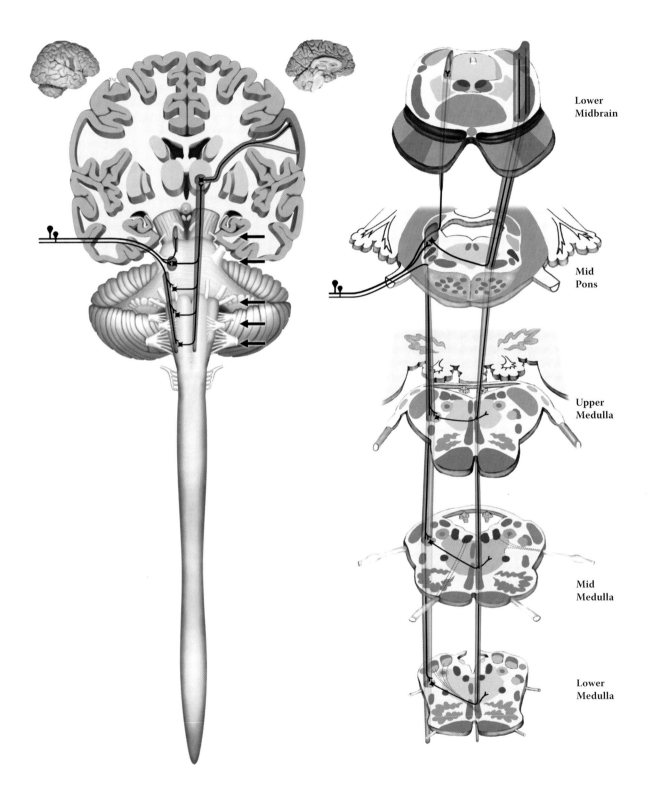

Lower Midbrain

Mid Pons

Upper Medulla

Mid Medulla

Lower Medulla

FIGURE 35: Trigeminal Pathways — Discriminative Touch, Pain, and Temperature

FIGURE 36
SENSORY SYSTEMS

SOMATOSENSORY AND TRIGEMINAL PATHWAYS

This diagram presents all the somatosensory pathways, the dorsal column-medial lemniscus, the anterolateral, and the trigeminal pathway as they pass through the midbrain region into the thalamus and onto the cortex. The view is a dorsal perspective (as in Figure 10 and Figure 40).

The pathway that carries discriminative touch sensation and information about joint position (as well as vibration) from the body is the medial lemniscus (see Figure 33). The equivalent pathway for the face comes from the principal nucleus of the trigeminal, which is located at the mid-pontine level (see Figure 8B and Figure 35). The anterolateral pathway conveying pain and temperature from the body has joined up with the medial lemniscus by this level (see Figure 34). The trigeminal pain and temperature fibers have likewise joined up with the other trigeminal fibers (see Figure 35).

The various sensory pathways are all grouped together at the level of the midbrain (see cross-section). At the level of the lower midbrain, these pathways are located near to the surface, dorsal to the substantia nigra; as they ascend they are found deeper within the midbrain, dorsal to the red nucleus (shown in cross-section in Figure 65A and Figure 65B).

The two pathways carrying the modalities of fine touch and position sense (and vibration) terminate in different specific relay nuclei of the thalamus (see Figure 12 and Figure 63):

*   The medial lemniscus in the VPL, ventral posterolateral nucleus
*   The trigeminal pathway in the VPM, ventral posteromedial nucleus

Sensory modality and topographic information is retained in these nuclei. There is physiologic processing of the sensory information, and some type of sensory "perception" likely occurs at the thalamic level.

After the synaptic relay, the pathways continue as the (superior) thalamo-cortical radiation through the posterior limb of the internal capsule, between the thalamus and lenticular nucleus (see Figure 26, Figure 27, Figure 28A, and Figure 28B). The fibers are then found within the white matter of the hemispheres. The somatosensory information is distributed to the cortex along the postcentral gyrus (see the small diagrams of the brain above the main illustration of Figure 36), also called S1. Precise localization and two-point discrimination are cortical functions.

The information from the face and hand is topographically located on the dorsolateral aspect of the hemispheres (see Figure 13 and Figure 14A). The information from the lower limb is localized along the continuation of this gyrus on the medial aspect of the hemispheres (see Figure 17). This cortical representation is called the sensory "homunculus," a distorted representation of the body and face with the trunk and lower limbs having very little area, whereas the face and fingers receive considerable representation.

Further elaboration of the sensory information occurs in the parietal association areas adjacent to the postcentral gyrus (see Figure 14A and Figure 60). This allows us to learn to recognize objects by tactile sensations (e.g., coins in the hand).

The pathways carrying pain and temperature from the body (the anterolateral system) and the face (spinal trigeminal system) terminate in part in the specific relay nuclei, ventral posterolateral and ventral posteromedial (VPL and VPM), respectively, but mainly in the intralaminar nuclei. These latter terminations may be involved with the emotional correlates that accompany many sensory experiences (e.g., pleasant or unpleasant).

The fibers that have relayed pain information project from these nuclei to several cortical areas, including the post-central gyrus, SI, and area SII (a secondary sensory area), which is located in the lower portion of the parietal lobe, as well as other cortical regions. The output from the intralaminar nuclei of the thalamus goes to widespread cortical areas.

CLINICAL ASPECT

Lesions of the thalamus may sometimes give rise to pain syndromes (also discussed with Figure 63).

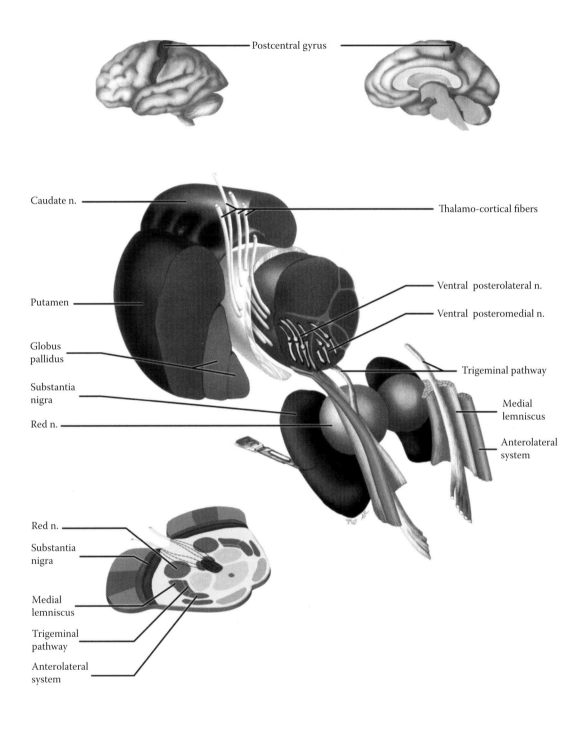

FIGURE 36: Somatosensory and Trigeminal Pathways

FIGURE 37
AUDITION 1

AUDITORY PATHWAY 1

The auditory pathway is somewhat more complex, firstly because it is bilateral, and secondly, because there are more synaptic stations (nuclei) along the way, with numerous connections across the midline. It also has a unique feature — a feedback pathway from the CNS to cells in the receptor organ, the cochlea.

The specialized hair cells in the cochlea respond maximally to certain frequencies (pitch) in a tonotopic manner; tones of a certain pitch cause patches of hair cells to respond maximally, and the distribution of this response is continuous along the cochlea. The peripheral ganglion for these sensory fibers is the **spiral ganglion**. The central fibers from the ganglion project to the first brainstem nuclei, the dorsal and ventral **cochlear nuclei**, at the level of entry of the VIIIth nerve at the uppermedullary level (see Figure 8B, Figure 40, and Figure 67A).

After this, the pathway can follow a number of different routes. In an attempt to make some semblance of order, these will be discussed in sequence, even though an axon may or may not synapse in each of these nuclei.

Most of the fibers leaving the cochlear nuclei will synapse in the **superior olivary complex**, either on the same side or on the opposite side. Crossing fibers are found in a structure known as the **trapezoid body**, a compact bundle of fibers that crosses the midline in the lower pontine region (see Figure 40 and Figure 67C). The main function of the superior olivary complex is sound localization; this is based on the fact that an incoming sound will not reach the two ears at the exact same moment.

Fibers from the superior olivary complex either ascend on the same side or cross (in the trapezoid body) and ascend on the other side. They form a tract, the **lateral lemniscus**, which begins just above the level of these nuclei (see Figure 40). The lateral lemniscus carries the auditory information upward through the pons (see Figure 66B) to the inferior colliculus of the midbrain. There are nuclei scattered along the way, and some fibers may ter-

minate or relay in these nuclei; the lateral lemnisci are interconnected across the midline (not shown).

Almost all the axons of the lateral lemniscus terminate in the **inferior colliculus** (see Figure 9A and Figure 65B). The continuation of this pathway to the medial geniculate nucleus of the thalamus is discussed in the following illustration.

In summary, audition is a complex pathway, with numerous opportunities for synapses. Even though named a "lemniscus," it does not transmit information in the efficient manner seen with the medial lemniscus. It is important to note that although the pathway is predominantly a crossed system, there is also a significant ipsilateral component. There are also numerous interconnections between the two sides.

The auditory pathway has a feedback system, from the higher levels to lower levels (e.g., from the inferior colliculus to the superior olivary complex). The final link in this feedback is somewhat unique in the mammalian CNS, for it influences the cells in the receptor organ itself. This pathway, known as the **olivo-cochlear bundle**, has its cells of origin in the vicinity of the superior olivary complex. It has both a crossed and an uncrossed component. Its axons reach the hair cells of the cochlea by traveling in the VIIIth nerve. This system changes the responsiveness of the peripheral hair cells.

NEUROLOGICAL NEUROANATOMY

The auditory system is shown at various levels of the brainstem, including the upper medulla, all three pontine levels, and the lower midbrain (inferior collicular) level.

The cochlear nuclei are the first CNS synaptic relays for the auditory fibers from the peripheral spiral ganglion; these nuclei are found along the incoming VIIIth nerve at the level of the upper medulla (see Figure 67A). The superior olivary complex, consisting of several nuclei, is located at the lower pontine level (see Figure 66C), along with the trapezoid body, containing the crossing auditory fibers. By the mid-pons (see Figure 66B), the lateral lemniscus can be recognized. These fibers move toward the outer margin of the upper pons and terminate in the inferior colliculus (see Figure 65B).

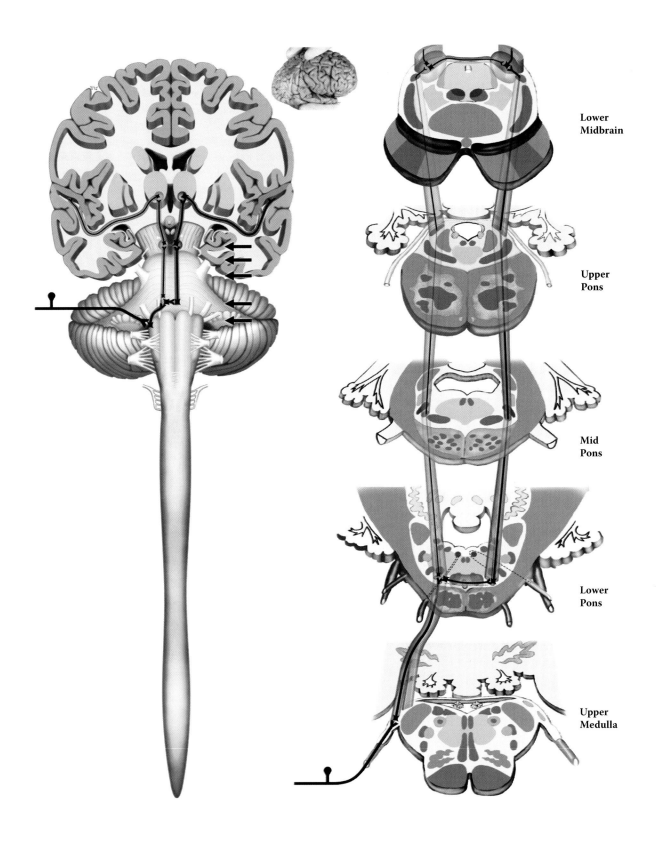

FIGURE 37: Auditory System 1 — Auditory Pathway 1

FIGURE 38
AUDITION 2

AUDITORY PATHWAY 2

This illustration shows the projection of the auditory system fibers from the level of the inferior colliculus, the lower midbrain, to the thalamus and then to the cortex.

Auditory information is carried via the lateral lemniscus to the inferior colliculus (see Figure 37 and Figure 40), after several synaptic relays. There is another synapse in this nucleus, making the auditory pathway overall somewhat different and more complex than the medial lemniscal and different than the visual pathways (see Figure 41A, Figure 41B, and Figure 41C). The inferior colliculi are connected to each other by a small commissure (not labeled).

The auditory information is next projected to a specific relay nucleus of the thalamus, the **medial geniculate** (nucleus) body (**MGB**, see Figure 12 and Figure 63). The tract that connects the two, the **brachium** of the inferior colliculus, can be seen on the dorsal aspect of the midbrain (see Figure 10; see also Figure 9A, not labeled); this is shown diagrammatically in the present figure.

From the medial geniculate nucleus the auditory pathway continues to the cortex. This projection, which courses beneath the lenticular (lentiform) nucleus of the basal ganglia (see Figure 22), is called the **sublenticular** pathway, the **inferior limb** of the internal capsule, or simply the **auditory radiation**. The cortical areas involved with receiving this information are the **transverse gyri of Heschl**, situated on the superior temporal gyrus, within the lateral fissure. The location of these gyri is shown in the inset as the primary auditory areas (also seen in a photographic view in the next illustration).

The medial geniculate nucleus is likely involved with some analysis and integration of the auditory information. More exact analysis occurs in the cortex. Further elaboration of auditory information is carried out in the adjacent cortical areas. On the dominant side for language, these cortical areas are adjacent to Wernicke's language area (see Figure 14A).

Sound frequency, known as **tonotopic** organization, is maintained all along the auditory pathway, starting in the cochlea. This can be depicted as a musical scale with high and low notes. The auditory system localizes the direction of a sound in the superior olivary complex (discussed with the previous illustration); this is done by analyzing the difference in the timing that sounds reach each ear and by the difference in sound intensity reaching each ear. The loudness of a sound would be represented physiologically by the number of receptors stimulated and by the frequency of impulses, as in other sensory modalities.

NEUROLOGICAL NEUROANATOMY

This view of the brain includes the midbrain level and the thalamus, with the lentiform nucleus lateral to it. The lateral ventricle is open (cut through its body) and the thalamus is seen to form the floor of the ventricle; the body of the caudate nucleus lies above the thalamus and on the lateral aspect of the ventricle.

The auditory fibers leave the inferior colliculus and course via the brachium of the inferior colliculus to the medial geniculate nucleus of the thalamus. From here the auditory radiation courses below the lentiform nucleus to the auditory gyri on the superior surface of the temporal lobe within the lateral fissure. The gyri are shown in the diagram above and in the next illustration.

This diagram also includes the lateral geniculate body (nucleus) which subserves the visual system and its projection, the optic radiation (to be discussed with Figure 41A and Figure 41B).

ADDITIONAL DETAIL

The temporal lobe structures are also shown, including the inferior horn of the lateral ventricle, the hippocampus proper, and adjoining structures relevant to the limbic system (Section D).

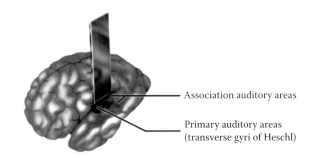

Association auditory areas

Primary auditory areas
(transverse gyri of Heschl)

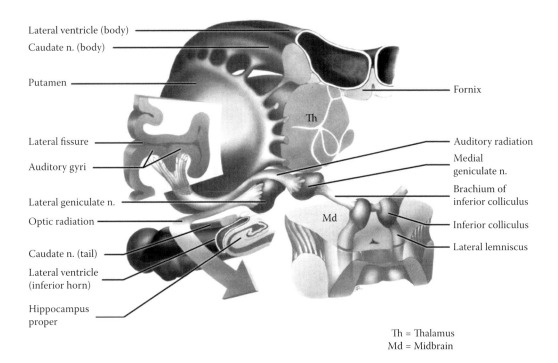

Lateral ventricle (body)

Caudate n. (body)

Putamen

Lateral fissure

Auditory gyri

Lateral geniculate n.

Optic radiation

Caudate n. (tail)

Lateral ventricle
(inferior horn)

Hippocampus
proper

Th

Md

Fornix

Auditory radiation

Medial
geniculate n.

Brachium of
inferior colliculus

Inferior colliculus

Lateral lemniscus

Th = Thalamus
Md = Midbrain

FIGURE 38: Auditory System 2 — Auditory Pathway 2

FIGURE 39
AUDITION 3

AUDITORY GYRI (PHOTOGRAPHIC VIEW)

This photographic view of the left hemisphere is shown from the lateral perspective (see Figure 14A). The lateral fissure has been opened, and this exposes two gyri, which are oriented transversely. These gyri are the areas of the cortex that receive the incoming auditory sensory information first. They are named the **transverse gyri of Heschl** (as was also shown in the previous illustration), the auditory gyri, areas 41 and 42 (see Figure 60).

The lateral fissure forms a complete separation between this part of the temporal lobe and the frontal and parietal lobes above. Looked at descriptively, the auditory gyri occupy the superior aspect of the temporal lobe, within the lateral fissure.

Cortical representation of sensory systems reflects the particular sensation (modality). The auditory gyri are organized according to pitch, giving rise to the term **tonotopic** localization. This is similar to the representation of the somatosensory system on the postcentral gyrus (somatotopic localization; the sensory "homunculus").

Further opening of the lateral fissure reveals some cortical tissue that is normally completely hidden from view. This area is the **insula** or insular cortex (see Figure 14B). The insula typically has five short gyri, and these are seen in the depth of the lateral fissure. It is important not to confuse the two areas, auditory gyri and insula. The position of the insula in the depth of the lateral fissure is also shown in a dissection of white matter bundles (see Figure 19B) and in the coronal slice of the brain (see Figure 29).

It should be noted that the lateral fissure has within it a large number of blood vessels, branches of the middle cerebral artery, which have been removed (see Figure 58). These branches emerge and then become distributed to the cortical tissue of the dorsolateral surface, including the frontal, temporal, parietal, and occipital cortex (discussed with Figure 58 and Figure 60). Other small branches to the internal capsule and basal ganglia are given off within the lateral fissure (discussed with Figure 62).

CLINICAL ASPECT

Since the auditory system has a bilateral pathway to the cortex, a lesion of the auditory pathway or cortex on one side will not lead to a total loss of hearing (deafness) of the opposite ear. Nonetheless, the pathway still has a strong crossed aspect; speech is directed to the dominant hemisphere.

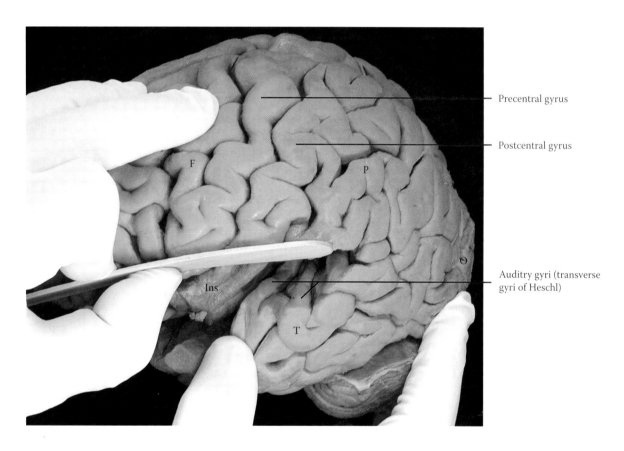

F = Frontal lobe
P = Parietal lobe
T = Temporal lobe
O = Occipital lobe

Ins = Insula

FIGURE 39: Auditory System 3 — Auditory Gyri (photograph)

FIGURE 40
SENSORY SYSTEMS

SENSORY NUCLEI AND ASCENDING TRACTS

This diagrammatic presentation of the internal structures of the brainstem is shown from the dorsal perspective (as in Figure 10 and Figure 36). The information concerning the various structures will be presented in an abbreviated manner, as most of the major points have been reviewed previously. The orientation of the cervical spinal cord representation should be noted.

The major sensory systems include:

- Dorsal column-medial lemniscus (discriminative touch, joint position, and vibration) and its nuclei
- Anterolateral system (pain and temperature)
- Trigeminal system and its nuclei (discriminative touch, pain, and temperature)
- Lateral lemniscus (audition), with its nuclei

THE DORSAL COLUMN-MEDIAL LEMNISCUS

The dorsal columns (gracile and cuneate tracts) of the spinal cord terminate (synapse) in the nuclei gracilis and cuneatus in the lowermost medulla (see Figure 9B). Axons from these nuclei then cross the midline (decussate) as the internal arcuate fibers (see Figure 67C), forming a new bundle called the medial lemniscus. These fibers ascend through the medulla, change orientation in the pons, and move laterally, occupying a lateral position in the midbrain.

THE ANTEROLATERAL SYSTEM

This tract, having already crossed in the spinal cord, ascends and continues through the brainstem. In the medulla it is situated posterior to the inferior olive. At the upper pontine level, this tract becomes associated with the medial lemniscus, and the two lie adjacent to each other in the midbrain region.

THE TRIGEMINAL PATHWAY

The sensory afferents for discriminative touch synapse in the principal nucleus of V; the fibers then cross at the level of the mid-pons and form a tract that joins the medial lemniscus. The pain and temperature fibers descend and form the descending trigeminal tract through the medulla with the nucleus adjacent to it. These fibers synapse and cross, over a wide area of the medulla, eventually joining the other trigeminal tract. The two tracts form the trigeminal pathway, which joins with the medial lemniscus in the uppermost pons (see Figure 36).

THE LATERAL LEMNISCUS

The auditory fibers (of CN VIII) enter the brainstem at the uppermost portion of the medulla. After the initial synapse in the cochlear nuclei, many of the fibers cross the midline, forming the trapezoid body. Some of the fibers synapse in the superior olivary complex. From this point, the tract known as the lateral lemniscus is formed. The fibers relay in the inferior colliculus.

CLINICAL ASPECT

This diagram allows the visualization of all the pathways together, which assists in understanding lesions of the brainstem. The cranial nerve nuclei affected help locate the level of the lesion.

One of the classic lesions of the brainstem is an infarct of the lateral medulla (see Figure 67B), known as the Wallenberg syndrome. (The blood supply of the brainstem is reviewed with Figure 58.) This lesion affects the pathways and cranial nerve nuclei located in the lateral area of the medulla, including the anterolateral tract and the lateral lemniscus, but not the medial lemniscus; the descending trigeminal system is also involved, as are the nuclei of CN IX and X. Additional deficits may include vestibular or cerebellar signs, as the vestibular nuclei are nearby and afferents to the cerebellum may be interrupted. Notwithstanding the fact that the lateral lemniscus is most likely involved in this lesion, auditory deficits are not commonly associated with this clinical syndrome, probably due to the fact that this is a bilateral pathway. The lateral meduallary syndrome is discussed with Figure 67B.

ADDITIONAL DETAIL

The superior cerebellar peduncles are shown in this diagram, although not part of the sensory systems. These will be described with the cerebellum (see Figure 57). This fiber pathway from the cerebellum to the thalamus decussates in the lower midbrain at the inferior collicular level (shown in cross-section, see Figure 65B).

The red nucleus is one of the prominent structures of the midbrain (see Figure 65A); its contribution to motor function in humans is not yet clear (discussed with Figure 47).

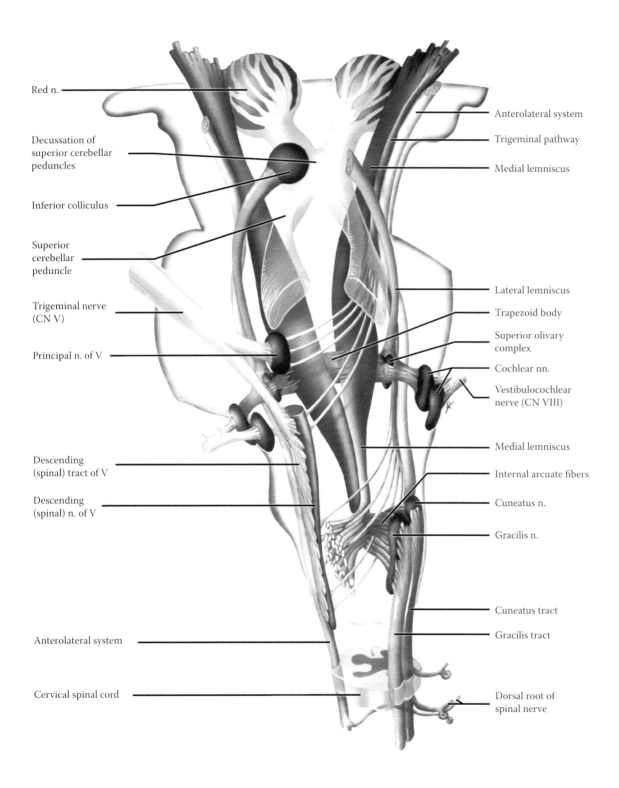

Red n.

Decussation of
superior cerebellar
peduncles

Inferior colliculus

Superior
cerebellar
peduncle

Trigeminal nerve
(CN V)

Principal n. of V

Descending
(spinal) tract of V

Descending
(spinal) n. of V

Anterolateral system

Cervical spinal cord

Anterolateral system

Trigeminal pathway

Medial lemniscus

Lateral lemniscus

Trapezoid body

Superior olivary
complex

Cochlear nn.

Vestibulocochlear
nerve (CN VIII)

Medial lemniscus

Internal arcuate fibers

Cuneatus n.

Gracilis n.

Cuneatus tract

Gracilis tract

Dorsal root of
spinal nerve

FIGURE 40: **Sensory Systems** — Sensory Nuclei and Ascending Tracts

FIGURE 41A
VISION 1

VISUAL PATHWAY 1

The visual image exists in the outside world, and is designated the visual field; there is a visual field for each eye. This image is projected onto the retina, where it is now termed the retinal field. Because of the lens of the eye, the visual information from the upper visual field is seen in the lower retina (and likewise for the lower visual field). The visual fields are also divided into temporal (lateral) and nasal (medial) portions. The temporal visual field of one eye is projected onto the nasal part of the retina of the ipsilateral eye, and onto the temporal part of the retina of the contralateral eye. The primary purpose of the visual apparatus (e.g., muscles) is to align the visual image on corresponding points of the retina of both eyes.

Visual processing begins in the retina with the photoreceptors, the highly specialized receptor cells, the **rods and cones**. The central portion of the visual field projects onto the **macular** area of the retina, composed of only cones, which is the area required for discriminative vision (e.g., reading) and color vision. Rods are found in the peripheral areas of the retina and are used for peripheral vision and seeing under conditions of low-level illumination. These receptors synapse with the **bipolar neurons** located in the retina, the first actual neurons in this system (functionally equivalent to DRG neurons). These connect with the **ganglion cells** (still in the retina) whose axons leave the retina at the optic disc to form the **optic nerve** (CN II). The optic nerve is in fact a tract of the CNS, as its myelin is formed by oligodendrocytes (the glial cell that forms and maintains CNS myelin).

After exiting from the orbit, the optic nerves undergo a partial crossing (decussation) in the **optic chiasm**. The fibers from both nasal retinas, representing the temporal visual fields, cross and then continue in the now-named **optic tract** (see Figure 15A and Figure 15B). The result of this rearrangement is to bring together the visual information from the visual field of one eye to the opposite side of the brain.

The visual fibers terminate in the **lateral geniculate nucleus (LGB)**, a specific relay nucleus of the thalamus (see Figure 12 and Figure 63). The lateral geniculate is a layered nucleus (see Figure 41C); the fibers of the optic nerve synapse in specified layers and, after processing, project to the **primary visual cortex, area 17**. The pro-

jection consists of two portions with some of the fibers projecting directly posteriorly, while others sweep forward alongside the inferior horn of the lateral ventricle in the temporal lobe, called **Meyer's loop** (see also Figure 41C); both then project to the visual cortex of the occipital lobe as the **geniculo-calcarine radiation**. The projection from thalamus to cortex eventually becomes situated behind the lenticular nucleus and is called the retro-lenticular portion of the internal capsule, or simply the **visual or optic radiation** (see also Figure 27, Figure 28B, and Figure 38).

The visual information goes to **area 17**, the **primary visual area**, also called the **calcarine cortex** (seen in the upper diagrams and also in the next illustration), and then to adjacent association areas 18 and 19.

CLINICAL ASPECT

The visual pathway is easily testable, even at the bedside. Lesions of the visual pathway are described as a deficit of the visual field, for example, loss of one-half of a field of vision is called **hemianopia** (visual loss is termed anopia). Loss of the visual field in both eyes is termed **homonymous** or **heteronymous**, as defined by the projection to the visual cortex on one side or both sides. Students should be able to draw the visual field defect in both eyes that would follow a lesion of the optic nerve, at the optic chiasm (i.e., bitemporal heteronymous hemianopia), and in the optic tract (i.e., homonymous hemianopia). (**Note to the Learner**: The best way of learning this is to do a sketch drawing of the whole visual pathway using colored pens or pencils.)

Lesions of the optic radiation are somewhat more difficult to understand:

- Loss of the fibers that project from the lower retinal field, those that sweep forward into the temporal lobe (Meyer's loop), results in a loss of vision in the upper visual field of both eyes on the side opposite the lesion, specifically the upper quadrant of both eyes, called superior (right or left) homonymous quadrantanopia.
- Loss of those fibers coming from the upper retinal field, which project directly posteriorly, passing deep within the parietal lobe, results in the loss of the lower visual field of both eyes on the side opposite the lesion, specifically the lower quadrant of both eyes, called inferior (right or left) homonymous quadrantanopia.

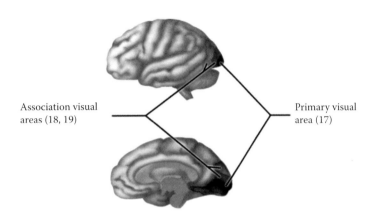

Association visual
areas (18, 19)

Primary visual
area (17)

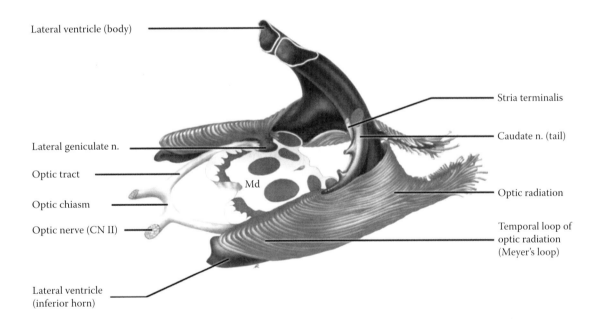

Lateral ventricle (body)

Stria terminalis

Caudate n. (tail)

Lateral geniculate n.

Optic tract

Optic chiasm

Optic radiation

Optic nerve (CN II)

Temporal loop of
optic radiation
(Meyer's loop)

Md

Lateral ventricle
(inferior horn)

Md = Midbrain

FIGURE 41A: Visual System 1 — Visual Pathway 1

FIGURE 41B
VISION 2

VISUAL PATHWAY 2 AND VISUAL CORTEX (PHOTOGRAPHS)

We humans are visual creatures. We depend on vision for access to information (the written word), the world of images (e.g., photographs, television), and the complex urban landscape. There are many cortical areas devoted to interpreting the visual world.

UPPER ILLUSTRATION (PHOTOGRAPHIC VIEW)

The visual fibers in the optic radiation terminate in **area 17**, the primary visual area, specifically the upper and lower gyri along the calcarine fissure. The posterior portion of area 17, extending to the occipital pole, is where macular vision is represented; the visual cortex in the more anterior portion of area 17 is the cortical region where the peripheral areas of the retina project.

The adjacent cortical areas, **areas 18 and 19**, are visual association areas; fibers are relayed here via the pulvinar of the thalamus (see below and Figure 12 and Figure 63). There are many other cortical areas for elaboration of the visual information, including a region on the inferior aspect of the hemisphere for face recognition.

LOWER ILLUSTRATION (PHOTOGRAPHIC VIEW)

This is a higher magnification of the medial aspect of the brain (shown in Figure 17). The interthalamic adhesion, fibers joining the thalamus of each side across the midline, has been cut (see Figure 6, not labeled). The optic chiasm is seen anteriorly; posteriorly, the tip of the pulvinar can be seen. The midbrain includes areas where fibers of the visual system synapse.

Fibers emerge from the **pulvinar**, the visually related association nucleus of the thalamus (see Figure 12 and Figure 63) and travel in the optic radiations to areas 18 and 19, the visual association areas of the cortex (shown in the previous diagram, alongside area 17). Some optic fibers terminate in the **superior colliculi** (see also Figure 9A and Figure 10), which are involved with coordinating eye movements (discussed with the next illustration). Visual fibers also end in the **pretectal "nucleus,"** an area in front of the superior colliculus, for the pupillary light reflex (reviewed with the next illustration). Some other fibers terminate in the **suprachiasmatic** nucleus of the hypothalamus (located above the optic chiasm), which is involved in the control of diurnal (day-night) rhythms.

The additional structures labeled in this illustration have been noted previously (see Figure 17 in Section A), except the superior medullary velum, located in the upper part of the roof of the fourth ventricle (see Figure 10); this band of white matter is associated with the superior cerebellar peduncles (discussed with the cerebellum, see Figure 57).

CLINICAL ASPECT

It is very important for the learner to know the visual system. The system traverses the whole brain and cranial fossa, from front to back, and testing the complete visual pathway from retina to cortex is an opportunity to sample the intactness of the brain from frontal pole to occipital pole.

Diseases of CNS myelin, such as multiple sclerosis (MS), affect the optic nerve or optic tract, causing visual loss. Sometimes this is the first manifestation of MS.

Visual loss can occur for many reasons, one of which is the loss of blood supply to the cortical areas. The visual cortex is supplied by the posterior cerebral artery (from the vertebro-basilar system, discussed with Figure 61). Part of the occipital pole, with the representation of the macular area of vision, may be supplied by the middle cerebral artery (from the internal carotid system, see Figure 60). In some cases, macular sparing is found after occlusion of the posterior cerebral artery, presumably because the blood supply to this area was coming from the carotid vascular supply.

ADDITIONAL DETAIL

The work on visual processing and its development has offered us remarkable insights into the formation of synaptic connections in the brain, critical periods in development, and the complex way in which sensory information is "processed" in the cerebral cortex. It is now thought that the primate brain has more than a dozen specialized visual association areas, including face recognition, color, and others. Neuroscience texts should be consulted for further details concerning the processing of visual information.

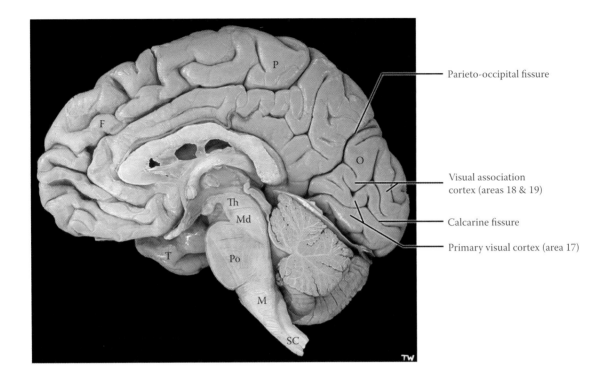

Parieto-occipital fissure

Visual association
cortex (areas 18 & 19)

Calcarine fissure

Primary visual cortex (area 17)

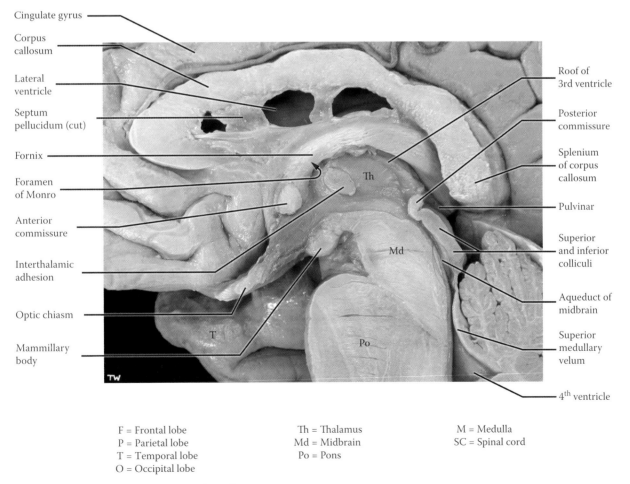

Cingulate gyrus

Corpus
callosum

Lateral
ventricle

Septum
pellucidum (cut)

Fornix

Foramen
of Monro

Anterior
commissure

Interthalamic
adhesion

Optic chiasm

Mammillary
body

Roof of
3rd ventricle

Posterior
commissure

Splenium
of corpus
callosum

Pulvinar

Superior
and inferior
colliculi

Aqueduct of
midbrain

Superior
medullary
velum

4th ventricle

F = Frontal lobe   Th = Thalamus   M = Medulla
P = Parietal lobe   Md = Midbrain   SC = Spinal cord
T = Temporal lobe   Po = Pons
O = Occipital lobe

FIGURE 41B: Visual System 2 — Visual Pathway 2 and Visual Cortex (photograph)

FIGURE 41C
VISION 3

VISUAL REFLEXES

The upper illustration shows the details of the optic radiation alongside the posterior horn of the lateral ventricle. The fibers end in the visual cortex along both banks of the calcarine fissure, the primary visual area, area 17 (see Figure 41A and Figure 41B).

This illustration also shows some fibers from the optic tract that project to the **superior colliculus** by-passing the lateral geniculate via the **brachium** of the superior colliculus (labeled in the lower illustration). This nucleus serves as an important center for visual reflex behavior, particularly involving eye movements. Fibers project to nuclei of the extra-ocular muscles (see Figure 8A and Figure 51A) and neck muscles via a small pathway, the **tecto-spinal tract**, which is found incorporated with the MLF, the medial longitudinal fasciculus (see Figure 51B).

Reflex adjustments of the visual system are also required for seeing nearby objects, known as the accommodation reflex. A small but extremely important group of fibers from the optic tract (not shown) project to the **pretectal area** for the pupillary light reflex.

- **Accommodation reflex** The accommodation reflex is activated when looking at a nearby object, as in reading. Three events occur simultaneously — convergence of both eyes (involving both medial recti muscles), a change (rounding) of the curvature of the lens, and pupillary constriction. This reflex requires the visual information to be processed at the cortical level. The descending cortico-bulbar fibers (see Figure 46 and Figure 48) go to the oculomotor nucleus and influence both the motor portion (to the medial recti muscles), and also to the parasympathetic (Edinger-Westphal) portion (to the smooth muscle of the lens and the pupil, via the ciliary ganglion) to effect the reflex.
- **Pupillary light reflex** Some of the visual information (from certain ganglion cells in the retina) is carried in the optic nerve and tract to the midbrain. A nucleus located in the area in front of the colliculi (the other name for the colliculi is the tectal area, see Figure 9A, Figure 10, and Figure 65), called the **pretectal area** (see also Figure 51B), is the site of synapse for the pupillary light reflex. Shining light on the retina causes a constriction of the pupil on the same side; this is the *direct* pupillary light reflex. Fibers also cross to the nucleus on the other side (via a commissure), and the pupil of the other eye reacts as well; this is the *consensual* light reflex. The efferent part of the reflex involves the parasympathetic nucleus (Edinger-Westphal) of the oculomotor nucleus (see Figure 8A and also Figure 65A); the efferent fibers course in CN III, synapsing in the ciliary ganglion (parasympathetic) in the orbit before innervating the smooth muscle of the iris, which controls the diameter of the pupil.

CLINICAL ASPECT

The pupillary light reflex is a critically important clinical sign, particularly in patients who are in a coma, or following a head injury. It is essential to ascertain the status of the reaction of the pupil to light, ipsilaterally and on the opposite side. The learner is encouraged to draw out this pathway and to work out the clinical picture of a lesion involving the afferent visual fibers, the midbrain area, and a lesion affecting the efferent fibers (CN III).

In a disease such as multiple sclerosis, or with diseases of the retina, there can be a reduced sensory input via the optic nerve, and this can cause a condition called a "relative afferent pupillary defect." A specific test for this is the swinging light reflex, which is performed in a dimly lit room. Both pupils will constrict when the light is shone on the normal side. As the light is shone in the affected eye, because of the diminished afferent input from the retina to the pretectal nucleus, the pupil of this eye will dilate in a paradoxical manner.

CN III, the oculomotor nerve, is usually involved in brain herniation syndromes, particularly uncal herniation (discussed with Figure 15B). This results in a fixed dilated pupil on one side, a critical sign when one is concerned about increased intracranial pressure from any cause. The significance and urgency of this situation must be understood by anyone involved in critical care.

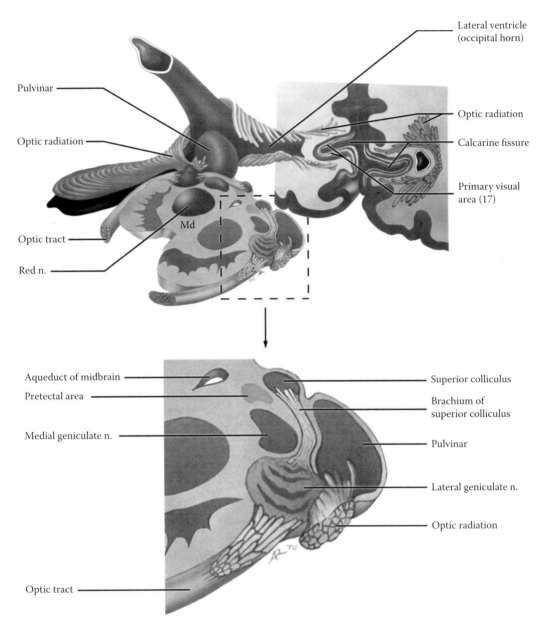

Pulvinar

Optic radiation

Optic tract

Red n.

Md

Lateral ventricle
(occipital horn)

Optic radiation

Calcarine fissure

Primary visual
area (17)

Aqueduct of midbrain

Pretectal area

Medial geniculate n.

Optic tract

Superior colliculus

Brachium of
superior colliculus

Pulvinar

Lateral geniculate n.

Optic radiation

Md = Midbrain

FIGURE 41C: **Visual System 3** — Visual Reflexes

PART II: RETICULAR FORMATION
FIGURE 42A
RETICULAR FORMATION 1

RETICULAR FORMATION: ORGANIZATION

The reticular formation, **RF**, is the name for a group of neurons found throughout the brainstem. Using the ventral view of the brainstem, the reticular formation occupies the central portion or core area of the brainstem from midbrain to medulla (see also brainstem cross-sections in Figure 65–Figure 67).

This collection of neurons is a phylogenetically old set of neurons that functions like a network or reticulum, from which it derives its name. The RF receives afferents from most of the sensory systems (see next illustration) and projects to virtually all parts of the nervous system.

Functionally, it is possible to localize different subgroups within the reticular formation:

- **Cardiac and respiratory "centers"**: Subsets of neurons within the medullary reticular formation and also in the pontine region are responsible for the control of the vital functions of heart rate and respiration. The importance of this knowledge was discussed in reference to the clinical emergency, tonsillar herniation (with Figure 9B).
- **Motor areas**: Both the pontine and medullary nuclei of the reticular formation contribute to motor control via the cortico-reticulo-spinal system (discussed in Section B, Part III, Introduction; also with Figure 49A and Figure 49B). In addition, these nuclei exert a very significant influence on muscle tone, which is very important clinically (discussed with Figure 49B).
- **Ascending projection system**: Fibers from the reticular formation ascend to the thalamus and project to various nonspecific thalamic nuclei. From these nuclei, there is a diffuse distribution of connections to all parts of the cerebral cortex. This whole system is concerned with consciousness and is known as the **ascending reticular activating system (ARAS)**.
- **Pre-cerebellar nuclei**: There are numerous nuclei in the brainstem that are located within the boundaries of the reticular formation that project to the cerebellum. These are not always included in discussions of the reticular formation.

It is also possible to describe the reticular formation topographically. The neurons appear to be arranged in three longitudinal sets; these are shown in the left-hand side of this illustration:

- The **lateral group** consists of neurons that are small in size. These are the neurons that receive the various inputs to the reticular formation, including those from the anterolateral system (pain and temperature, see Figure 34), the trigeminal pathway (see Figure 35), as well as auditory and visual input.
- The next group is the **medial group**. These neurons are larger in size and project their axons upward and downward. The ascending projection from the midbrain area is particularly involved with the consciousness system. Nuclei within this group, notably the nucleus gigantocellularis of the medulla, and the pontine reticular nuclei, caudal (lower) and oral (upper) portions, give origin to the two reticulo-spinal tracts (discussed with the next illustration, also Figure 49A and Figure 49B).
- Another set of neurons occupy the midline region of the brainstem, the **raphe nuclei**, which use the catecholamine serotonin for neurotransmission. The best-known nucleus of this group is the nucleus raphe magnus, which plays an important role in the descending pain modulation system (to be discussed with Figure 43).

In addition, both the locus ceruleus (shown in the upper pons) and the periaqueductal gray (located in the midbrain, see next illustration and also Figure 65 and Figure 65A) are considered part of the reticular formation (discussed with the next illustration).

In summary, the reticular formation is connected with almost all parts of the CNS. Although it has a generalized influence within the CNS, it also contains subsystems that are directly involved in specific functions. The most clinically significant aspects are:

- Cardiac and respiratory centers in the medulla
- Descending systems in the pons and medulla that participate in motor control and influence muscle tone
- Ascending pathways in the upper pons and midbrain that contribute to the consciousness system

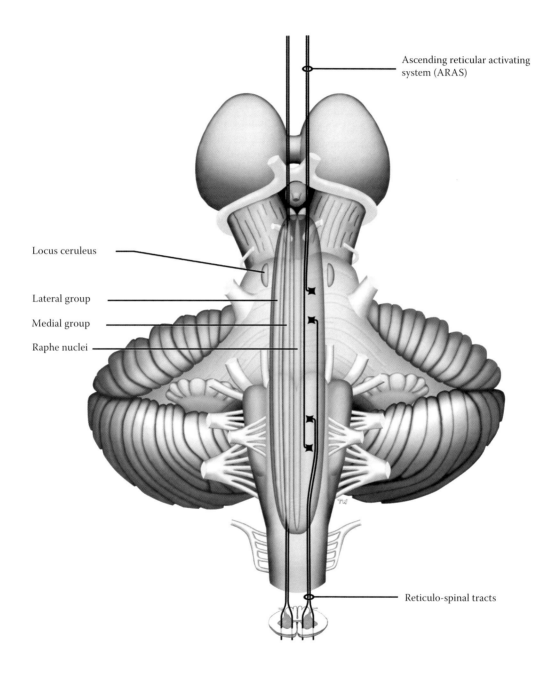

FIGURE 42A: Reticular Formation 1 — Organization

FIGURE 42B
RETICULAR FORMATION 2

RETICULAR FORMATION: NUCLEI

In this diagram, the reticular formation is being viewed from the dorsal (posterior) perspective (see Figure 10 and Figure 40). Various nuclei of the reticular formation, **RF**, which have a significant (known) functional role, are depicted, as well as the descending tracts emanating from some of these nuclei.

Functionally, there are afferent and efferent nuclei in the reticular formation and groups of neurons that are distinct because of the catecholamine neurotransmitter used, either serotonin or noradrenaline. The afferent and efferent nuclei of the RF include:

- Neurons that receive the various inputs to the RF are found in the lateral group (as discussed with the previous illustration). In this diagram, these neurons are shown receiving collaterals (or terminal branches) from the ascending antero-lateral system, carrying pain and temperature (see Figure 34; also Figure 35).
- The neurons of the medial group are larger in size, and these are the output neurons of the reticular formation, at various levels. These cells project their axons upward or downward. The **nucleus gigantocellularis** of the medulla, and the **pontine reticular nuclei, caudal**, and **oral** portions, give rise to the descending tracts that emanate from these nuclei — the medial and lateral reticulo-spinal pathways, part of the indirect voluntary and nonvoluntary motor system (see Figure 49A and Figure 49B).
- Raphe nuclei use the neurotransmitter serotonin and project to all parts of the CNS. Recent studies indicate that serotonin plays a significant role in emotional equilibrium, as well as in the regulation of sleep. One special nucleus of this group, the **nucleus raphe magnus**, located in the upper part of the medulla, plays a special role in the descending pain modulation pathway (described with the next illustration).

There are other nuclei in the brainstem that appear to functionally belong to the reticular formation yet are not located within the core region. These include the periaqueductal gray and the locus ceruleus.

The **periaqueductal gray** of the midbrain (for its location see Figure 65 and Figure 65A) includes neurons that are found around the aqueduct of the midbrain (see also Figure 20B). This area also receives input (illustrated but not labeled in this diagram) from the ascending sensory systems conveying pain and temperature, the anterolateral pathway; the same occurs with the trigeminal system. This area is part of a descending pathway to the spinal cord, which is concerned with pain modulation (as shown in the next illustration).

The **locus ceruleus** is a small nucleus in the upper pontine region (see Figure 66 and Figure 66A). In some species (including humans), the neurons of this nucleus accumulate a pigment that can be seen when the brain is sectioned (prior to histological processing, see photograph of the pons, Figure 66). Output from this small nucleus is distributed widely throughout the brain to virtually every part of the CNS, including all cortical areas, subcortical structures, the brainstem and cerebellum, and the spinal cord. The neurotransmitter that is used by these neurons is noradrenaline and its electrophysiological effects at various synapses are still not clearly known. Although the functional role of this nucleus is still not completely understood, the locus ceruleus has been thought to act like an "alarm system" in the brain. It has been implicated in a wide variety of CNS activities, such as mood, the reaction to stress, and various autonomic activities.

The cerebral cortex sends fibers to the RF nuclei, including the periaqueductal gray, forming part of the cortico-bulbar system of fibers (see Figure 46). The nuclei that receive this input and then give off the pathways to the spinal cord form part of an indirect voluntary motor system — the cortico-reticulo-spinal pathways (discussed in Section B, Part III, Introduction; see Figure 49A and Figure 49B). In addition, this system is known to play an extremely important role in the control of muscle tone (discussed with Figure 49B).

CLINICAL ASPECT

Lesions of the cortical input to the reticular formation in particular have a very significant impact on muscle tone. In humans, the end result is a state of increased muscle tone, called spasticity, accompanied by hyper-reflexia, an increase in the responsiveness of the deep tendon reflexes (discussed with Figure 49B).

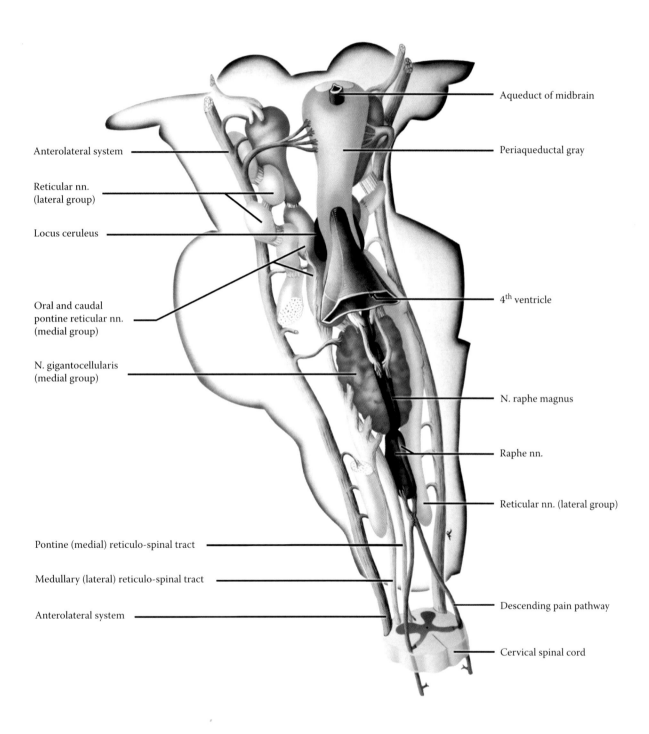

Aqueduct of midbrain

Periaqueductal gray

Anterolateral system

Reticular nn.
(lateral group)

Locus ceruleus

4th ventricle

Oral and caudal
pontine reticular nn.
(medial group)

N. gigantocellularis
(medial group)

N. raphe magnus

Raphe nn.

Reticular nn. (lateral group)

Pontine (medial) reticulo-spinal tract

Medullary (lateral) reticulo-spinal tract

Anterolateral system

Descending pain pathway

Cervical spinal cord

FIGURE 42B: Reticular Formation 2 — Nuclei

FIGURE 43
RETICULAR FORMATION 3

PAIN MODULATION SYSTEM

Pain, both physical and psychic, is recognized by the nervous system at multiple levels. Localization of pain, knowing which parts of the limbs and body wall are involved, requires the cortex of the postcentral gyrus (SI); SII is also likely involved in the perception of pain (discussed with Figure 36). There is good evidence that some "conscious" perception of pain occurs at the thalamic level.

We have a built-in system for dampening the influences of pain from the spinal cord level — the **descending pain modulation pathway**. This system apparently functions in the following way: The neurons of the periaqueductal gray can be activated in a number of ways. It is known that many ascending fibers from the anterolateral system and trigeminal system activate neurons in this area (only the anterolateral fibers are being shown in this illustration), either as collaterals or direct endings of these fibers in the midbrain. This area is also known to be rich in opiate receptors, and it seems that neurons of this region can be activated by circulating endorphins. Experimentally, one can activate these neurons by direct stimulation or by a local injection of morphine. In addition, descending cortical fibers (cortico-bulbar) may activate these neurons (see Figure 46).

The axons of some of the neurons of the periaqueductal gray descend and terminate in one of the serotonin-containing raphe nuclei in the upper medulla, the **nucleus raphe magnus**. From here, there is a descending, crossed, pathway, which is located in the dorsolateral white matter (funiculus) of the spinal cord. The serotonergic fibers terminate in the substantia gelatinosa of the spinal cord, a nuclear area of the dorsal horn of the spinal cord where the pain afferents synapse (see Figure 32). The descending serotonergic fibers are thought to terminate on small interneurons, which contain enkephalin. There is evidence that these enkephalin-containing spinal neurons inhibit the transmission of the pain afferents entering the spinal cord from peripheral pain receptors. Thus, descending influences are thought to modulate a local circuit.

There is a proposed mechanism that these same interneurons in the spinal cord can be activated by stimulation of other sensory afferents, particularly those from the touch receptors in the skin and the mechanoreceptors in the joints; these give rise to anatomically large well-myelinated peripheral nerve fibers, which send collaterals to the dorsal horn (see Figure 32). This is the physiological basis for the **gate theory of pain**. In this model, the same circuit is activated at a segmental level.

It is useful to think about multiple gates for pain transmission. We know that mental states and cognitive processes can affect, positively and negatively, the experience of pain and our reaction to pain. The role of the limbic system and the "emotional reaction" to pain will be discussed in Section D.

CLINICAL ASPECT

In our daily experience with local pain, such as a bump or small cut, the common response is to vigorously rub and/or shake the limb or the affected region. What we may be doing is activating the local segmental circuits via the touch- and mechano-receptors to decrease the pain sensation.

Some of the current treatments for pain are based upon the structures and neurotransmitters being discussed here. The gate theory underlies the use of transcutaneous stimulation, one of the current therapies offered for the relief of pain. More controversial and certainly less certain is the postulated mechanism(s) for the use of acupuncture in the treatment of pain.

Most discussions concerning pain refer to ACUTE pain, or short-term pain caused by an injury or dental procedure. CHRONIC pain should be regarded from a somewhat different perspective. Living with pain on a daily basis, caused, for example, by arthritis, cancer, or diabetic neuropathy, is an unfortunately tragic state of being for many people. Those involved with pain therapy and research on pain have proposed that the CNS actually rewires itself in reaction to chronic pain and may in fact become more sensitized to pain the longer the pain pathways remain active; some of this may occur at the receptor level. Many of these people are now being referred to "pain clinics," where a team of physicians and other health professionals (e.g., anesthetists, neurologists, psychologists) try to assist people, using a variety of therapies, to alleviate their disabling condition.

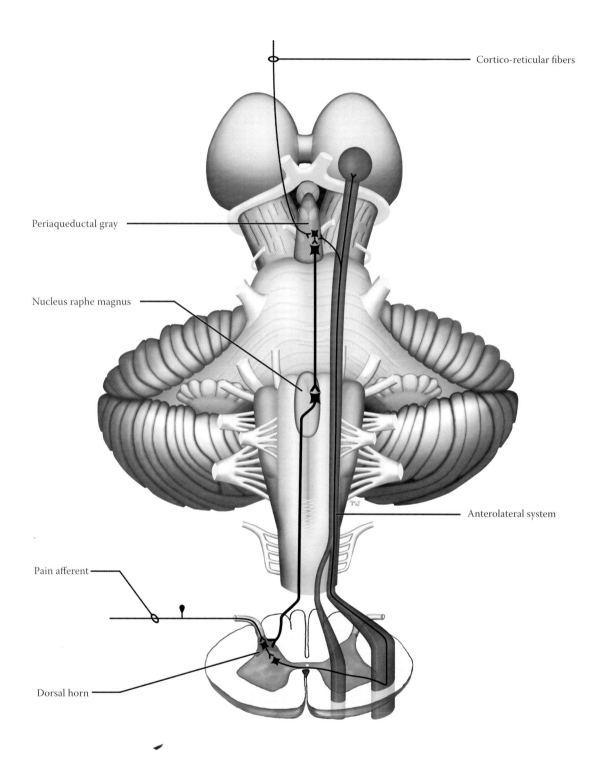

Cortico-reticular fibers

Periaqueductal gray

Nucleus raphe magnus

Anterolateral system

Pain afferent

Dorsal horn

FIGURE 43: Reticular Formation 3 — Pain Modulation System

PART III: MOTOR SYSTEMS INTRODUCTION

There are multiple areas involved in motor control, which is the reason for the title Motor Systems (plural). The parts of the CNS that regulate the movement of our muscles include: motor areas of the cerebral cortex, the basal ganglia (including the substantia nigra and the subthalamic nucleus), the cerebellum (with its functional subdivisions), nuclei of the brainstem including portions of the reticular formation, and finally the output motor neurons of the cranial nerve motor nuclei and the spinal cord (the anterior horn cells, also known as the lower motor neurons).

One way of approaching this complexity is to separate motor activity into a voluntary system and a nonvoluntary system.

- **Voluntary motor control** involves both direct and indirect pathways:
  - The **direct voluntary pathway**, for the control of fine motor movements, includes the cortico-bulbar fibers to cranial nerve nuclei and the cortico-spinal fibers and its pathway continuation in the spinal cord, the lateral cortico-spinal tract.
  - The **indirect voluntary pathway**, an older system for the control of proximal joint movements and axial musculature, involves the motor cortex acting through the reticular formation of the brainstem.
- **Nonvoluntary motor regulation** is an older system for adjustment of the body to vestibular and gravitational changes, as well as visual input. The various nuclei of the brainstem (the red nucleus, the vestibular nuclei, and the reticular formation) are regulated by older functional parts of the cerebellum but may be influenced by the cerebral cortex. This system also controls muscle tone and the deep tendon reflex, the reactivity of the muscle (stretch) reflex.

There are three areas of the cerebral cortex directly involved in motor control (see Figure 14A, Figure 17, Figure 53, and Figure 60):

- The **motor cortex** is the precentral gyrus anatomically, **area 4**, also called the motor strip. The various portions of the body are functionally represented along this gyrus; the fingers and particularly the thumb, as well as the tongue and lips are heavily represented on the dorsolateral surface, with the lower limb on the medial surface of the hemisphere. This motor "homunculus" is not unlike the sensory homun-

culus. The large neurons of the motor strip (in the deeper cortical layers) send their axons as projection fibers to form the cortico-bulbar and cortico-spinal tracts. It is this cortical strip that contributes most to voluntary movements.

- Anterior to this is another wedge-shaped cortical area, the **premotor cortex, area 6**, with a less definite body representation. This cortical area sends its axons to the motor cortex as well as to the cortico-spinal tract, and its function likely has more to do with proximal joint control and postural adjustments needed for movements.
- The **supplementary motor cortex** is located on the dorsolateral surface and mostly on the medial surface of the hemisphere, anterior to the motor areas. This is an organizing area for movements and its axons are sent to the premotor and motor cortex.

These motor areas of the cerebral cortex are regulated by the basal ganglia and certain (newer) parts of the cerebellum. These two important large areas of the brain are "working behind the scenes" to adjust and calibrate the neuronal circuits of the cerebral cortex involved in motor control. All these areas also receive input from other parts of the cerebral cortex, particularly from the sensory postcentral gyrus, as well as from the parietal lobe.

The voluntary and nonvoluntary motor systems act directly or indirectly upon the motor neurons in the spinal cord and the cranial nerve motor nuclei, whose axons innervate the muscles. Therefore, there are several pathways that "descend" through the spinal cord — each with its own crossing (decussation) and each of which may result in a functional loss of the control of movement, with a change in responsiveness of the stretch (deep tendon) reflexes.

The motor pathways (tracts) are called descending because they commence in the cortex or brainstem and influence motor cells lower down in the neuraxis, either in the brainstem or spinal cord. Those neurons in the cortex or brainstem (including the reticular formation) giving rise to these pathways are collectively called the **upper motor neurons**. The motor neurons in the spinal cord or brainstem that give origin to the peripheral efferent fibers (spinal and cranial nerves) are often called collectively the **lower motor neuron** (discussed with Figure 44).

LEARNING PLAN

This section will consider the motor areas of the cerebral cortex, the basal ganglia, the cerebellum, the motor nuclei of the thalamus, and the nuclei of the brainstem and reticular formation involved in motor regulation. The same standardized diagram of the nervous system will be used as with the sensory systems, as well as the inclusion of select X-sections of the spinal cord and brainstem.

The descending tracts or pathways that will be considered include:

- **Cortico-spinal tract**: This pathway originates in motor areas of the cerebral cortex. The cortico-spinal tract, from cortex to spinal cord, is a relatively new tract and the most important for voluntary movements in humans, particularly of the hand and digits — the direct voluntary motor pathway.
- **Cortico-bulbar fibers**: This is a descriptive term that is poorly defined and includes all fibers that go to the brainstem, both cranial nerve nuclei and other brainstem nuclei. The fibers that go to the reticular formation include those that form part of the indirect voluntary motor pathway. The cortico-pontine fibers are described with the cerebellum.
- **Rubro-spinal tract**: The red nucleus of the midbrain gives rise to the rubro-spinal tract. Its connections are such that it may play a role in voluntary and nonvoluntary motor activity; this may be the case in higher primates, but its precise role in humans is not clear.
- **Reticulo-spinal tracts**: These tracts are involved in the indirect voluntary pathways and in nonvoluntary motor regulation, as well as in the underlying control of muscle tone and reflex responsiveness. Two tracts descend from the reticular formation, one from the pontine region, the medial reticulo-spinal tract, and one from the medulla, the lateral reticulo-spinal tract.
- **Lateral vestibulo-spinal tract**: The lateral vestibular nucleus of the pons gives rise to the lateral vestibulo-spinal tract. This nucleus plays an important role in the regulation of our responses to gravity (vestibular afferents). It is therefore a nonvoluntary pathway. It is under control of the cerebellum, not the cerebral cortex.
- **Medial longitudinal fasciculus (MLF)**: This is a complex pathway of the brainstem and upper spinal cord that serves to coordinate various eye and neck reflexes. There are both ascending and descending fibers within the MLF, from vestibular and other nuclei.

Broca's area for the motor control of speech is situated on the dominant side on the dorsolateral surface, a little anterior to the lower portions of the motor areas (see Figure 14A). The frontal eye field, in front of the premotor area, controls voluntary eye movements (see Figure 14A).

CLINICAL ASPECT

The conceptual approach to the motor system as comprising an upper motor neuron and a lower motor neuron is most important for clinical neurology. A typical human lesion of the brain (e.g., vascular, trauma, tumor) usually affects cortical and subcortical areas, and several of the descending systems, resulting in a mixture of deficits of movement, as well as a change in muscle tone (flaccidity or spasticity) and an alteration of the stretch reflexes (discussed with Figure 49B).

There is one abnormal reflex that indicates, in the human, that there has been a lesion interrupting the cortico-spinal pathway — at any level (cortex, white matter, internal capsule, brainstem, spinal cord). The reflex involves stroking the lateral aspect of the bottom of the foot (a most uncomfortable sensation for most people). Normally, the response involves flexion of the toes, the **plantar reflex**, and oftentimes an attempt to withdraw the limb. Testing this same reflex after a lesion interrupts the cortico-spinal pathway results in an upward movement of the big toe (extension) and a fanning apart of the other toes. The abnormal response is called a **Babinski sign — not reflex** — and it can be elicited almost immediately after any lesion that interrupts any part of the cortico-spinal pathway, from cortex through to spinal cord (except spinal shock, see Figure 5).

Most interestingly, this Babinski sign is normally present in the infant and disappears somewhere in the second year of life, concurrent with the myelination that occurs in this pathway.

FIGURE 44
SPINAL CORD CROSS-SECTION

MOTOR-ASSOCIATED NUCLEI

UPPER ILLUSTRATION

The motor regions of the spinal cord in the ventral horn are shown in this diagram. The lateral motor nuclei supply the distal musculature (e.g., the hand), and as would be expected this area is largest in the region of the limb plexuses (brachial and lumbosacral, see Figure 69). The medial group of neurons supplies the axial musculature.

LOWER ILLUSTRATION

In the spinal cord, the neurons that are located in the ventral or anterior horn, and are (histologically) the anterior horn cells, are usually called the **lower motor neurons**. Physiologists call these neurons the **alpha motor neurons**. In the brainstem, these neurons include the motor neurons of the cranial nerves (see Figure 8A). Since all of the descending influences converge upon the lower motor neurons, these neurons have also been called, in a functional sense, **the final common pathway**. The lower motor neuron and its axon and the muscle fibers that it activates are collectively called the **motor unit**. The intactness of the motor unit determines muscle strength and muscle function.

MOTOR REFLEXES

The **myotatic reflex** is elicited by stretching a muscle (e.g., by tapping on its tendon), and this causes a contraction of the same muscle that was stretched; thus the reflex is also known as the **stretch reflex**, the **deep tendon reflex**, often simply **DTR**. In this reflex arc (shown on the left side), the information from the muscle spindle (afferent) ends directly on the anterior horn cell (efferent); there is only one synapse (i.e., a monosynaptic reflex).

All other reflexes, even a simple withdrawal reflex (e.g., touching a hot surface) involves some central processing (more than one synapse, multisynaptic) in the spinal cord, prior to the response (shown on the right side). All these reflexes involve hard-wired circuits of the spinal cord but are influenced by information descending from higher levels of the nervous system.

Recent studies indicate that complex motor patterns are present in the spinal cord, such as stepping movements with alternating movements of the limbs, and that influences from higher centers provide the organization for these built-in patterns of activity.

CLINICAL ASPECT

The deep tendon reflex is a monosynaptic reflex and perhaps the most important for a neurological examination. The degree of reactivity of the lower motor neuron is influenced by higher centers, also called descending influences, particularly by the reticular formation (to be discussed with Figure 49B). An increase in this reflex responsiveness is called *hyperreflexia*, a decrease *hyporeflexia*. The state of activity of the lower motor neuron also influences **muscle tone** — the "feel" of a muscle at rest and the way in which the muscles react to passive stretch (by the examiner); again, there be may be *hypertonia* or *hypotonia*.

Disease or destruction of the anterior horn cells results in weakness or paralysis of the muscles supplied by those neurons. The extent of the weakness depends upon the extent of the neuronal loss and is rated on a clinical scale, called the MRC (Medical Research Council). There is also a decrease in muscle tone, and a decrease in reflex responsiveness (hyporeflexia) of the affected segments; the plantar response is normal.

The specific disease that affects these neurons is poliomyelitis, a childhood infectious disease carried in fecal-contaminated water. This disease entity has almost been totally eradicated in the industrialized world by immunization of all children.

In adults, the disease that affects these neurons specifically (including cranial nerve motor neurons) is amyotrophic lateral sclerosis, ALS, also known as Lou Gehrig's disease. In this progressive degenerative disease there is also a loss of the motor neurons in the cerebral cortex (the upper motor neurons). The clinical picture depends upon the degree of loss of the neurons at both levels. People afflicted with this devastating disease suffer a continuous march of loss of function, including swallowing and respiratory function, leading to their death. Researchers are actively seeking ways to arrest the destruction of these neurons.

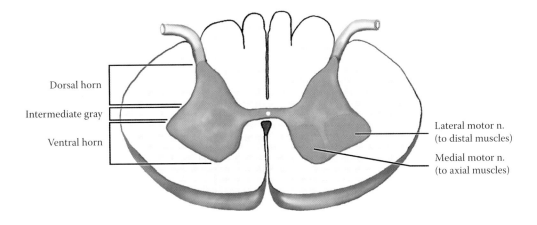

Dorsal horn

Intermediate gray

Ventral horn

Lateral motor n.
(to distal muscles)

Medial motor n.
(to axial muscles)

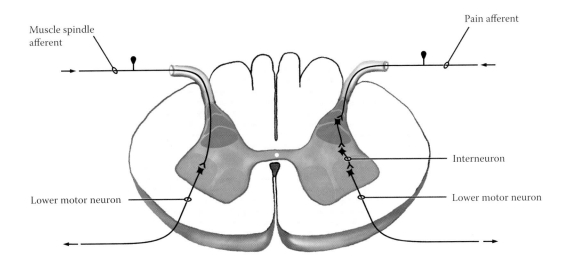

Muscle spindle
afferent

Pain afferent

Interneuron

Lower motor neuron

Lower motor neuron

FIGURE 44: Spinal Cord Nuclei — Motor

FIGURE 45
CORTICO-SPINAL TRACT —
THE PYRAMIDAL SYSTEM

DIRECT VOLUNTARY PATHWAY

The cortico-spinal tract, a direct pathway linking the cortex with the spinal cord, is the most important one for voluntary motor movements in humans.

This pathway originates mostly from the motor areas of the cerebral cortex, areas 4 and 6 (see Figure 14A, Figure 17, and Figure 60; discussed in Section B, Part III, Introduction and with Figure 48). The well-myelinated axons descend through the white matter of the hemispheres, through the posterior limb of the internal capsule (see Figure 26, Figure 27, Figure 28A, and Figure 28B), continue through the midbrain and pons (see below) and are then found within the medullary pyramids (see Figure 6 and Figure 7). Hence, the cortico-spinal pathway is often called the **pyramidal tract**, and clinicians may sometimes refer to this pathway as the pyramidal system. At the lowermost part of the medulla, most (90%) of the cortico-spinal fibers decussate (cross) in the **pyramidal decussation** (see Figure 7) and form the **lateral cortico-spinal tract** in the spinal cord (see Figure 68).

Many of these fibers end directly on the lower motor neuron, particularly in the cervical spinal cord. This pathway is involved with controlling the individualized movements, particularly of our fingers and hands (i.e., the distal limb musculature). Experimental work with monkeys has shown that, after a lesion is placed in the medullary pyramid, there is muscle weakness and a loss of ability to perform fine movements of the fingers and hand (on the opposite side); the animals were still capable of voluntary gross motor movements of the limb. There was no change in the deep tendon reflexes, and a decrease in muscle tone was reported. The innervation for the lower extremity is similar but clearly involves less voluntary activity.

Those fibers that do not cross in the pyramidal decussation form the **anterior (or ventral) cortico-spinal tract**. Many of the axons in this pathway will cross before terminating, while others supply motor neurons on both sides. The ventral pathway is concerned with movements of the proximal limb joints and axial movements, similar to other pathways of the nonvoluntary motor system.

Other areas of the cortex contribute to the cortico-spinal pathway; these include the sensory cortical areas, the postcentral gyrus (also discussed with the next illustration).

NEUROLOGICAL NEUROANATOMY

The cross-sectional levels for following this pathway include the upper midbrain, the mid-pons, the mid-medulla, and cervical and lumbar spinal cord levels.

After emerging from the internal capsule, the cortico-spinal tract is found in the midportion of the cerebral peduncles in the midbrain (see Figure 6, Figure 7, next illustration, and Figure 48). The cortico-spinal fibers are then dispersed in the pontine region and are seen as bundles of axons among the pontine nuclei (see Figure 66B). The fibers collect again in the medulla as a single tract, in the pyramids on each side of the midline (see Figure 6, Figure 7, Figure 67, and Figure 67B). At the lowermost level of the medulla, 90% of the fibers decussate and form the lateral cortico-spinal tract, situated in the lateral aspect of the spinal cord (see Figure 68). The ventral cortico-spinal tract is found in the anterior portion of the white matter of the spinal cord (see Figure 68).

CLINICAL ASPECT

Lesions involving the cortico-spinal tract in humans are quite devastating, as they rob the individual of voluntary motor control, particularly the fine skilled motor movements. This pathway is quite commonly involved in strokes, as a result of vascular lesions of the cerebral arteries or of the deep arteries to the internal capsule (reviewed with Figure 60 and Figure 62). This lesion results in a weakness (paresis) or paralysis of the muscles on the opposite side. The clinical signs in humans will reflect the additional loss of cortical input to the brainstem nuclei, particularly to the reticular formation.

Damage to the tract in the spinal cord is seen after traumatic injuries (e.g., automobile and diving accidents). In this case, other pathways would be involved and the clinical signs will reflect this damage, with the loss of the nonvoluntary tracts (discussed with Figure 68). If one-half of the spinal cord is damaged, the loss of function is ipsilateral to the lesion.

A Babinski sign (discussed in Section B, Part III, Introduction) is seen with all lesions of the cortico-spinal tract (except spinal shock, see Figure 5).

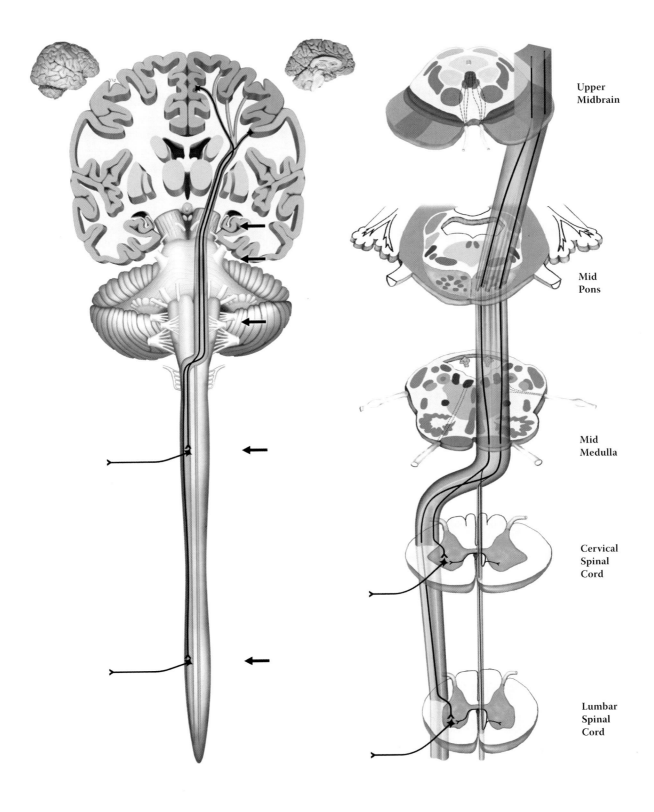

FIGURE 45: Cortico-Spinal Tract — Pyramidal System

FIGURE 46
CORTICO-BULBAR FIBERS

NUCLEI OF THE BRAINSTEM

The word "bulb" (i.e., bulbar) is descriptive and refers to the brainstem. The cortico-bulbar fibers do not form a single pathway. The fibers end in a wide variety of nuclei of the brainstem; those fibers ending in the pontine nuclei are considered separately (see Figure 48).

Wide areas of the cortex send fibers to the brainstem as projection fibers (see Figure 16). These axons course via the internal capsule and continue into the cerebral peduncles of the midbrain (see Figure 26). The fibers involved with motor control occupy the middle third of the cerebral peduncle along with the cortico-spinal tract (described with the previous illustration; see Figure 48), supplying the motor cranial nerve nuclei of the brainstem (see Figure 8A and Figure 48), the reticular formation and other motor-associated nuclei of the brainstem.

- **Cranial Nerve Nuclei**: The motor neurons of the cranial nerves of the brainstem are lower motor neurons (see Figure 8A and Figure 48); the cortical motor cells are the upper motor neurons. These motor nuclei are generally innervated by fibers from both sides, i.e., each nucleus receives input from both hemispheres.

There are two exceptions to this rule, which are very important in the clinical setting:
  - The major exception is the cortical input to the **facial nucleus**. The portion of the facial nucleus supplying the upper facial muscles is supplied from both hemispheres, whereas the part of the nucleus supplying the lower facial muscles is innervated only by the opposite hemisphere (crossed).
  - The cortical innervation to the **hypoglossal** nucleus is not always bilateral. In some individuals, there is a predominantly crossed innervation.
- **Brainstem motor control nuclei**: Cortical fibers influence all the brainstem motor nuclei, particularly the reticular formation, including the red nucleus and the substantia nigra, but not the lateral vestibular nucleus (see Figure 49A, Figure 49B, and Figure 50). The cortico-retic-

ular fibers are extremely important for voluntary movements of the proximal joints (indirect voluntary pathway) and for the regulation of muscle tone.
- **Other brainstem nuclei**: The cortical input to the sensory nuclei of the brainstem is consistent with cortical input to all relay nuclei; this includes the somatosensory nuclei, the nuclei cuneatus and gracilis (see Figure 33). There is also cortical input to the periaqueductal gray, as part of the pain modulation system (see Figure 43).

CLINICAL ASPECT

Loss of cortical innervation to the cranial nerve motor nuclei is usually associated with a weakness, not paralysis, of the muscles supplied. For example, a lesion on one side may result in difficulty in swallowing or phonation, and often these problems dissipate in time.

Facial movements: A lesion of the facial area of the cortex or of the cortico-bulbar fibers affects the muscles of the face differentially. A patient with such a lesion will be able to wrinkle his or her forehead normally on both sides when asked to look up, but will not be able to show the teeth or smile symmetrically on the side opposite the lesion. Because of the marked weakness of the muscles of the lower face, there will be a drooping of the lower face on the side opposite the lesion. This will also affect the muscle of the cheek (the buccinator muscle) and cause some difficulties with drinking and chewing (the food gets stuck in the cheek and oftentimes has to be manually removed); sometimes there is also drooling.

This clinical situation must be distinguished from a lesion of the **facial nerve** itself, a lower motor neuron lesion, most often seen with Bell's palsy (a lesion of the facial nerve as it emerges from the skull); in this case, the movements of the muscles of both the upper and lower face are lost on one (affected) side.

Tongue movements: The fact that the hypoglossal nucleus may or my not receive innervation from the cortex of both sides or only from the opposite side makes interpretation of tongue deviation not a reliable sign in the clinical setting. A lesion affecting the hypoglossal nucleus or nerve is a lower motor lesion of one-half of the tongue (on the same side) and will lead to paralysis and atrophy of the side affected.

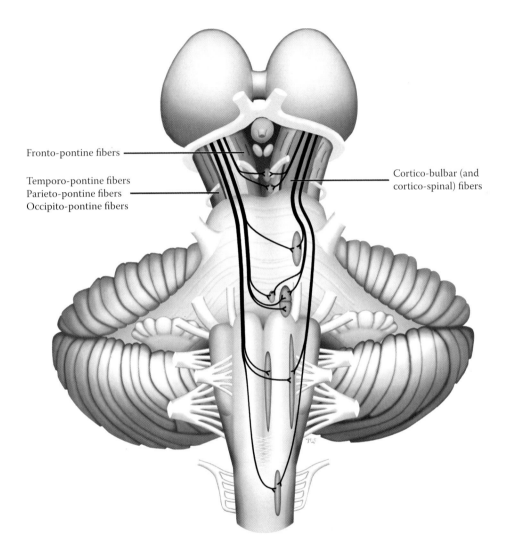

Fronto-pontine fibers

Temporo-pontine fibers
Parieto-pontine fibers
Occipito-pontine fibers

Cortico-bulbar (and
cortico-spinal) fibers

FIGURE 46: Cortico-Bulbar Tracts — Nuclei of the Brainstem

FIGURE 47
RUBRO-SPINAL TRACT

VOLUNTARY/NONVOLUNTARY MOTOR CONTROL

The **red nucleus** is a prominent nucleus of the midbrain. It gets its name from a reddish color seen in fresh dissections of the brain, presumably due to its high vascularity. The nucleus (see Figure 48, Figure 51B, and Figure 65A) has two portions, a small-celled upper division and a portion with large neurons more ventrally located. The rubro-spinal pathway originates, at least in humans, from the larger cells.

The red nucleus receives its input from the motor areas of the cerebral cortex and from the cerebellum (see Figure 53). The cortical input is directly onto the projecting cells, thus forming a potential two-step pathway from motor cortex to spinal cord.

The rubro-spinal tract is also a crossed pathway, with the decussation occurring in the ventral part of the midbrain (see also Figure 48 and Figure 51B). The tract descends within the central part of the brainstem (the tegmentum), and is not clearly distinguishable from other fiber systems. The fibers then course in the lateral portion of the white matter of the spinal cord, just anterior to and intermingled with the lateral cortico-spinal tract (see Figure 68 and Figure 69).

The rubro-spinal tract is a well-developed pathway in some animals. In monkeys, it seems to be involved in flexion movements of the limbs. Stimulation of this tract in cats produces an increase in tone of the flexor muscles.

NEUROLOGICAL NEUROANATOMY

The location of this tract within the brainstem is shown at cross-sectional levels of the upper midbrain, the mid-pons, the mid-medulla, and cervical and lumbar spinal cord levels. The tract is said to continue throughout the length of the spinal cord in primates but probably only extends into the cervical spinal cord in humans.

The fibers of CN III (oculomotor) exit through the medial aspect of this nucleus at the level of the upper midbrain (see Figure 65A).

CLINICAL ASPECT

The functional significance of this pathway in humans is not well known. The number of large cells in the red nucleus in humans is significantly less than in monkeys. Motor deficits associated with a lesion involving only the red nucleus or only the rubro-spinal tract have not been adequately described. Although the rubro-spinal pathway may play a role in some flexion movements, it seems that the cortico-spinal tract predominates in the human.

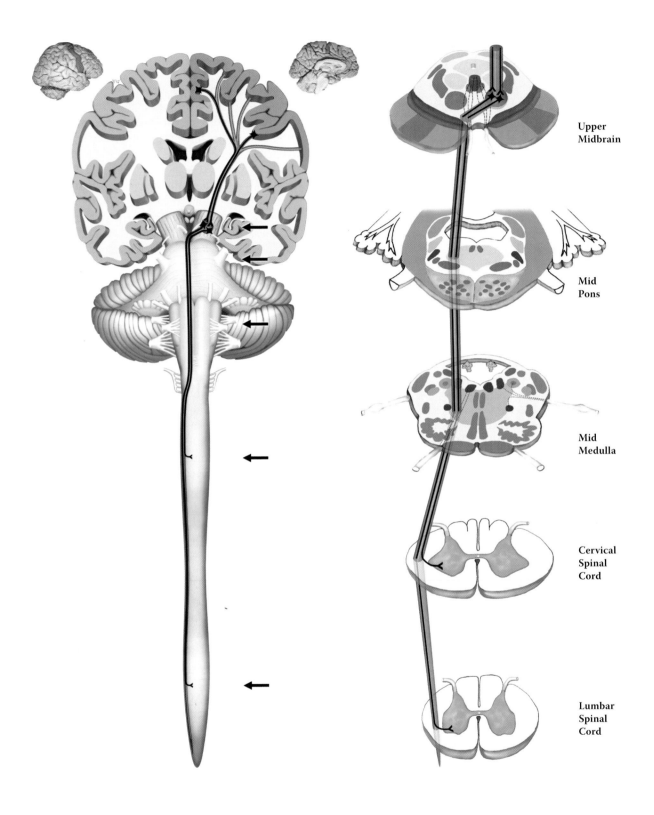

Upper
Midbrain

Mid
Pons

Mid
Medulla

Cervical
Spinal
Cord

Lumbar
Spinal
Cord

FIGURE 47: Rubro-Spinal Tract

FIGURE 48
MOTOR TRACTS AND
CRANIAL NERVE NUCLEI

DESCENDING TRACTS AND CORTICO-PONTINE FIBERS

The descending pathways that have been described are shown, using the somewhat oblique posterior view of the brainstem (see Figure 10 and Figure 40), along with those cranial nerve nuclei that have a motor component. These pathways will be presented in summary form:

- **Cortico-spinal tract** (see Figure 45): These fibers course in the middle third of the cerebral peduncle, are dispersed in the pontine region between the pontine nuclei, and regroup as a compact bundle in the medulla, situated within the pyramids. At the lowermost part of the medulla (Figure 7), most of the fibers decussate to form the lateral cortico-spinal tract of the spinal cord (see Figure 68 and Figure 69). A small portion of the tract continues ipsilaterally, mostly into the cervical spinal cord region, as the anterior (ventral) cortico-spinal tract.
- **Cortico-bulbar fibers** (see Figure 46): The cortical fibers that project to the cranial nerve nuclei of the brainstem are shown in this diagram. The term also includes those cortical fibers that project to the reticular formation and other brainstem nuclei. These are also located in the middle third of the cerebral peduncle and are given off at various levels within the brainstem.
- **Rubro-spinal tract** (see Figure 47): This tract from the lower portion of the red nucleus decussates in the midbrain region and descends through the brainstem. In the spinal cord, the fibers are located anterior to the lateral cortico-spinal tract (see Figure 68).

CORTICO-PONTINE FIBERS

The cortico-pontine fibers are part of a circuit that involves the cerebellum. The cortical fibers arise from the motor areas as well as from widespread parts of the cerebral cortex. The fibers are located in the outer and inner thirds of the cerebral peduncle (see also Figure 46): the fronto-pontine fibers in the inner third, and fibers from the other lobes in the outer third. They terminate in the nuclei of the pons proper (see Figure 6), and the information is then relayed (after crossing) to the cerebellum via the massive middle cerebellar peduncle (discussed with Figure 55; see also Figure 6 and Figure 7). The role of this circuit in motor control will be explained with the cerebellum (see Figure 54–Figure 57).

The motor cranial nerve nuclei and their function have been discussed (see Figure 7 and Figure 8A), and their location within the brainstem will be described (see Figure 64–Figure 67). Only topographical aspects will be described here:

- **CN III — Oculomotor** (to most extra-ocular muscles and parasympathetic): These fibers traverse through the medial portion of the red nucleus, before exiting in the fossa between the cerebral peduncles, the interpeduncular fossa (see Figure 65A).
- **CN IV — Trochlear** (to the superior oblique muscle): The fibers from this nucleus cross in the posterior aspect of the lower midbrain before exiting posteriorly (see Figure 10 and Figure 66A). The slender nerve then wraps around the lower border of the cerebral peduncles in its course anteriorly.
- **CN V — Trigeminal** (to muscles of mastication): The motor fibers pierce the middle cerebellar peduncle in the mid-pontine region, along with the sensory component.
- **CN VI — Abducens** (to the lateral rectus muscle): The anterior course of the exiting fibers could not be depicted from this perspective.
- **CN VII — Facial** (to muscles of facial expression): The fibers to the muscles of facial expression have an internal loop before exiting. The nerve loops over the abducens nucleus, forming a bump called the facial colliculus in the floor of the fourth ventricle (see Figure 10). It should be noted that the nerve of only one side is being shown in this illustration.
- **CN IX — Glossopharyngeal and CN X-Vagus** (motor and parasympathetic): The fibers exit on the lateral aspect of the medulla, behind the inferior olive.
- **CN XI — Spinal Accessory** (to neck muscles): The fibers that supply the large muscles of the neck (sternomastoid and trapezius) originate in the upper spinal cord and ascend into the skull before exiting.
- **CN XII — Hypoglossal** (to muscles of the tongue): These fibers actually course anteriorly, exiting from the medulla between the inferior olive and the cortico-spinal (pyramidal) tract.

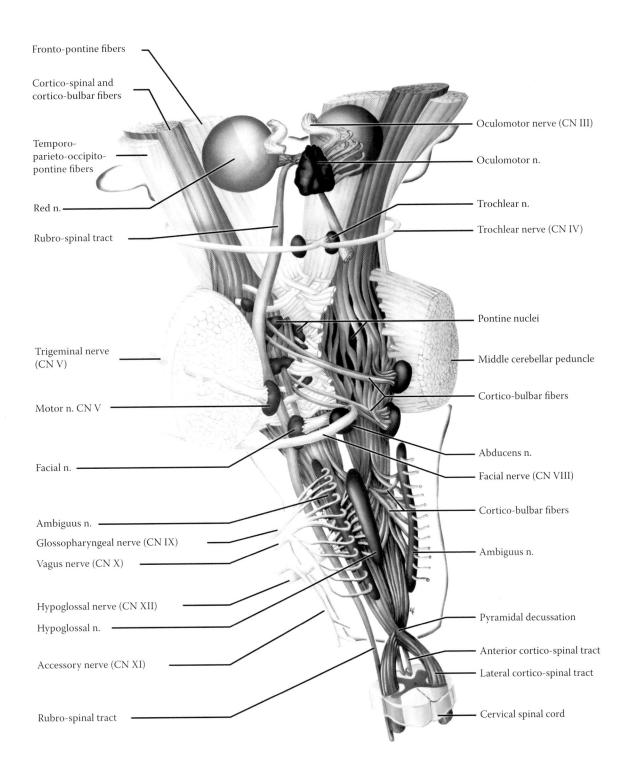

Fronto-pontine fibers

Cortico-spinal and cortico-bulbar fibers

Temporo-parieto-occipito-pontine fibers

Red n.

Rubro-spinal tract

Trigeminal nerve (CN V)

Motor n. CN V

Facial n.

Ambiguus n.

Glossopharyngeal nerve (CN IX)

Vagus nerve (CN X)

Hypoglossal nerve (CN XII)

Hypoglossal n.

Accessory nerve (CN XI)

Rubro-spinal tract

Oculomotor nerve (CN III)

Oculomotor n.

Trochlear n.

Trochlear nerve (CN IV)

Pontine nuclei

Middle cerebellar peduncle

Cortico-bulbar fibers

Abducens n.

Facial nerve (CN VIII)

Cortico-bulbar fibers

Ambiguus n.

Pyramidal decussation

Anterior cortico-spinal tract

Lateral cortico-spinal tract

Cervical spinal cord

FIGURE 48: Descending Tracts and Cortico-Pontine Fibers

FIGURE 49A AND FIGURE 49B RETICULO-SPINAL TRACTS

INDIRECT VOLUNTARY AND NONVOLUNTARY MOTOR REGULATION

As has been noted (see Figure 42A and Figure 42B), the reticular formation is a collection of nuclei that participates in a number of functions, some quite general (e.g., "arousal") and others more specific (e.g., respiratory control). These nuclei of the reticular formation are also part of the indirect voluntary motor pathway, as well as nonvoluntary motor regulation (see Section B, Part III, Introduction).

The indirect voluntary pathway, the cortico-reticulo-spinal pathway, is thought to be an older pathway for the control of movements, particularly of proximal joints and the axial musculature. Therefore, some voluntary movements can still be performed after destruction of the cortico-spinal pathway (discussed with Figure 45). Muscle tone and reflex responsiveness are greatly influenced by activity in the reticular formation as part of the nonvoluntary motor system; it is important to note that cortical input to the reticular formation is part of this regulation.

The reticular formation receives input from many sources, including most sensory pathways (anterolateral, trigeminal, auditory, and visual). At this point, the focus is on the input from the cerebral cortex, from both hemispheres. These axons form part of the "cortico-bulbar system of fibers" (discussed with Figure 46).

Note to the Learner: Understanding the complexity of the various parts of the motor system and the role of the reticular formation in particular is not easy. One approach is to start with the basic reflex arc — the reticular formation assumes a significant role in the modification of this response, i.e., hyperreflexia or hyporeflexia, as well as muscle tone. In addition, there is the role of the reticular formation and other motor brainstem nuclei in the nonvoluntary response of the organism to gravitational changes. The next step would be the role of the reticular formation in motor control, particularly for axial musculature, as part of the indirect voluntary motor system. It now becomes important to understand that the cortex has an important role in controlling this system.

There are two pathways from the reticular formation to the spinal cord: one originates in the pontine region (this illustration) and one in the medullary region (next illustration).

FIGURE 49A — PONTINE (MEDIAL) RETICULO-SPINAL TRACT

This tract originates in the pontine reticular formation from two nuclei: the upper one is called the **oral portion** of the pontine reticular nuclei (nucleus reticularis pontis oralis), and the lower part is called the **caudal portion** (see Figure 42B). The tract descends to the spinal cord and is located in the medial region of the white matter (see Figure 68 and Figure 69); this pathway therefore is called the medial reticulo-spinal tract.

Functionally, this pathway exerts its action on the extensor muscles, both movements and tone. The area in the pons is known as the reticular extensor facilitatory area. The fibers terminate on the anterior horn cells controlling the axial muscles, likely via interneurons (see Figure 44). This system is complementary to that from the lateral vestibular nucleus (see Figure 50).

NEUROLOGICAL NEUROANATOMY

The location of the tract in the brainstem is shown at cross-sectional levels of the mid-pons, the lower pons, the mid-medulla, and cervical and lumbar spinal cord levels. The tract is intermingled with others in the white matter of the spinal cord.

CLINICAL ASPECT

Lesions involving the cortico-bulbar fibers including the cortico-reticular fibers will be discussed with the medullary reticular formation (next illustration).

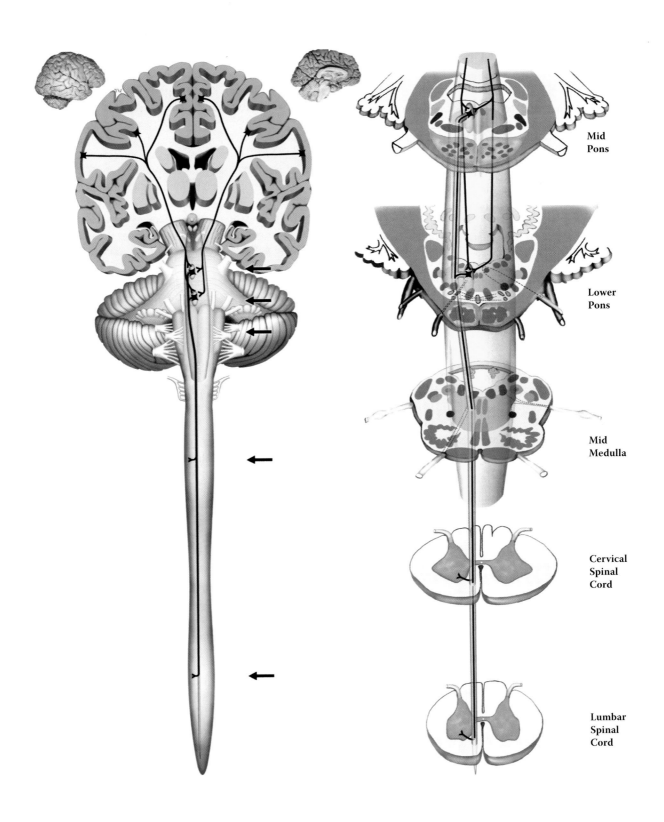

Mid
Pons

Lower
Pons

Mid
Medulla

Cervical
Spinal
Cord

Lumbar
Spinal
Cord

FIGURE 49A: Pontine (Medial) Reticulo-Spinal Tract

FIGURE 49B
MEDULLARY (LATERAL)
RETICULO-SPINAL TRACT

This tract originates in the medullary reticular formation, mainly from the **nucleus gigantocellularis** (meaning very large cells, see Figure 42A, Figure 42B, and Figure 67C). The tract descends more laterally in the spinal cord than the pontine pathway, and is thus named the lateral reticulo-spinal tract (see Figure 68 and Figure 69); some of the fibers are crossed. The tract lies beside the lateral vestibulo-spinal pathway.

The pathway also has its greatest influence on axial musculature. This part of the reticular formation is functionally the reticular extensor inhibitory area, opposite to that of the pontine reticular formation. This area depends for its normal activity on influences coming from the cerebral cortex.

NEUROLOGICAL NEUROANATOMY

The location of the tract in the brainstem is shown at the cross-sectional levels of the mid-pons, the lower pons, the mid-medulla, and cervical and lumbar spinal cord levels, intermingled with other tracts in the white matter of the spinal cord (see Figure 68 and Figure 69).

CLINICAL ASPECT: SPASTICITY

A lesion destroying the cortico-bulbar fibers, an **upper motor neuron lesion**, results in an increase in the tone of the extensor/anti-gravity muscles, which develops over a period of days. This increase in tone, called **spasticity**, tested by passive flexion and extension of a limb, is velocity dependent, meaning that the joint of the limb has to be moved quickly. It is the anti-gravity muscles that are affected in spasticity; in humans, for reasons that are difficult to explain, these muscles are the flexors of the upper limb and the extensors of the lower limb. There is also an increase in responsiveness of the stretch reflex, called **hyperreflexia**, as tested using the deep tendon reflex, DTR (discussed with Figure 44), which also develops over a period of several days.

There are two hypotheses for the increase in the stretch (monosynaptic) reflex responsiveness:

- **Denervation supersensitivity**: One possibility is a change of the level of responsivity of the neurotransmitter receptors of the motor neurons themselves caused by the loss of the descending input, leading to an increase in excitability.
- **Collateral sprouting**: Another possibility is that axons adjacent to an area that has lost synaptic input will sprout branches and occupy the vacated synaptic sites of the lost descending fibers. In this case, the sprouting is thought to be of the incoming muscle afferents (called 1A afferents, from the muscle spindles).

There is experimental evidence (in animals) for both mechanisms. Spasticity and hyperreflexia usually occur in the same patient. Another feature accompanying hyperreflexia is **clonus**. This can be elicited by grasping the foot and jerking the ankle upward; in a person with hyperreflexia, the response is a short burst of flexion-extension responses of the ankle, which the tester can feel and which also can be seen.

Lesions involving parts of the motor areas of the cerebral cortex, large lesions of the white matter of the hemispheres or of the posterior limb of the internal capsule, and certain lesions of the upper brainstem all may lead to a similar clinical state in which a patient is paralyzed or has marked weakness, with spasticity and hyperreflexia (with or without clonus) on the contralateral side some days after the time of the damage. The cortico-spinal tract would also be involved in most of these lesions, with loss of voluntary motor control, and with the appearance of the Babinski sign in most cases immediately after the lesion (see Introduction to this section).

A similar situation occurs following large lesions of the spinal cord in which all the descending motor pathways are disrupted, both voluntary and nonvoluntary. Destruction of the whole cord would lead to paralysis below the level of the lesion (paraplegia), bilateral spasticity, and hyperreflexia (usually with clonus), a severely debilitating state.

It is most important to distinguish this state from that seen in a Parkinsonian patient who has a change of muscle tone called **rigidity** (discussed with Figure 24), with no change in reflex responsiveness and a normal plantar response.

This state should be contrasted with a **lower motor neuron lesion** of the anterior horn cell, with hypotonia and hyporeflexia as well as weakness (e.g., polio, discussed with Figure 44).

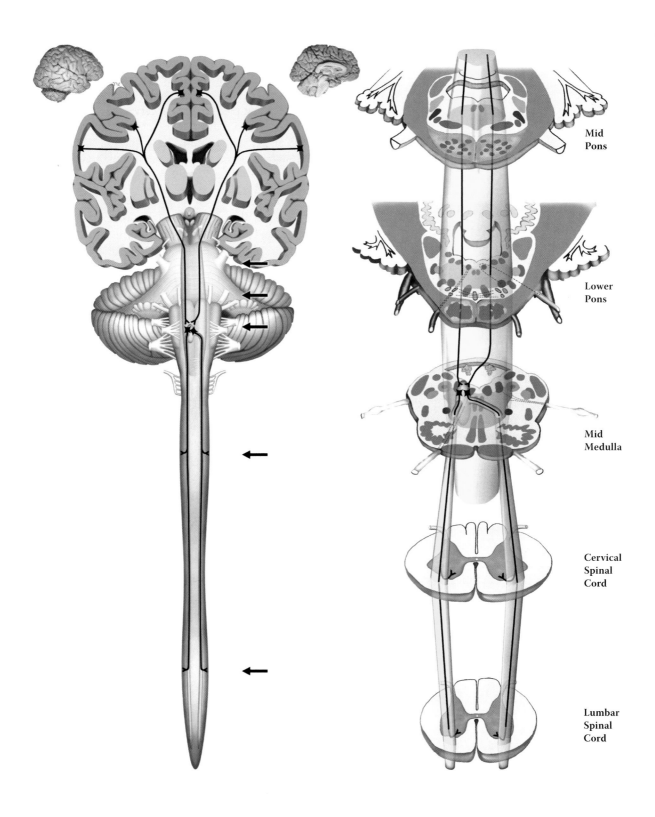

FIGURE 49B: Medullary (Lateral) Reticulo-Spinal Tract

Mid
Pons

Lower
Pons

Mid
Medulla

Cervical
Spinal
Cord

Lumbar
Spinal
Cord

FIGURE 50
LATERAL VESTIBULO-SPINAL TRACT

NONVOLUNTARY MOTOR REGULATION

This pathway is very important in that it provides a link between the vestibular influences (i.e., gravity and balance) and the control of axial musculature, via the spinal cord. The main function is to provide corrective muscle activity when the body (and head) tilt or change orientation in space (activation of the vestibular system, CN VIII, see Figure 8B).

This tract originates in the **lateral vestibular nucleus**, which is located in the lower pontine region (see next illustration and Figure 66C). The nucleus is found at the lateral edge of the fourth ventricle and is characterized by extremely large neurons. (This nucleus is also called Deiter's nucleus in some texts and the large neurons are often called by the same name.)

The lateral vestibular nucleus receives its major inputs from the vestibular system and from the cerebellum; there is no cerebral cortical input. This tract descends through the medulla and traverses the entire spinal cord in the ventral white matter (see Figure 68 and Figure 69). It does not decussate. The fibers terminate in the medial portion of the anterior horn, namely on those motor cells that control the axial musculature (see Figure 44).

Functionally, this pathway increases extensor muscle tone and activates extensor muscles. It is easier to think of these muscles as anti-gravity muscles in a four-legged animal; in humans, one must translate these muscles in functional terms, which are the flexors of the upper extremity and the extensors of the lower extremity.

NEUROLOGICAL NEUROANATOMY

The same cross-sectional levels have been used as with the reticular formation, starting at the mid-pons. The vestibular nuclei are found at the lower pontine level and are seen through the mid-medulla; the tract descends throughout the spinal cord, as seen at cervical and lumbar levels. In the spinal cord the tract is positioned anteriorly, just in front of the ventral horn (see Figure 68 and Figure 69) and innervates the medial group of motor nuclei.

CLINICAL ASPECT

A lesion of this pathway would occur with spinal cord injuries and this would be one of the "upper motor neuron" pathways involved, leading to spasticity and hyperreflexia.

> **Decorticate rigidity**: Humans with severe lesions of the cerebral hemispheres but whose brainstem circuitry is intact often exhibit a postural state known as decorticate rigidity. In this condition, there is a state of flexion of the forearm and extension of the legs.
>
> **Decerebrate Rigidity**: Humans with massive cerebral trauma, anoxic damage, or midbrain destructive lesions exhibit a postural state in which all four limbs are rigidly extended. The back is arched and this may be so severe as to cause a posture known as opisthotonus, in which the person is supported by the back of the neck and the heels.

Physiologically, these conditions are not related to Parkinsonian rigidity but to the abnormal state of spasticity (see discussion with the previous illustration). The postulated mechanism involves the relative influence of the pontine and medullary reticular formations, along with the vestibulo-spinal pathway, with and without the input from the cerebral cortex.

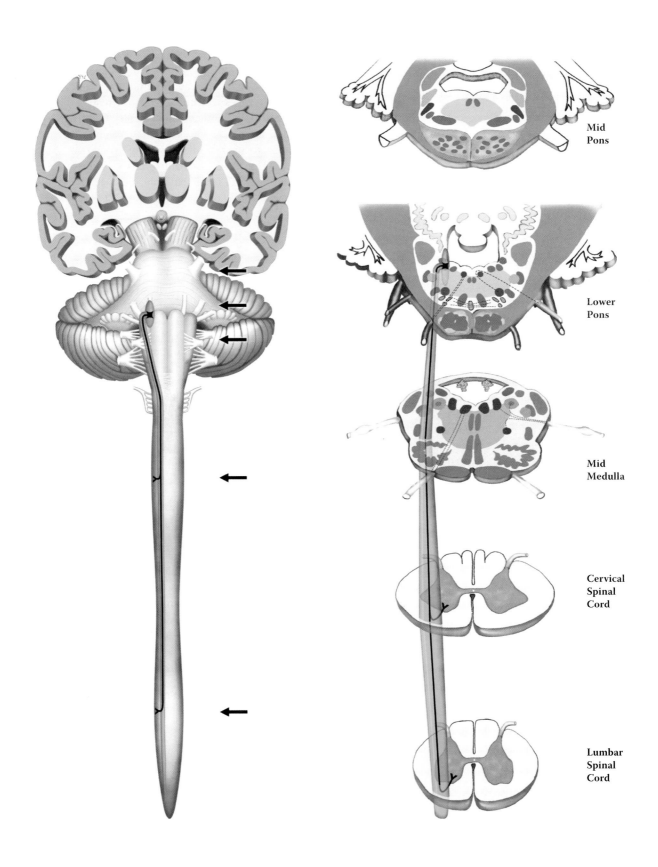

Mid Pons

Lower Pons

Mid Medulla

Cervical Spinal Cord

Lumbar Spinal Cord

FIGURE 50: Lateral Vestibulo-Spinal Tract

FIGURE 51A
VESTIBULAR SYSTEM

VESTIBULAR NUCLEI AND EYE MOVEMENTS

The vestibular system carries information about our position in relation to gravity and changes in that position. The sensory system is located in the inner ear and consists of three **semicircular canals** and other sensory organs in a bony and membranous labyrinth. There is a peripheral ganglion (the spiral ganglion), and the central processes of these cells, CN VIII, enter the brainstem at the cerebellar-pontine angle, just above the cerebellar flocculus (see Figure 6, Figure 7, and Figure 8B).

The vestibular information is carried to **four vestibular nuclei**, which are located in the upper part of the medulla and lower pons: superior, lateral, medial, and inferior (see Figure 8B; also Figure 66C, Figure 67A, and Figure 67B). The **lateral vestibular nucleus** gives rise to the lateral vestibulo-spinal tract (as described in the previous illustration; see also the following illustration). This is the pathway that serves to adjust the postural musculature to changes in relation to gravity.

The **medial and inferior vestibular nuclei** give rise to both ascending and descending fibers, which join a conglomerate bundle called the **medial longitudinal fasciculus (MLF)** (described more fully with the next illustration). The descending fibers from the medial vestibular nucleus, if considered separately, could be named the **medial vestibulo-spinal tract** (see Figure 68). This system is involved with postural adjustments to positional changes, using the axial musculature.

The ascending fibers adjust the position of the eyes and coordinate eye movements of the two eyes by interconnecting the three cranial nerve nuclei involved in the control of eye movements — CN III (oculomotor) in the upper midbrain, CN IV (trochlear) in the lower midbrain, and CN VI (abducens) in the lower pons (see Figure 8A, Figure 48, and also Figure 51B). If one considers lateral gaze, a movement of the eyes to the side (in the horizontal plane), this requires the coordination of the lateral rectus muscle (abducens nucleus) of one side and the medial rectus (oculomotor nucleus) of the other side; this eye movement is called conjugate. These fibers for coordinating the eye movements are carried in the MLF.

There is a "gaze center" within the pontine reticular formation for **saccadic** eye movements. These are extremely rapid (ballistic) movements of both eyes, yoked together, usually in the horizontal plane so that we can shift our focus extremely rapidly from one object to another. The fibers controlling this movement originate from the cortex, from the frontal eye field (see Figure 14A), and also likely course in the MLF.

CLINICAL ASPECT

A not uncommon tumor, called an acoustic neuroma, can occur along the course of the acoustic nerve, usually at the cerebello-pontine angle. This is a slow-growing benign tumor, composed of Schwann cells, the cell responsible for myelin in the peripheral nervous system. Initially, there will be a complaint of loss of hearing, or perhaps a ringing noise in the ear (called tinnitus). Because of its location, as it grows it will begin to compress the adjacent nerves (including CN VII). Eventually, if left unattended, there would be additional symptoms due to further compression of the brainstem and an increase in intracranial pressure. Modern imaging techniques allow early detection of this tumor. Surgical removal, though, still requires considerable skill so as not to damage CN VIII itself (which would produce a loss of hearing), or CN VII (which would produce a paralysis of facial muscles) and adjacent neural structures.

ADDITIONAL DETAIL

There is a small nucleus in the periaqueductal gray region of the midbrain that is associated with the visual system and is involved in the coordination of eye and neck movements. This nucleus is called the interstitial nucleus (of Cajal). It is located near the oculomotor nucleus. This nucleus (see also the next illustration) receives input from various sources and contributes fibers to the MLF. Some have named this pathway the interstitio-spinal "tract."

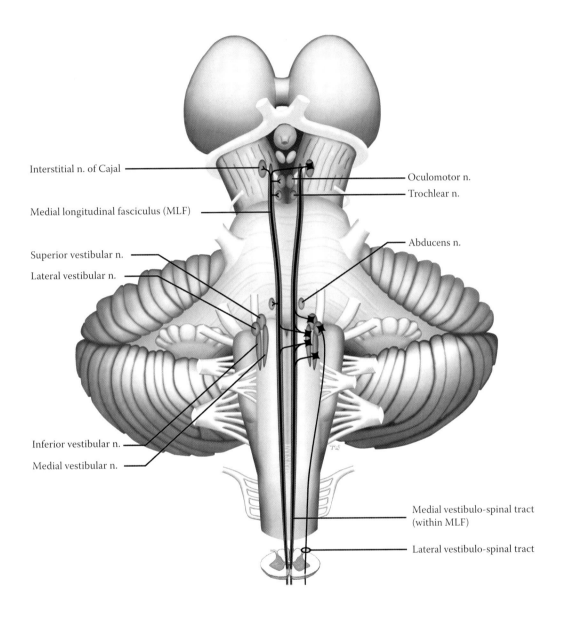

Interstitial n. of Cajal

Medial longitudinal fasciculus (MLF)

Superior vestibular n.

Lateral vestibular n.

Inferior vestibular n.

Medial vestibular n.

Oculomotor n.

Trochlear n.

Abducens n.

Medial vestibulo-spinal tract (within MLF)

Lateral vestibulo-spinal tract

FIGURE 51A: Vestibular Nuclei and Eye Movements

FIGURE 51B
MEDIAL LONGITUDINAL
FASCICULUS (MLF)

MLF AND ASSOCIATED TRACTS

This diagram shows the brainstem from the posterior perspective (as in Figure 10 and Figure 40). Note the orientation of the spinal cord (with the ventral horn away from the viewer).

The MLF is a tract within the brainstem and upper spinal cord that links the visual world and vestibular events with the movements of the eyes and the neck, as well as linking up the nuclei that are responsible for eye movements. The tract runs from the midbrain level to the upper thoracic level of the spinal cord. It has a rather constant location near the midline, dorsally, just anterior to the aqueduct of the midbrain and the fourth ventricle (see brainstem cross-sections, e.g., Figure 65A, Figure 66A, and Figure 67A).

The MLF is, in fact, composed of several tracts running together:

- **Vestibular fibers**: Of the four vestibular nuclei (see previous illustration), descending fibers originate from the medial vestibular nuclei and become part of the MLF; this can be named separately the medial vestibulo-spinal tract. There are also ascending fibers that come from the medial, inferior, and superior vestibular nuclei that also are carried in the MLF. Therefore, the MLF carries both ascending and descending vestibular fibers.
- **Visuomotor fibers**: The interconnections between the various nuclei concerned with eye movements are carried in the MLF (as described in the previous illustration).
- **Vision-related fibers**: Visual information is received by various brainstem nuclei.
  - The superior colliculus is a nucleus for the coordination of visual-related reflexes, including eye movements (see Figure 9A). The superior colliculus coordinates the movements of the eyes and the turning of the neck in response to visual information. It also receives input from the visual association cortical areas, areas 18 and 19 (see Figure 17 and Figure 41B). The descending fibers from the superior colliculus, called the

tecto-spinal tract, are closely associated with the MLF and can be considered part of this system (although in most books it is discussed separately). As shown in the upper inset, these fibers cross in the midbrain. (Note that the superior colliculus [SC] of only one side is shown in order not to obscure the crossing fiber systems at that level.)
- The small interstitial nucleus and its contribution have already been noted and discussed with the previous illustration.

The lower inset shows the MLF in the ventral funiculus (white matter) of the spinal cord, at the cervical level (see Figure 68 and Figure 69). The three components of the tract are identified, those coming from the medial vestibular nucleus, the fibers from the interstitial nucleus, and the tecto-spinal tract. These fibers are mingled together in the MLF.

In summary, the MLF is a complex fiber bundle that is necessary for the proper functioning of the visual apparatus. The MLF interconnects the three cranial nerve nuclei responsible for movements of the eyes, with the motor nuclei controlling the movements of the head and neck. It allows the visual movements to be influenced by vestibular, visual, and other information, and carries fibers (upward and downward) that coordinate the eye movements with the turning of the neck.

The diagram also shows the posterior commissure (not labeled). This small commissure carries fibers connecting the superior colliculi. In addition, it carries the important fibers for the consensual pupillary light reflex coordinated in the pretectal "nucleus" (discussed with Figure 41C).

CLINICAL ASPECT

A lesion of the MLF interferes with the normal conjugate movements of the eyes. When a person is asked to follow an object (e.g., the tip of a pencil moving to the right) with the head steady, the two eyes move together in the horizontal plane. With a lesion of the MLF (such as demyelination in multiple sclerosis), the abducting eye (the right eye) moves normally but the adducting eye (the left eye) fails to follow; yet, adduction is preserved on convergence. Clearly the nuclei and the nerves are intact; the lesion, then, is in the fibers coordinating the movement. This condition is known as **internuclear ophthalmoplegia**. Sometimes there is also monocular horizontal nystagmus (rapid side-to-side movements) of the abducting eye.

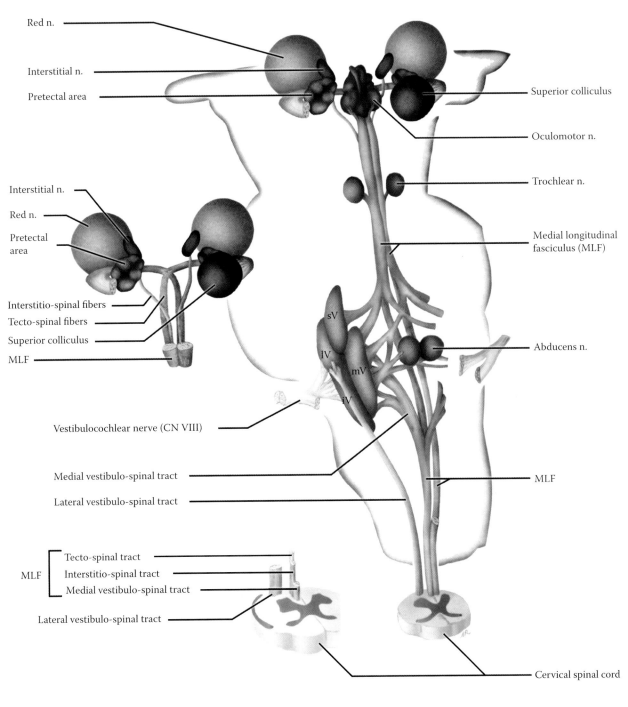

Red n.

Interstitial n.

Pretectal area

Superior colliculus

Oculomotor n.

Trochlear n.

Medial longitudinal
fasciculus (MLF)

Interstitial n.

Red n.

Pretectal
area

sV

lV

mV

iV

Interstitio-spinal fibers

Tecto-spinal fibers

Superior colliculus

MLF

Abducens n.

Vestibulocochlear nerve (CN VIII)

Medial vestibulo-spinal tract

Lateral vestibulo-spinal tract

MLF

Tecto-spinal tract

Interstitio-spinal tract

Medial vestibulo-spinal tract

Lateral vestibulo-spinal tract

MLF

Cervical spinal cord

sV = Superior vestibular n.
lV = Lateral vestibular n.
mV = Medial vestibular n.
iV = Inferior vestibular n.

FIGURE 51B: Medial Longitudinal Fasciculus (MLF)

FIGURE 52
MOTOR REGULATORY
SYSTEM A

BASAL GANGLIA CIRCUITRY

UPPER ILLUSTRATION

This is the same view of the basal ganglia as shown previously (see Figure 24), with the head of the caudate nucleus removed. The illustration includes the two other parts of the basal ganglia as a functional "system" — the subthalamic nucleus and the substantia nigra.

- **The subthalamic nucleus (S)** is situated in a small region below the level of the thalamus.
- **The substantia nigra (SN)** is a flattened nucleus located in the midbrain region. It is composed of two parts (see Figure 65A).
  - The **pars compacta** has the pigment-containing cells (see Figure 15B and Figure 65). These neurons project their fibers to the caudate and putamen (the striatum or neostriatum). This is called the nigro-striatal "pathway," although the fibers do not form a compact bundle; the neurotransmitter involved is dopamine.
  - The **pars reticulata** is situated more ventrally. It receives fibers from the striatum and is also an output nucleus from the basal ganglia to the thalamus, like the internal segment of the globus pallidus (see below).

LOWER ILLUSTRATION: BASAL GANGLIA CIRCUITRY

Information flows into the caudate (C) and putamen (P) from all areas of the cerebral cortex (in a topographic manner, see next illustration), from the substantia nigra (dopaminergic from the pars compacta), and from the centromedian nucleus of the thalamus (see below). This information is processed and passed through to the globus pallidus, internal segment (GPi), and the pars reticulata of the substantia nigra; these are the output nuclei of the basal ganglia.

Most of this information is relayed to the specific relay nuclei of the thalamus, the ventral anterior (**VA**) and **ventral lateral** (**VL**) nuclei (see Figure 12 and Figure 63). These project to the premotor and supplementary motor cortical areas (see Figure 14A, Figure 17, and Figure 60).

(This is to be contrasted with the projection of the cerebellum to the cortex, discussed with Figure 57.) The circuitry involving the basal ganglia, the thalamus, and the motor cortical areas will be described in detail with the next illustration.

In addition, there is a subcircuit involving the subthalamic nucleus (S): the external segment of the globus pallidus sends fibers to the subthalamic nucleus, and this nucleus sends fibers to the internal segment of the globus pallidus, the output portion.

Another subloop of the basal ganglia involves the centromedian nucleus of the thalamus, a nonspecific nucleus (see Figure 12). The loop starts in the striatum (only the caudate nucleus is shown here), to both segments of the globus pallidus; then fibers from the globus pallidus internal segment are sent to the centromedian nucleus, which then sends its fibers back to the striatum (see Figure 63).

CLINICAL ASPECT (SEE ALSO FIGURE 24)

Parkinson's disease: The degeneration of the dopamine-containing neurons of the pars compacta of the substantia nigra, with the consequent loss of their dopamine input to the basal ganglia (the striatum) leads to this clinical entity. Those afflicted with this disease have slowness of movement (bradykinesia), reduced facial expressiveness ("mask-like" face), and a tremor at rest, typically a "pill-rolling" type of tremor. On examination, there is **rigidity**, manifested as an increased resistance to passive movement of both flexors and extensors, which is not velocity-dependent. (This is to be contrasted with spasticity, discussed with Figure 49B.) In addition, there is *no* change in reflexes.

The medical treatment of Parkinson's disease has limitations, although various medications and combinations (as well as newer drugs) can be used for many years. For these patients, as well as in other select clinical cases, a surgical approach for the alleviation of the symptoms of the Parkinson's disease has been advocated, including placing lesions in the circuitry or using stimulating electrodes (with external control devices). To date, the theory has been that these surgical approaches are attempting to restore the balance of excitation and inhibition to the thalamus, thereby restoring the appropriate influence to the cortical areas involved in motor control.

The motor abnormality associated with a lesion of the subthalamic nucleus is called **hemiballismus**. The person is seen to have sudden flinging movements of a limb, on the side of the body opposite to the lesion. The likely cause for this is usually a vascular lesion.

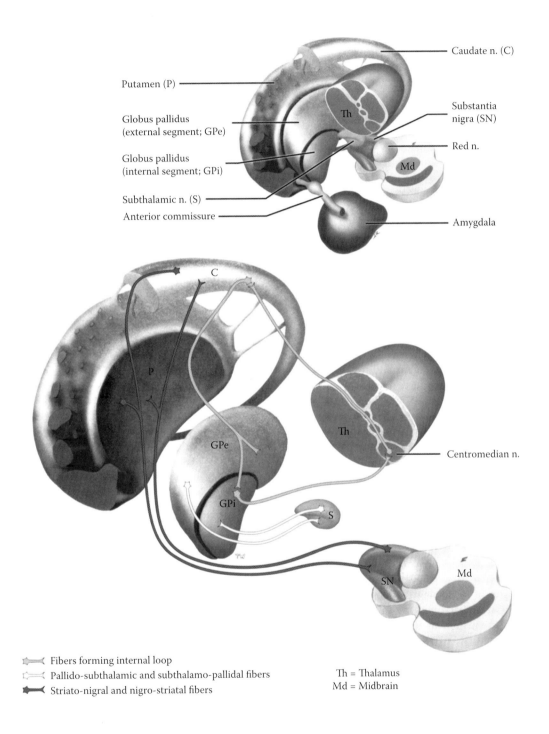

Caudate n. (C)

Putamen (P)

Globus pallidus
(external segment; GPe)

Globus pallidus
(internal segment; GPi)

Subthalamic n. (S)

Anterior commissure

Substantia
nigra (SN)

Red n.

Amygdala

Th

Md

C

P

Th

GPe

GPi

S

SN

Md

Centromedian n.

Fibers forming internal loop
Pallido-subthalamic and subthalamo-pallidal fibers
Striato-nigral and nigro-striatal fibers

Th = Thalamus
Md = Midbrain

FIGURE 52: Basal Ganglia Circuitry

FIGURE 53
MOTOR REGULATORY
SYSTEM B

THALAMUS: MOTOR CIRCUITS

The specific relay nuclei of the thalamus that are linked with the motor systems, the basal ganglia and the cerebellum, are the **ventral lateral (VL) and the ventral anterior (VA)** nuclei (see Figure 12 and Figure 63). These project to the different cortical areas involved in motor control, the motor strip, the premotor area, and the supplementary motor area (as shown in the upper insets). These thalamic nuclei also receive input from these cortical areas, in line with the reciprocal connections of the thalamus and cortex. One of the intralaminar nuclei, the centromedian nucleus, is also linked with the circuitry of the basal ganglia (described in the previous illustration).

Basal Ganglia: The neostriatum receives input from wide areas of the cerebral cortex, as well as from the dopamergic neurons of the substantia nigra. Fibers are then sent to the globus pallidus. The major outflow from the basal ganglia, from the internal (medial) segment of the globus pallidus, follows two slightly different pathways to the thalamus, as pallido-thalamic fibers. One group of fibers passes around, and the other passes through the fibers of the internal capsule (represented on the diagram by large stippled arrows). These merge and end in the ventral anterior (VA) and ventral lateral (VL) nuclei of the thalamus (see Figure 63). (The ventral anterior nucleus is not seen on this section through the thalamus.) The other outflow from the basal ganglia via the pars reticulata of the substantia nigra generally follows the same projection to these thalamic nuclei (not shown). The projection from these thalamic nuclei to the cerebral cortex goes to the premotor and supplementary motor areas, as shown in the small insets (in the upper figures; see Figure 14A and Figure 17; also Figure 60), cortical areas concerned with motor regulation and planning.

The pathway from thalamus to cortex is excitatory. The basal ganglia influence is to modulate the level of excitation of the thalamic nuclei. Too much inhibition leads to a situation that the motor cortex has insufficient activation, and the prototypical syndrome for this is Parkinson's (discussed with Figure 24 and Figure 52). Too little inhibition leads to a situation that the motor cortex receives too much stimulation and the prototypical syndrome for this is Huntington's chorea (discussed with Figure 24). The analogy that has been used to understand these diseases is to a motor vehicle, in which a balance is needed between the brake and the gas pedal for controlled forward motion in traffic.

The MOTOR areas of the cerebral cortex that receive input from these two subsystems of the motor system are shown diagrammatically in the small insets, both on the dorsolateral surface and on the medial surface of the hemispheres (see Figure 14 and Figure 17).

Cerebellum (to be reviewed *after* study of the cerebellum): The other part of the motor regulatory systems, the cerebellum, also projects (via the superior cerebellar peduncles) to the thalamus. The major projection is to the VL nucleus, but to a different portion of it than the part that receives the input from the basal ganglia. From here, the fibers project to the motor areas of the cerebral cortex, predominantly the precentral gyrus as well as the premotor area, areas 4 and 6, respectively (see Figure 57).

CLINICAL ASPECT

Many years ago it was commonplace to refer to the basal ganglia as part of the **extrapyramidal motor system** (in contrast to the pyramidal motor system — discussed with Figure 45, the cortico-spinal tract). It is now known that the basal ganglia exert their influence through the appropriate parts of the cerebral cortex, which then acts either directly, i.e., using the cortico-spinal (pyramidal) tract, or indirectly via certain brainstem nuclei (cortico-bulbar pathways, see Figure 46) to alter motor activity. The term extrapyramidal should probably be abandoned, but it is still frequently encountered in a clinical setting.

Tourette's syndrome is a motor disorder manifested by tics, uncontrolled sudden movements; occasionally, these individuals have bursts of uncontrolled language, which rarely contains vulgar expletives. This disorder starts in childhood and usually has other associated behavioral problems, including problems with attention. There is growing evidence that this disorder is centered in the basal ganglia. The condition may persist into adulthood.

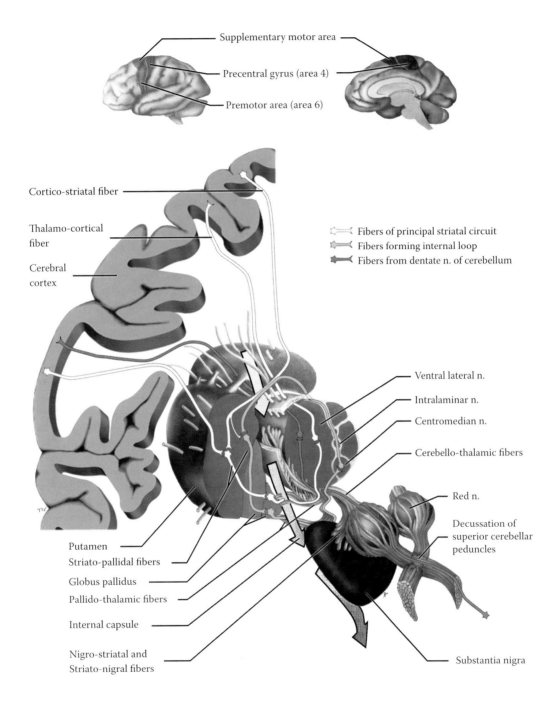

Supplementary motor area

Precentral gyrus (area 4)

Premotor area (area 6)

Cortico-striatal fiber

Thalamo-cortical fiber

Cerebral cortex

▷▭◁ Fibers of principal striatal circuit
◁▭◁ Fibers forming internal loop
◀▭◀ Fibers from dentate n. of cerebellum

Ventral lateral n.

Intralaminar n.

Centromedian n.

Cerebello-thalamic fibers

Red n.

Decussation of superior cerebellar peduncles

Putamen
Striato-pallidal fibers
Globus pallidus
Pallido-thalamic fibers
Internal capsule
Nigro-striatal and Striato-nigral fibers

Substantia nigra

FIGURE 53: Thalamus — Motor Circuits

FIGURE 54
CEREBELLUM 1

FUNCTIONAL LOBES

The cerebellum has been subdivided anatomically according to some constant features and fissures (see Figure 9A and Figure 9B). In the midline, the worm-like portion is the **vermis**; the lateral portions are the **cerebellar hemispheres**. The horizontal fissure lies approximately at the division between the superior and the inferior surfaces. The deep primary fissure is found on the superior surface and the area in front of it is the **anterior lobe** of the cerebellum. The only other parts to be noted are the nodulus and lingula of the vermis, as well as the tonsil.

In order to understand the functional anatomy of the cerebellum and its contribution to the regulation of motor control, it is necessary to subdivide the cerebellum into operational units. The three functional lobes of the cerebellum are

A. Vestibulocerebellum
B. Spinocerebellum
C. Neo- or cerebrocerebellum

These lobes of the cerebellum are defined by the areas of the cerebellar cortex involved, the related deep cerebellar nucleus, and the connections (afferents and efferents) with the rest of the brain.

There is a convention of portraying the functional cerebellum as if it is found in a single plane, using the lingula and the nodulus of the vermis as fixed points (see also Figure 17).

Note to the Learner: The best way to visualize this is to use the analogy of a book, with the binding toward you — representing the horizontal fissure. Place the fingers of your left hand on the edge of the front cover (the superior surface of the cerebellum) and the fingers of your right hand on the edges of the back cover (the inferior surface of the cerebellum), then (gently) open up the book so as to expose both the front and back covers. Both are now laid out in a single plane; now, the lingula is at the "top" of the cerebellum and the nodulus is at the bottom of the diagram. This same "flattening" can be done with an isolated brainstem and attached cerebellum in the laboratory.

Having done this, as is shown in the upper part of this figure, it is now possible to discuss the three functional lobes of the cerebellum.

- The **vestibulocerebellum** is the functional part of the cerebellum responsible for balance and gait. It is composed of two cortical components, the flocculus and the nodulus; hence, it is also called the **flocculonodular lobe**. The flocculus is a small lobule of the cerebellum located on its inferior surface and oriented in a transverse direction, below the middle cerebellar peduncle (see Figure 6 and Figure 7); the nodulus is part of the vermis. The vestibulocerebellum sends its fibers to the **fastigial** nucleus, one of the deep cerebellar nuclei (discussed with Figure 56 and Figure 57).
- The **spinocerebellum** is concerned with coordinating the activities of the limb musculature. Part of its role is to act as a *comparator* between the intended and the actual movements. It is made up of three areas:
  - The **anterior lobe** of the cerebellum, the cerebellar area found on the superior surface, in front of the primary fissure (see Figure 9A)
  - Most of the **vermis** (other than the parts mentioned above, see Figure 9A and Figure 9B)
  - A strip of tissue on either side of the vermis called the **paravermal** or **intermediate zone** — there is no anatomical fissure demarcating this functional area

  The output deep cerebellar nuclei for this functional part of the cerebellum are mostly the **interposed** nuclei, the globose and emboliform nuclei (see Figure 56A and Figure 56B) and, in part, the fastigial nucleus.
- The **neocerebellum** includes the remainder of the cerebellum, the areas behind the primary fissure and the inferior surface of the cerebellum (see Figure 9A and Figure 9B), with the exception of the vermis itself and the adjacent strip, the paravermal zone. This is the largest part of the cerebellum and the newest from an evolutionary point of view. It is also known as the **cerebrocerebellum**, since most if its connections are with the cerebral cortex. The output nucleus of this part of the cerebellum is the **dentate** nucleus (see Figure 56 and Figure 57). The neocerebellum is involved with the overall coordination of voluntary motor activities and is also involved in motor planning.

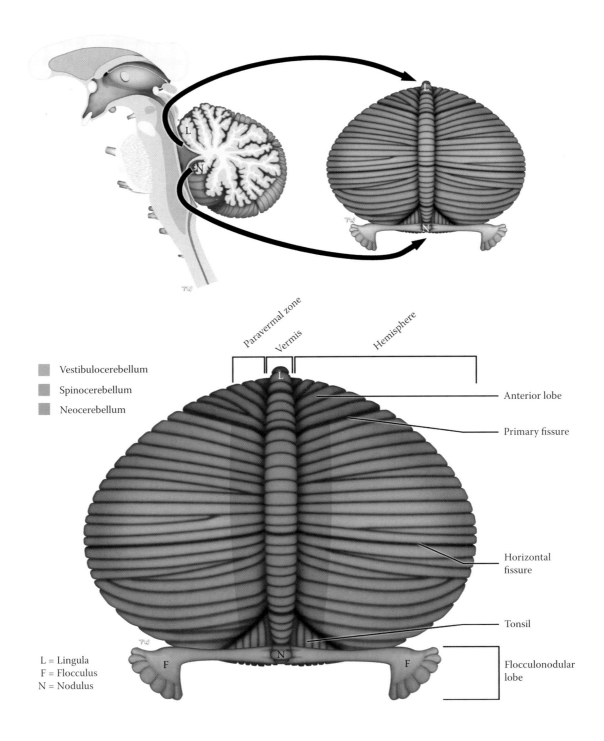

Vestibulocerebellum

Spinocerebellum

Neocerebellum

Paravermal zone

Vermis

Hemisphere

Anterior lobe

Primary fissure

Horizontal
fissure

Tonsil

Flocculonodular
lobe

L = Lingula
F = Flocculus
N = Nodulus

FIGURE 54: Cerebellum 1 — Functional Lobes

FIGURE 55
CEREBELLUM 2

CEREBELLAR AFFERENTS

Information relevant to the role of the cerebellum in motor regulation comes from the cerebral cortex, the brainstem, and from the muscle receptors in the periphery. The information is conveyed to the cerebellum mainly via the middle and inferior cerebellar peduncles.

- **Inferior Cerebellar Peduncle**: The inferior cerebellar peduncle goes from the medulla to the cerebellum. It lies behind the inferior olivary nucleus and can sometimes be seen on the ventral view of the brainstem (as in Figure 7). This peduncle conveys a number of fiber systems to the cerebellum. These are shown schematically in this diagram of the ventral view of the brainstem and cerebellum. They include the following:
  - **The posterior (dorsal) spino-cerebellar** pathway conveys proprioceptive information from most of the body. This is one of the major tracts of the inferior peduncle. These fibers, carrying information from the muscle spindles, relay in the dorsal nucleus of Clarke in the spinal cord (see Figure 32). They ascend ipsilaterally in a tract that is found at the edge of the spinal cord (see Figure 68). The dorsal spino-cerebellar fibers terminate ipsilaterally; these fibers are distributed to the spino-cerebellar areas of the cerebellum.
  - The homologous tract for the upper limb is the **cuneo-cerebellar** tract. These fibers relay in the accessory (external) cuneate nucleus, located in the lower medulla (see Figure 67B and Figure 67C). This pathway is not shown in the diagram.
  - The **olivo-cerebellar** tract is also carried in this peduncle. The fibers originate from the inferior olivary nucleus (see Figure 6, Figure 7, Figure 67, and Figure 67B), cross in the medulla, and are distributed to all parts of the cerebellum. These axons have been shown to be the climbing fibers to the main dendritic branches of the Purkinje neurons.
  - Other cerebellar afferents from other nuclei of the brainstem, including the reticular formation, are conveyed to the cerebellum via this peduncle. Most important are those from the medial (and inferior) vestibular nuclei to the vestibulocerebellum. Afferents from the visual and auditory system are also known to be conveyed to the cerebellum.
- **Middle Cerebellar Peduncle**: All parts of the cerebral cortex contribute to the massive **cortico-pontine** system of fibers (also described with Figure 48). These fibers descend via the anterior and posterior limbs of the internal capsule, then the inner and outer parts of the cerebral peduncle, and terminate in the pontine nuclei. The fibers synapse and cross, and go to all parts of the cerebellum via the middle cerebellar peduncle (see Figure 6 and Figure 7). This input provides the cerebellum with the cortical information relevant to motor commands and the planned (intended) motor activities.
- **Superior Cerebellar Peduncle**: Only one afferent tract enters via the superior cerebellar peduncle (see below). This peduncle carries the major efferent pathway from the cerebellum (discussed with Figure 57).

ADDITIONAL DETAIL

One group of cerebellar afferents, those carried in the **ventral (anterior) spino-cerebellar tract**, enters the cerebellum via the superior cerebellar peduncle. These fibers cross in the spinal cord, ascend (see Figure 68), enter the cerebellum, and cross again, thus terminating on the same side from which they originated.

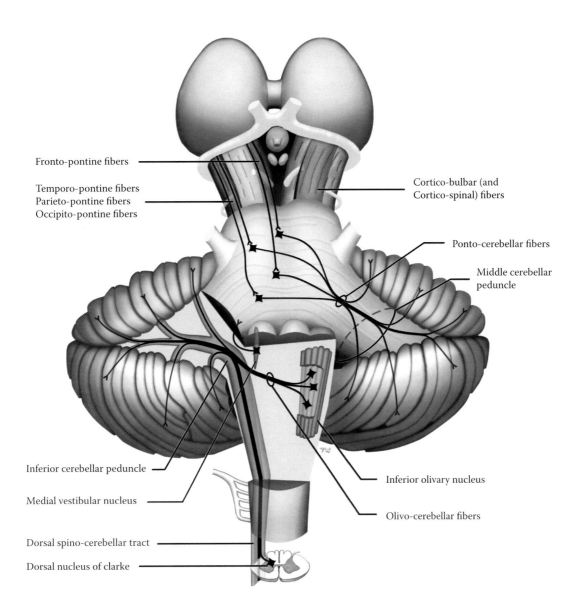

Fronto-pontine fibers

Temporo-pontine fibers
Parieto-pontine fibers
Occipito-pontine fibers

Cortico-bulbar (and
Cortico-spinal) fibers

Ponto-cerebellar fibers

Middle cerebellar
peduncle

Inferior cerebellar peduncle

Medial vestibular nucleus

Inferior olivary nucleus

Olivo-cerebellar fibers

Dorsal spino-cerebellar tract

Dorsal nucleus of clarke

FIGURE 55: Cerebellum 2 — Cerebellar Afferents

FIGURE 56A
CEREBELLUM 3

INTRACEREBELLAR (DEEP CEREBELLAR) NUCLEI

The brainstem is presented from the anterior perspective, with the cerebellum attached (as in Figure 6, Figure 7, Figure 8A, and Figure 8B). This diagram shows the **intracerebellar** nuclei — also called the **deep cerebellar nuclei** — within the cerebellum.

There are four pairs of deep cerebellar nuclei — the **fastigial** nucleus, the **globose** and **emboliform** nuclei (together called the intermediate or interposed nucleus), and the lateral or **dentate** nucleus. Each belongs to a different functional part of the cerebellum. These nuclei are the output nuclei of the cerebellum to other parts of the central nervous system.

- The fastigial (medial) nucleus is located next to the midline.
- The globose and emboliform nuclei are slightly more lateral; often these are grouped together and called the intermediate or interposed nucleus.
- The dentate nucleus, with its irregular margin, is most lateral. This nucleus is sometimes called the lateral nucleus and is by far the largest.

The nuclei are located within the cerebellum at the level of the junction of the medulla and the pons. Therefore, the cross-sections shown at this level (see Figure 66C) may include these deep cerebellar nuclei. Usually, only the dentate nucleus can be identified in sections of the gross brainstem and cerebellum done at this level (see Figure 67).

Two of the afferent fiber systems are shown on the left side — representing cortico-ponto-cerebellar fibers and spino-cerebellar fibers. All afferent fibers send collaterals to the deep cerebellar nuclei en route to the cerebellar cortex, and these are excitatory. Therefore, these neurons are maintained in a chronic state of activity.

The lateral vestibular nucleus functions as an additional deep cerebellar nucleus, because its main input is from the vestibulocerebellum (shown in the next illustration); its output is to the spinal cord (see Figure 50).

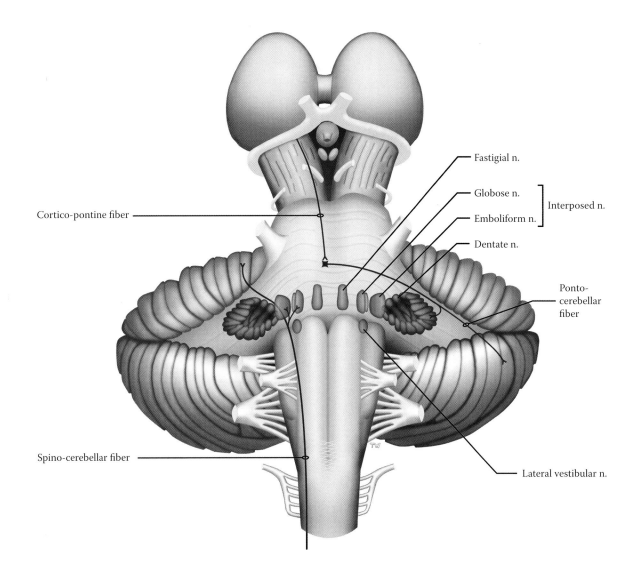

Cortico-pontine fiber

Fastigial n.

Globose n.
Emboliform n.
} Interposed n.

Dentate n.

Ponto-
cerebellar
fiber

Spino-cerebellar fiber

Lateral vestibular n.

FIGURE 56A: Cerebellum 3 — Intracerebellar (Deep Cerebellar) Nuclei

FIGURE 56B
CEREBELLUM 4

INTRACEREBELLAR CIRCUITRY

The cerebellum is being presented from the dorsal perspective (as in Figure 9A). The third ventricle is situated between the two diencephala; the pineal gland is seen attached to the posterior aspect of the thalamus. Below are the colliculi, superior and inferior. On the right side of the illustration, the cerebellar hemisphere has been cut away, revealing the "interior" on this side.

The cerebellum is organized with cortical tissue on the outside, the **cerebellar cortex**. The cortex consists of three layers, and all areas of the cerebellum are histologically alike. The most important cell of the cortex is the **Purkinje neuron**, which forms a layer of cells; their massive dendrites receive the input to the cerebellum. Various interneurons are also located in the cortex. The axon of the Purkinje neuron is the only axonal system to leave the cerebellar cortex.

Deep within the cerebellum are the intracerebellar nuclei or the deep cerebellar nuclei, now shown from the posterior view (see Figure 56A).

Overall, the circuitry is as follows: All (excitatory) afferents to the cerebellum go to both the deep cerebellar nuclei (via collaterals) and the cerebellar cortex. After processing in the cortex, the Purkinje neuron sends its axon on to the neurons of the deep cerebellar nuclei — all Purkinje neurons are inhibitory. Their influence modulates the activity of the deep cerebellar neurons, which are tonically active (described in more detail below). The output of the deep cerebellar neurons, which is excitatory, influences neurons in the brainstem and cerebral cotex via the thalamus (discussed with the next illustration).

The connections of the cortical areas with the intracerebellar nuclei follow the functional divisions of the cerebellum:

- The vestibulocerebellum is connected to the fastigial nucleus, as well as to the lateral vestibular nucleus.
- The spinocerebellum connects with the interposed nucleus (the globose and emboliform).
- The neocerebellum connects to the dentate nucleus.

Axons from the deep nuclei neurons project from the cerebellum to many areas of the CNS, including brainstem motor nuclei (e.g., vestibular, reticular formation) and thalamus (to motor cortex). In this way, the cerebellum exerts its influence on motor performance. This will be discussed with the next illustration.

DETAILS OF CEREBELLAR CIRCUITRY

The cerebellum receives information from many parts of the nervous system, including the spinal cord, the vestibular system, the brainstem, and the cerebral cortex. Most of this input is related to motor function, but some is also sensory. These afferents are excitatory in nature and influence the ongoing activity of the neurons in the intracerebellar nuclei, as well as projecting to the cerebellar cortex.

The incoming information to the cerebellar cortex is processed by various interneurons of the cerebellar cortex and eventually influences the Purkinje neuron. This will lead to either increased or decreased firing of this neuron. Its axon is the only one to leave the cerebellar cortex, and these axons project, in an organized manner, to the deep cerebellar nuclei.

The Purkinje neurons are inhibitory, and their influence modulates the activity of the deep cerebellar nuclei. Increased firing of the Purkinje neuron increases the ongoing inhibition onto these deep cerebellar nuclei, while decreased Purkinje cell firing results in a decrease in the inhibitory effect on the deep cerebellar cells, i.e., this results in the increased firing of the deep cerebellar neurons (called disinhibition).

It is interesting to note that the cerebellar cortex projects fibers directly to the lateral vestibular nucleus (see Figure 50, not illustrated). As would be anticipated, these are inhibitory. The lateral vestibular nucleus could therefore, in some sense, be considered one of the intracerebellar nuclei. This nucleus also receives input from the vestibular system, and then projects to the spinal cord (see Figure 50 and Figure 51A).

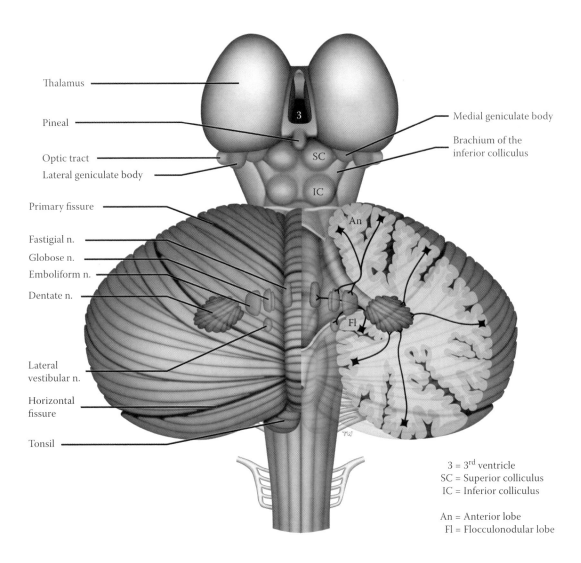

Thalamus

Pineal

Optic tract

Lateral geniculate body

Primary fissure

Fastigial n.

Globose n.

Emboliform n.

Dentate n.

Lateral
vestibular n.

Horizontal
fissure

Tonsil

Medial geniculate body

Brachium of the
inferior colliculus

SC

IC

An

Fl

3 = 3rd ventricle
SC = Superior colliculus
IC = Inferior colliculus

An = Anterior lobe
Fl = Flocculonodular lobe

FIGURE 56B: Cerebellum 4 — Intracerebellar Circuitry

FIGURE 57
CEREBELLUM 5

CEREBELLAR EFFERENTS

This is again a dorsal view of the diencephalon, brainstem, and cerebellum, with the deep cerebellar (intracerebellar) nuclei. The cerebellar tissue has been removed in the midline, revealing the fourth ventricle (as in Figure 10); the three cerebellar peduncles are also visualized from this posterior perspective (see Figure 10).

The output from the cerebellum will be described, following the functional divisions of the cerebellum:

- **Vestibulocerebellum**: Efferents from the fastigial nuclei go to brainstem motor nuclei (e.g., vestibular nuclei and reticular formation), influencing balance and gait. They exit in a bundle that is found adjacent to the inferior cerebellar peduncle (named the juxtarestiform body).
- **Spinocerebellum**: The emboliform and globose, the interposed nucleus, also project to brainstem nuclei, including the red nucleus of the midbrain. They also project to the appropriate limb areas of the motor cortex via the thalamus (see below); these are the fibers involved in the comparator function of this part of the cerebellum.
- **Neocerebellum**: The dentate nucleus is the major outflow from the cerebellum via the superior cerebellar peduncle (see Figure 10 and Figure 40). This peduncle connects the cerebellar efferents, through the midbrain, to the thalamus on their way to the motor cortex. Some of the fibers terminate in the red nucleus of the midbrain, particularly those from the interposed nucleus. The majority of the fibers, those from the dentate nucleus, terminate in the ventral lateral (VL) nucleus of the thalamus (see Figure 53 and Figure 63). From here they are relayed to the motor cortex, predominantly area 4, and also to the premotor cortex, area 6. The neocerebellum is involved in motor coordination and planning. (This is to be compared with the influence of the basal ganglia on motor activity, see Figure 53.)

DETAILED PATHWAY

The outflow fibers of the superior cerebellar peduncles originate mainly from the dentate nucleus, with some from the interposed nucleus (as shown). The axons start laterally and converge toward the midline (see Figure 10 and Figure 40), passing in the roof of the upper half of the fourth ventricle (see Figure 21 and Figure 41B). The fibers continue to "ascend" through the upper part of the pons (see the cross-sections in Figure 66 and Figure 66A). In the lower midbrain there is a complete decussation of the peduncles (see Figure 65B).

CORTICAL LOOP

The cerebral cortex is linked to the neocerebellum by a circuit that forms a loop. Fibers are relayed from the cerebral cortex via the pontine nuclei to the cerebellum. The ponto-cerebellar fibers cross and go to the neocerebellum of the opposite side. After cortical processing in the cerebellar cortex, the fibers project to the dentate nucleus. These efferents project to the thalamus, after crossing (decussating) in the lower midbrain. From the thalamus, fibers are relayed mainly to the motor areas of the cerebral cortex. Because of the two crossings, the messages are returned to the same side of the cerebral cortex from which the circuit began.

CLINICAL ASPECT

Lesions of the neocerebellum (of one side) cause motor deficits to occur on the same side of the body, that is, ipsilaterally for the cerebellum. The explanation for this lies in the fact that the cortico-spinal tract is also a crossed pathway (see Figure 45). For example, the errant messages from the left cerebellum that are delivered to the right cerebral cortex cause the symptoms to appear on the left side — contralaterally for the cerebral cortex but ipsilaterally from the point of view of the cerebellum.

The cerebellar symptoms associated with lesions of the neocerebellum are collectively called **dyssynergia**, in which the range, direction, and amplitude of voluntary muscle activity are disturbed. The specific symptoms include the following:

- Distances are improperly gauged when pointing, called dysmetria, and include pastpointing.
- Rapid alternating movements are poorly performed, called dysdiadochokinesis.
- Complex movements are performed as a series of successive movements, which is called a decomposition of movement.
- There is a tremor seen during voluntary movement, an **intention tremor**. (This is in contrast to the Parkinsonian tremor, which is present at rest.)
- Disturbances also occur in the normally smooth production of words, resulting in slurred and explosive speech.

In addition, cerebellar lesions in humans are often associated with hypotonia and sluggish deep tendon reflexes.

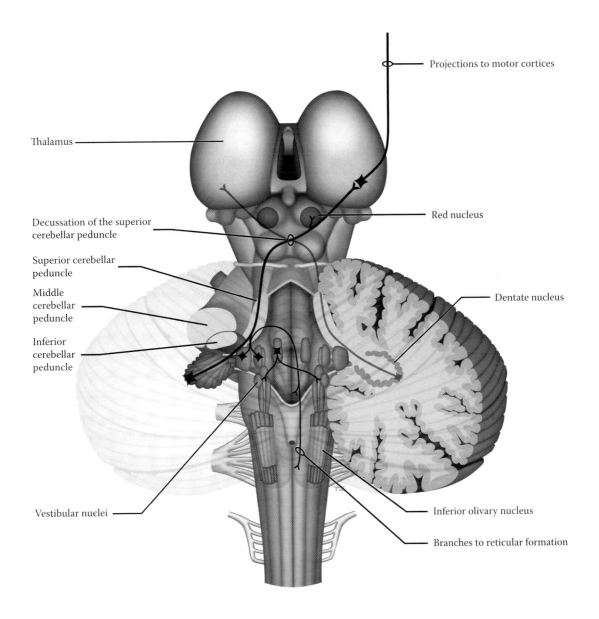

Projections to motor cortices

Thalamus

Red nucleus

Decussation of the superior
cerebellar peduncle

Superior cerebellar
peduncle

Middle
cerebellar
peduncle

Inferior
cerebellar
peduncle

Dentate nucleus

Vestibular nuclei

Inferior olivary nucleus

Branches to reticular formation

FIGURE 57: Cerebellum 5 — Cerebellar Efferents

Section C

NEUROLOGICAL NEUROANATOMY

INTRODUCTION

A thorough understanding of the structure and function of the nervous system is the foundation for clinical neurology. The neurologist's task is to analyze the history of the illness and the symptoms and signs of the patient, decide whether the problem is in fact neurological, configure the patient's complaints and the physical findings to establish where in the nervous system the problem is located (localization), and then to ascertain a cause for the disease (etiology). At some stage, laboratory investigations and imaging studies are used to confirm the localization of the disease and to assist in establishing the diagnosis. An appropriate therapeutic plan would then be proposed, and the patient can be advised of the long-term outlook of his or her disease (the prognosis), and its impact on life, family, and employment (psychosocial issues).

A simple mnemonic using the letter "w" helps to recall the basic steps necessary to establish a neurological diagnosis:

- **Whether** the signs and symptoms are consistent with involvement of the nervous system, based upon a detailed history and a complete neurological examination
- **Where** the nervous system problem is located (i.e., **localization**)
- **What** is the etiology of the disease, its pathophysiological mechanism(s)

Diseases can be recognized by skilled and knowledgeable expert clinicians based upon their presentation (for example, vascular lesions have a sudden onset vs. a slow onset for tumors), the age of the patient, the part(s) of the nervous system involved, and the evolution of the disease process. The task is more complex in children, depending on the age, because the nervous system continues to develop through infancy and childhood; diseases interfere with and interrupt this developmental pattern. Knowledge of normal growth and development is necessary to practice pediatric neurology.

LEARNING PLAN

The learning objective of this section is to synthesize the structural and functional aspects of the nervous system.

This should enable the student to localize the disease process within the nervous system.

The vascular supply of the brain will be studied at this point, allowing the learner to integrate the vascular information with the functioning of the nervous system. The thalamus will be presented once again, permitting a synthesis of the connections of the thalamic nuclei, both on the input side and with the cerebral cortex.

There is an emphasis in this section on the brainstem microanatomy; often the brainstem presents an overwhelming challenge to students struggling to learn the nervous system (see below). Finally, the spinal cord will be presented with all the ascending (sensory) and descending (motor) tracts.

VASCULAR SUPPLY

The CNS is dependent upon a continuous supply of blood; viability of the neurons depends upon the immediate and constant availability of both oxygen and glucose. Interruption of this lifeline causes sudden loss of function. Study of the nervous system must include a complete knowledge of the blood supply and which structures (nuclei and tracts) are situated in the vascular territory of the various arteries. Failure of the blood supply to a region, either because of occlusion or hemorrhage, will lead to death of the neurons and axons, leading to functional deficits.

Areas of gray matter, where the neurons are located, have a greater blood supply than white matter. Loss of oxygen and glucose supply to these neurons will lead to loss of electrical activity after a few minutes (in adults), and if continued for several minutes, to neuronal death. Although white matter requires less blood supply, loss of adequate supply leads to destruction of the axons in the area of the infarct and an interruption of pathways. After loss of the cell body or interruption of the axon, the distal portion of the axon (the part on the other side of the lesion separated from the cell body) and the synaptic connections will degenerate, leading to a permanent loss of function.

Every part of the nervous system lies within the vascular territory of an artery, sometimes with an overlap from adjacent arteries. Visualization of the arterial (and venous) branches can be accomplished using:

- **Arteriogram**: By injecting a radiopaque substance into the arteries (this is a procedure that is done by a neuroradiologist) and following its course through a rapid series of x-rays (called an **arteriogram**), a detailed view of the vasculature of the brain is obtained; either the carotid or vertebral artery is usually injected, according to which arterial tree is under investigation. This is an invasive procedure carrying a certain degree of risk.
- **MR Angiogram**: Using neuroradiology imaging with MRI (discussed with Figure 59), the major blood vessels (such as the circle of Willis) can be visualized; this is called a magnetic resonance angiogram (**MRA**).

CLINICAL ASPECT

It is extremely important to know which parts of the brain are located in the territory supplied by each of the major cerebral and brainstem blood vessels, and to understand the functional contribution of these parts. This is fundamental for clinical neurology.

A clinical syndrome involving the arteries of the brain is often called a cerebrovascular accident (**CVA**) or "**a stroke**." The nature of the process, blood vessel occlusion through infarction or embolus, or hemorrhage, is not specified by the use of this term; nor does the term indicate which blood vessel is involved. The clinical event is a sudden loss of function; the clinical deficit will depend upon where the occlusion or hemorrhage occurred.

Occlusion is more common than hemorrhage, often caused by an embolus (e.g., from the heart). Hemorrhage may occur into the brain substance (parenchymal), causing destruction of the brain tissue and at the same time depriving areas distally of blood.

HISTOLOGICAL NEUROANATOMY

This section presents the detailed neuroanatomy that is needed for localization of lesions in the brainstem. A series of illustrations is presented through the brainstem to enable the learner to integrate the nuclei, both cranial nerve and other important nuclei, and the tracts passing through that region. Accompanying these schematics are photographs of the brainstem from the human brain — at the same levels. The same approach is used for the spinal cord, a common site for clinical disease and traumatic injuries.

INTRACRANIAL PRESSURE (ICP)

In addition to knowledge of the brain and the function of the various parts and the blood supply, many disease processes exert their effect because of a rise in intracranial pressure (ICP). This may lead to a displacement of brain tissue within the skull. The adult skull is a rigid container filled with the brain, the cerebrospinal fluid (CSF), and blood. The interior of the skull is divided into compartments by folds of dura: the **falx cerebri** in the midline between the hemispheres (see Figure 16) and the **tentorium cerebelli**, which partially separates the hemispheres from the contents of the posterior cranial fossa, the brainstem and cerebellum (see Figure 17 and Figure 30). The opening in the tentorium for the brainstem, called the tentorial notch or incisura, is at the level of the upper midbrain (see Figure 30). (**Note to the Learner**: Anatomy texts should be consulted for a visual understanding of these structures.)

CLINICAL ASPECT

Any increase in volume inside the skull — due to brain swelling, tumor, abscess, hemorrhage, abnormal amount of CSF — causes a rise in pressure inside the skull (i.e., ICP). Although brain tissue itself has no pain fibers, the blood vessels and meninges do, hence any pulling on the meninges may give rise to a headache. This process may be acute, subacute, or chronic. A prolonged increase in ICP can be detected clinically by examining the optic disc; its margins will become blurred and the disc itself engorged, called **papilledema**.

Any space-occupying lesion (e.g., sudden hemorrhage, slow-growing tumor), depending upon the lesion and its progression, will sooner or later cause a displacement of brain tissue from one compartment to another. This pathological displacement causes damage to the brain. This is called a **brain herniation** syndrome and typically occurs:

- Through the foramen magnum, **tonsillar herniation** (discussed with Figure 9B)
- Through the tentorial notch, **uncal herniation** (discussed with Figure 15B)
- Under the falx cerebri

These shifts are life-threatening and require emergency management. (**Note to the Learner**: This would be an opportune time to review the signs and symptoms associated with these clinical emergencies, such as testing of the pupillary light reflex and the pathway involved.)

FIGURE 58
BLOOD SUPPLY 1

THE ARTERIAL CIRCLE OF WILLIS (PHOTOGRAPHIC VIEW WITH OVERLAY)

The arterial circle (of Willis) is a set of arteries interconnecting the two sources of blood supply to the brain, the vertebral and internal carotid arteries. It is located at the base of the brain, surrounding the optic chiasm and the hypothalamus (the mammillary nuclei) (review Figure 15A and Figure 15B). Within the skull, it is situated above the pituitary fossa (and gland). The major arteries to the cerebral cortex of the hemispheres are branches of this arterial circle. This illustration is a photographic view of the inferior aspect of the brain, including brainstem and cerebral hemispheres, with the blood vessels (as in Figure 15A). Branches from the major arteries have been added to the photographic image.

The cut end of the **internal carotid** arteries is a starting point. Each artery divides into the **middle cerebral** artery (MCA) and the **anterior cerebral** artery (ACA). The MCA courses within the lateral fissure. The anterior portion of the temporal lobe has been removed on the left side of this illustration in order to follow the course of the MCA in the lateral fissure. Within the fissure, small arteries are given off to the basal ganglia, called the **striate** arteries (not labeled; see Figure 62). The artery emerges at the surface (see Figure 14A) and courses upward, dividing into branches that are distributed onto the dorsolateral surface of the hemispheres (see Figure 60).

By removing (or lifting) the optic chiasm, the ACA can be followed anteriorly. This artery heads into the interhemispheric fissure (see Figure 16) and will be followed when viewing the medial surface of the brain (see Figure 17 and Figure 61). A very short artery connects the ACAs of the two sides, the **anterior communicating** artery.

The vertebro-basilar system supplies the brainstem and cerebellum, and the posterior part of the hemispheres. The two **vertebral** arteries unite at the lower border of the pons to form the midline **basilar** artery, which courses in front of the pons. The basilar artery terminates at the midbrain level by dividing into two **posterior cerebral** arteries. These supply the inferior aspect of the brain and particularly the occipital lobe (see Figure 61).

The arterial circle is completed by the **posterior communicating** artery (normally one on each side), which connects the internal carotid (or middle cerebral) artery, often called the anterior circulation, with the posterior cerebral artery, the posterior circulation.

Small arteries directly from the circle (not shown) provide the blood supply to the diencephalon (thalamus and hypothalamus), some parts of the internal capsule, and part of the basal ganglia. The major blood supply to these regions is from the striate arteries (see Figure 62).

The branches from the vertebral and basilar artery supply the brainstem. Small branches directly from the vertebral and basilar arteries (not shown), known as **paramedian** arteries, supply the medial structures of the brainstem (further discussed with Figure 67B). There are three major branches from this arterial tree to the cerebellum — the **posterior inferior cerebellar** artery (**PICA**), **the anterior inferior cerebellar** artery (AICA), and the **superior cerebellar** artery. All supply the lateral aspects of the brainstem, including nuclei and tracts, en route to the cerebellum; these are often called the **circumferential** branches.

The blood supply to the spinal cord is shown in Figure 2B and is discussed with Figure 68.

CLINICAL ASPECT

The vascular territories of the various cerebral blood vessels are shown in color in this diagram. The most common clinical lesion involving the cerebral blood vessels is occlusion, often due to an embolus originating from the heart or the carotid bifurcation in the neck. These clinical deficits will be described with each of the major branches to the cerebral cortex (with Figure 60 and Figure 61).

In the eventuality of a slow occlusion of one of the major blood vessels of the circle, sometimes one of the communicating branches becomes large enough to provide sufficient blood to be shunted to the area deprived (see Figure 59B). One of the vascular syndromes of the brainstem, the lateral medullary syndrome (of Wallenberg) is discussed with Figure 67B.

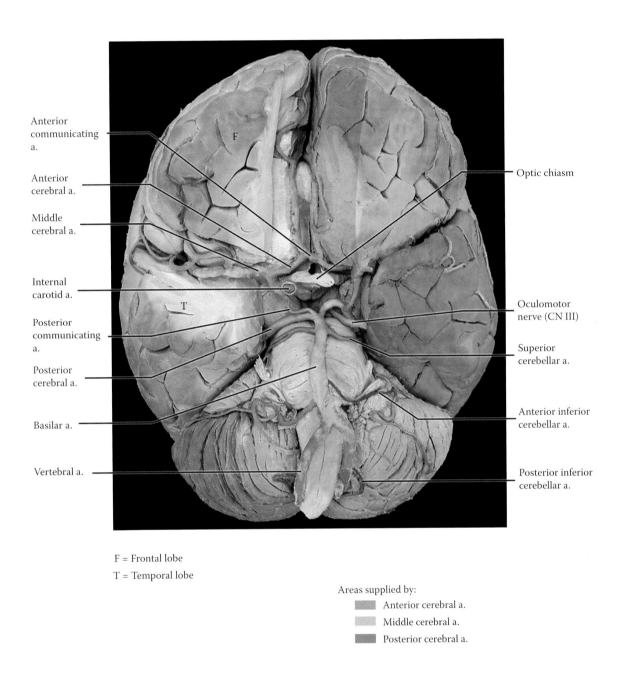

Anterior
communicating
a.

Anterior
cerebral a.

Middle
cerebral a.

Internal
carotid a.

Posterior
communicating
a.

Posterior
cerebral a.

Basilar a.

Vertebral a.

F

T

Optic chiasm

Oculomotor
nerve (CN III)

Superior
cerebellar a.

Anterior inferior
cerebellar a.

Posterior inferior
cerebellar a.

F = Frontal lobe

T = Temporal lobe

Areas supplied by:

Anterior cerebral a.

Middle cerebral a.

Posterior cerebral a.

FIGURE 58: Blood Supply 1 — Arterial Circle of Willis (photograph with overlay)

FIGURE 59A
BLOOD SUPPLY 2

MR ANGIOGRAM — MRA

Recent advances in technology have allowed for a visualization of the major blood vessels supplying the brain, notably the arterial circle of Willis. This investigation does not require an invasive procedure (described with the next illustration), although an injection intravenously of a contrast substance called gadolinium maybe used (discussed with Figure 28B). Although the quality of such images cannot match the detail seen after an angiogram of select blood vessels (shown in the next illustration), the noninvasive nature of this procedure, and the fact that the patient is not exposed to any risk, clearly establishes this investigation as desirable to provide some information about the state of the cerebral vasculature.

UPPER RADIOGRAPH

This arteriogram shows the **circle of Willis** as seen as if looking at the brain from below (as in the previous illustration). The internal carotid artery goes through the cavernous (venous) sinus of the skull, forming a loop that is called the carotid siphon. It then divides into the anterior cerebral artery, which goes anteriorly, and the middle cerebral artery, which goes laterally. The basilar artery is seen at its termination, as it divides into the posterior cerebral arteries. The anterior communicating artery is present, and there are two posterior communicating arteries completing the circle, joining the internal carotid with the posterior cerebral on each side.

LOWER RADIOGRAPH

This is the same angiogram, displayed at a different orientation, as though you are looking at the patient "face-on,' but, wtih his/her head tilted forward slightly. The two vertebral arteries can be seen, joining to form the basilar artery; it is not uncommon to see the asymmetry in these vessels. The posterior inferior cerebellar artery (PICA) can be seen, a branch of the vertebral (it is also labeled in the upper radiograph), but not the anterior inferior cerebellar artery, a branch of the basilar (see Figure 58). The basilar artery gives off the superior cerebellar arteries and then ends by dividing into the posterior cerebral arteries. The internal carotid artery can be followed through its curvature in the petrous temporal bone of the skull, before dividing into the anterior and middle cerebral arteries.

CLINICAL ASPECT

One of the characteristic vascular lesions in the arteries that make up the arterial circle of Willis is a type of aneurysm, called a **Berry aneurysm**. This is caused by a weakness of part of the wall of the artery, causing a local ballooning of the artery. Often these aneurysms rupture spontaneously, particularly if there is accompanying hypertension. This sudden rupture occurs into the subarachnoid space and may also involve nervous tissue of the base of the brain. The whole event is known as a **subarachnoid hemorrhage**, and this diagnosis must be considered when one is faced clinically with an acute major cerebrovascular event, without trauma, accompanied by intensely severe headache and often a loss of consciousness.

Sometimes these aneurysms leak a little blood, which causes an irritation of the meninges and accompanying symptoms of headache. An MRA can, at the minimum, visualize whether there is an aneurysm on one of the vessels of the circle, and whether the major blood vessels are patent.

Note to the Learner: One of the best ways of learning the arterial supply to the brain and the circle of Willis is to actually make a sketch drawing, accompanied by a list of the areas supplied and the major losses that would follow a sudden occlusion. The blood supply to the brainstem and the most common vascular lesions affecting this area will be discussed with the illustrations to follow.

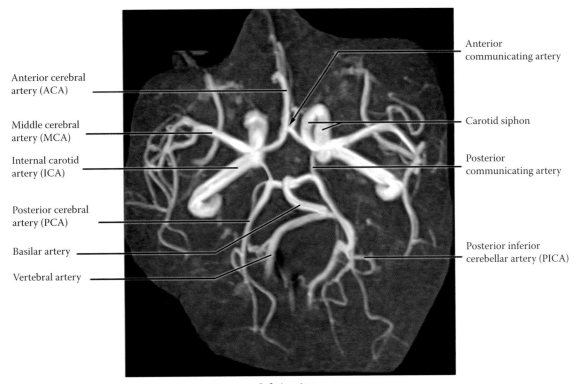

Anterior cerebral
artery (ACA)

Middle cerebral
artery (MCA)

Internal carotid
artery (ICA)

Posterior cerebral
artery (PCA)

Basilar artery

Vertebral artery

Anterior
communicating artery

Carotid siphon

Posterior
communicating artery

Posterior inferior
cerebellar artery (PICA)

Inferior view

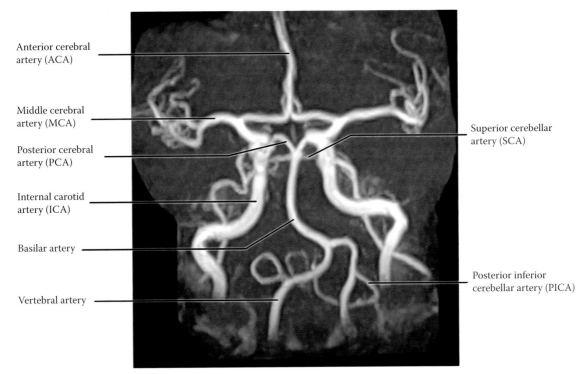

Anterior cerebral
artery (ACA)

Middle cerebral
artery (MCA)

Posterior cerebral
artery (PCA)

Internal carotid
artery (ICA)

Basilar artery

Vertebral artery

Superior cerebellar
artery (SCA)

Posterior inferior
cerebellar artery (PICA)

Tilted anterior view

FIGURE 59A: Blood Supply 2 — MR Angiogram: MRA (radiograph)

FIGURE 59B
BLOOD SUPPLY 3

CEREBRAL ANGIOGRAM

This radiograph was done by injecting a radiopaque dye into the left internal carotid artery. The usual procedure involves threading a catheter from the groin up the aorta and into the internal carotid artery, under fluoroscopic guidance, a procedure not without risk; then a radiopaque dye is injected within the artery.

In this particular case, there had been a slow occlusion of the right internal carotid, allowing time for the anterior communicating artery of the circle of Willis to become widely patent; therefore, blood was shunted into the anterior and middle cerebral arteries on the affected side. This is not usual, and in fact, this radiogram was chosen for this reason.

The middle cerebral artery goes through the lateral fissure and breaks up into various branches on the dorsolateral surface of the hemisphere (shown in the next illustration). The lenticulostriate (striate) arteries given off en route supply the interior structures of the hemisphere (to be discussed with Figure 62).

This radiograph shows the profuseness of the blood supply to the brain, the hemispheres, and is presented to give the student that visual image, as well as to show the appearance of an angiogram.

CLINICAL ASPECT

Visualization of the blood supply to the brain is required for the accurate diagnosis of aneurysms and occlusions affecting these blood vessels. Procedures are now done within the blood vessels (intravascular), using specialized catheters to destroy an identified blood clot, or to insert a metal "coil" into an aneurysm (thereby "curing" the problem). These procedures are done by interventional neuroradiologists.

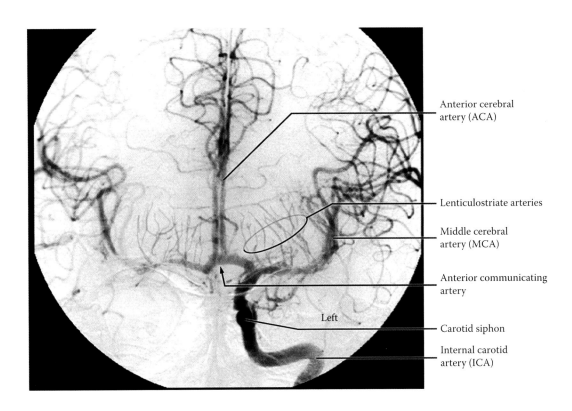

Anterior cerebral
artery (ACA)

Lenticulostriate arteries

Middle cerebral
artery (MCA)

Anterior communicating
artery

Left

Carotid siphon

Internal carotid
artery (ICA)

FIGURE 59B: Blood Supply 3 — Cerebral Angiogram (radiograph)

FIGURE 60
BLOOD SUPPLY 4

CORTICAL: DORSOLATERAL
(PHOTOGRAPHIC VIEW WITH
OVERLAY)

This illustration shows the blood supply to the cortical areas of the dorsolateral aspect of the hemispheres; it has been created by superimposing the blood vessels onto the photographic view of the brain (the same brain as in Figure 14A).

After coursing through the depths of the lateral fissure (see Figure 58 and Figure 59B), the **middle cerebral artery** emerges and breaks into a number of branches that supply different parts of the dorsolateral cortex — the frontal, parietal, and temporal areas of the cortex. Each branch supplies a different territory, as indicated; branches supply the precentral and post-central gyri, the major motor and sensory areas for the face and head and the upper limbs. On the dominant side, this includes the language areas (see Figure 14A).

The vascular territories of the various cerebral blood vessels are shown in color in this diagram. The branches of the middle cerebral artery extend toward the midline sagittal fissure, where branches from the other cerebral vessels (anterior and posterior cerebral) are found, coming from the medial aspect of the hemispheres (see next illustration). A zone remains between the various arterial territories — the arterial **borderzone** region (a watershed area). This area is poorly perfused and prone to infarction, particularly if there is a sudden loss of blood pressure (e.g., with cardiac arrest or after a major hemorrhage).

CLINICAL ASPECT

The most common clinical lesion involving these blood vessels is occlusion, often due to an embolus originating from the heart or the carotid bifurcation in the neck. This results in infarction of the nervous tissue supplied by that branch — the clinical deficit will depend upon which branch or branches are involved. For example, loss of sensory or motor function to the arm and face region will be seen after the blood vessel to the central region is occluded. The type of language loss that occurs will depend upon the branch affected, in the dominant hemisphere — a deficit in expressive language will be seen with a lesion affecting Broca's area, whereas a comprehension deficit is found with a lesion affecting Wernicke's area.

Acute strokes are now regarded as an emergency with a narrow therapeutic window. According to current evidence, if the site of the blockage can be identified and the clot (or embolus) removed within three hours, there is a good chance that the individual will have significant if not complete recovery of function. The therapeutic measures include a substance that will dissolve the clot, or interventional neuroradiology whereby a catheter is threaded through the vasculature and into the brain and the clot is removed. Major hospitals now have a "stroke protocol," including a CT scan, to investigate these people immediately when brought to emergency so that therapeutic measures can be instituted.

A clinical syndrome has been defined in which there is a temporary loss of blood supply affecting one of the major blood vessels. Some would limit this temporary loss to less than one hour, whereas others suggest that this period could extend to several hours. This syndrome is called **a transient ischemic attack (TIA)**. Its cause could be blockage of a blood vessel that resolves spontaneously, or perhaps an embolus that breaks up on its own. Regardless, people are being educated to look at this event as a *brain attack*, much like a heart attack, and to seek medical attention immediately. The statistics indicate that many of these people would go on to suffer a significant stroke.

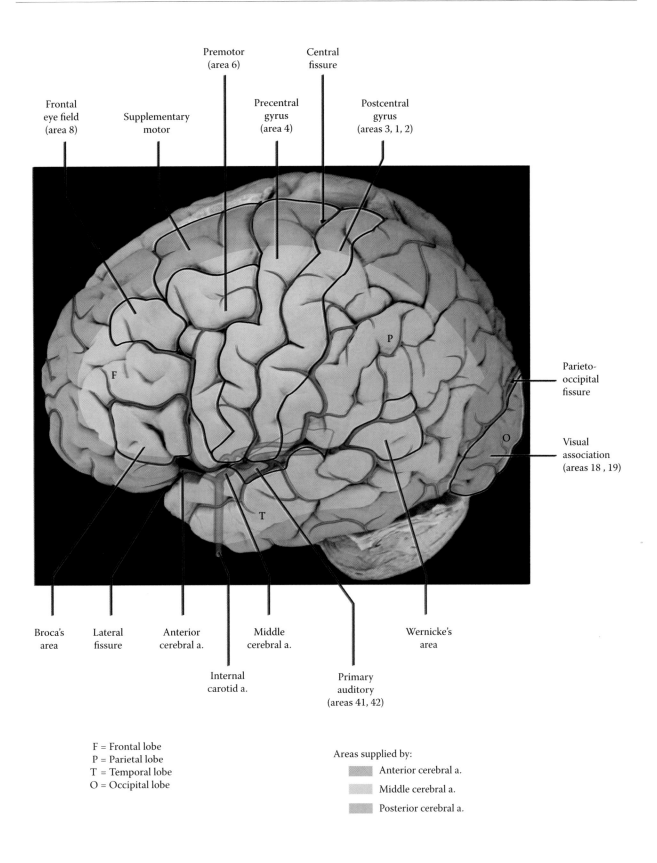

Premotor
(area 6)

Central
fissure

Frontal
eye field
(area 8)

Supplementary
motor

Precentral
gyrus
(area 4)

Postcentral
gyrus
(areas 3, 1, 2)

Parieto-
occipital
fissure

Visual
association
(areas 18 , 19)

Broca's
area

Lateral
fissure

Anterior
cerebral a.

Middle
cerebral a.

Wernicke's
area

Internal
carotid a.

Primary
auditory
(areas 41, 42)

F = Frontal lobe
P = Parietal lobe
T = Temporal lobe
O = Occipital lobe

Areas supplied by:

Anterior cerebral a.

Middle cerebral a.

Posterior cerebral a.

FIGURE 60: Blood Supply 4 — Cortical Dorsolateral Surface (photograph with overlay)

FIGURE 61
BLOOD SUPPLY 5

CORTICAL: MEDIAL (PHOTOGRAPHIC VIEW WITH OVERLAY)

In this illustration, the blood supply to the medial aspect of the hemispheres has been superimposed onto this view of the brain (see Figure 17). Two arteries supply this part — the anterior cerebral artery and the posterior cerebral artery. The vascular territories of the various cerebral blood vessels are shown in color in this diagram.

The **anterior cerebral artery** (ACA) is a branch of the internal carotid artery from the circle of Willis (see Figure 58, Figure 59A, and Figure 59B). It runs in the interhemispheric fissure, above the corpus callosum (see Figure 16) and supplies the medial aspects of both the frontal lobe and the parietal lobe; this includes the cortical areas responsible for sensory-motor function of the lower limb.

The **posterior cerebral artery** (PCA) supplies the occipital lobe and the visual areas of the cortex, areas 17, 18, and 19 (see Figure 41A and Figure 41B). The posterior cerebral arteries are the terminal branches of the basilar artery from the vertebral or posterior circulation (see Figure 58). The demarcation between these arterial territories is the parieto-occipital fissure.

Both sets of arteries have branches that spill over to the dorsolateral surface. As noted (in the previous illustration), there is a potential gap between these and the territory supplied by the middle cerebral artery, known as the arterial borderzone or watershed region.

BRAINSTEM

The blood supply to the brainstem and cerebellum is shown from this perspective, and should be reviewed with Figure 58. The three cerebellar arteries — posterior inferior, anterior inferior, and superior — are branches of the vertebro-basilar artery, supplying the lateral aspects of the brainstem en route to the cerebellum.

CLINICAL ASPECT

The deficit most characteristic of an occlusion of the ACA is selective loss of function of the lower limb. Clinically, the control of micturition seems to be located on this medial area of the brain, perhaps in the supplementary motor area (see Figure 53), and symptoms related to voluntary bladder control may also occur with lesions in this area.

The clinical deficit found after occlusion of the posterior cerebral artery on one side is a loss of one-half of the visual field of both eyes — a contralateral homonymous hemianopia. The blood supply to the calcarine cortex, the visual cortex, area 17, is discussed with Figure 41B. (**Note to the Learner**: This is an opportune time to review the optic pathway and to review the visual field deficits that are found after a lesion in different parts of the visual system.)

Recent studies indicate that the core of tissue that has lost its blood supply is surrounded by a region where the blood supply is marginal, but which is still viable and may be rescued — the "**penumbra**," as it is now called. In this area surrounding the infarcted tissue, the blood supply is reduced below the level of nervous tissue functionality and the area is therefore "silent," but the neurons are still viable.

These studies have led to a rethinking of the therapy of strokes:

- In the acute stage, if the patient can be seen quickly and investigated immediately, the site of the lesion might be identified. This is the basis for the immediate treatment of strokes with powerful drugs to dissolve the clot or the use of interventional neuroradiology (in large centers). If done soon enough after the "stroke," it may be possible to avert any clinical deficit.
- There may be an additional period beyond this timeframe when damaged neurons in the penumbra can be rescued through the use of neuroprotective agents — specific pharmacological agents that protect the neurons from the damaging consequences of loss of blood supply.

As loss of function and diminished quality of life are the end result of strokes, and with our aging population, it is clear that this is a most active area of neuroscience research.

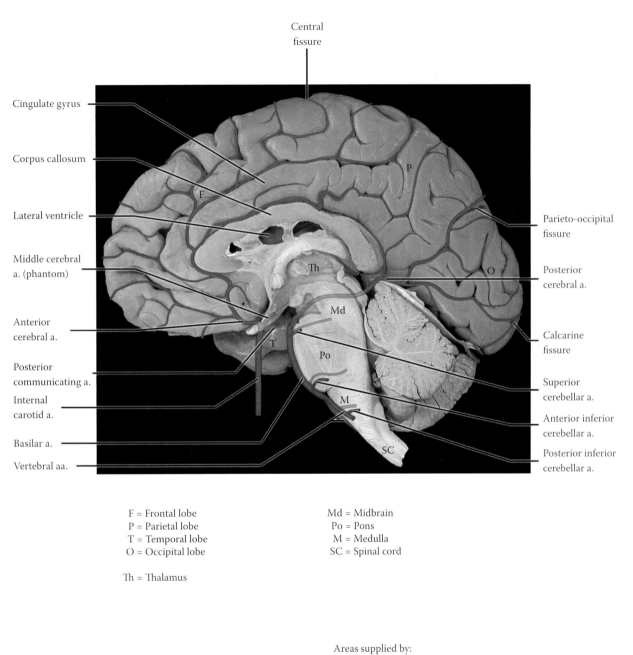

Central fissure

Cingulate gyrus

Corpus callosum

Lateral ventricle

Middle cerebral a. (phantom)

Anterior cerebral a.

Posterior communicating a.

Internal carotid a.

Basilar a.

Vertebral aa.

Parieto-occipital fissure

Posterior cerebral a.

Calcarine fissure

Superior cerebellar a.

Anterior inferior cerebellar a.

Posterior inferior cerebellar a.

F = Frontal lobe
P = Parietal lobe
T = Temporal lobe
O = Occipital lobe

Th = Thalamus

Md = Midbrain
Po = Pons
M = Medulla
SC = Spinal cord

Areas supplied by:

 Anterior cerebral a.

Posterior cerebral a.

FIGURE 61: Blood Supply 5 — Cortical Medial Surface (photograph with overlay)

FIGURE 62
BLOOD SUPPLY 6

INTERNAL CAPSULE (PHOTOGRAPHIC VIEW WITH OVERLAY)

One of the most important sets of branches of the middle cerebral artery is found within the lateral fissure (this artery has been dissected in Figure 58). These are known as the striate arteries, also called lenticulostriate arteries (see Figure 59B). These branches supply most of the internal structures of the hemispheres, including the internal capsule and the basal ganglia (discussed with Figure 26; see also Figure 27 and Figure 29).

In this illustration, a coronal section of the brain (see Figure 29), the middle cerebral artery is shown traversing the lateral fissure. The artery begins as a branch of the circle of Willis (see Figure 58; also Figure 59B). Several small branches are given off, which supply the area of the lenticular nucleus and the internal capsule, as well as the thalamus. The artery then emerges, after passing through the lateral fissure, to supply the dorsolateral cortex (see Figure 60).

These small blood vessels are the major source of blood supply to the internal capsule and the adjacent portions of the basal ganglia (head of caudate nucleus and putamen), as well as the thalamus (see Figure 26). Some of these striate arteries enter the brain through the anterior perforated space (area) which is located where the olfactory tract divides (see Figure 15B and Figure 79; also shown in Figure 80B). Additional blood supply to these structures comes directly from small branches of the circle of Willis (discussed with Figure 58).

CLINICAL ASPECT

These small-caliber arteries are functionally different from the cortical (cerebral) vessels. Firstly, they are end-arteries, and do not anastomose. Secondly, they react to a chronic increase of blood pressure (hypertension) by a necrosis of the muscular wall of the blood vessels, called fibrinoid necrosis. Following this there are two possibilities:

- These blood vessels may occlude, causing small infarcts in the region of the internal capsule. As these small infarcts resolve, they leave small "holes" called lacunes (lakes), which can be visualized radiographically. Hence, they are known as **lacunar infarcts**, otherwise called a "stroke."

The extent of the clinical deficit with this type of infarct depends upon its location and size in the internal capsule. A relatively small lesion may cause major motor and/or sensory deficits on the contralateral side. This may result in a devastating incapacity of the person, with contralateral paralysis. (**Note to the Leaner:** The learner should review the major ascending and descending tracts at this time and their course through the internal capsule.)

- The other possibility is that these weakened blood vessels can rupture, leading to hemorrhage deep in the hemispheres. (Brain hemorrhage can be visualized by CT, computed tomography; reviewed with Figure 28A).

Although the blood supply to the white matter of the brain is significantly less (because of the lower metabolic demand), this nervous tissue is also dependent upon a continuous supply of oxygen and glucose. A loss of blood supply to the white matter will result in the loss of the axons (and myelin) and, hence, interruption of the transmission of information. This type of stroke may result in a more extensive clinical deficit, due to the fact that the hemorrhage itself causes a loss of brain tissue, as well as a loss of the blood supply to areas distal to the site of the hemorrhage.

ADDITIONAL DETAIL

Choroidal arteries, branches from the circle, supply the choroid plexus of the lateral venrricles.

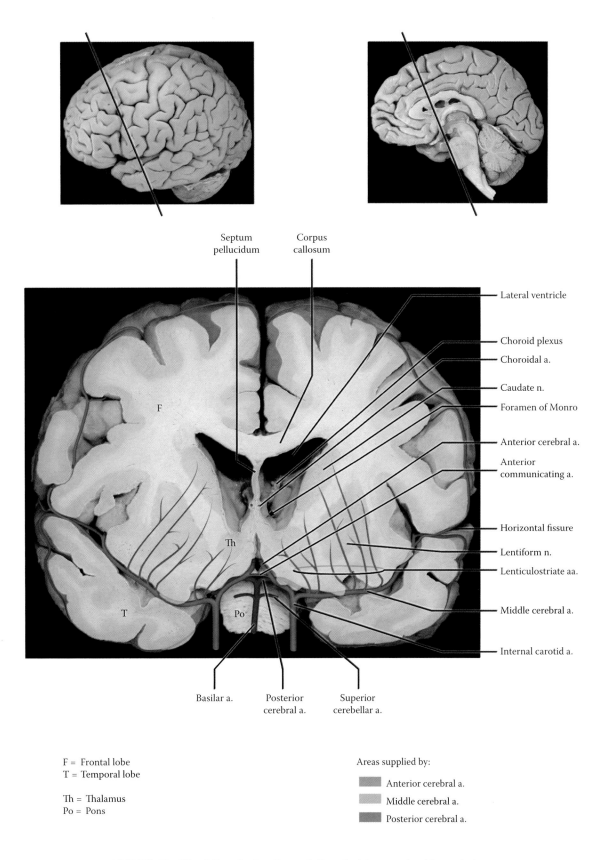

FIGURE 62: Blood Supply 6 — Internal Capsule (photograph with overlay)

FIGURE 63
THALAMUS

NUCLEI AND CONNECTIONS

The Thalamus was introduced previously in Section A (Orientation) with a schematic perspective, as well as an introduction to the nuclei and their functional aspects (see Figure 11 and Figure 12). At this stage, it is important to integrate knowledge of the thalamic nuclei with the inputs, both sensory and motor, and the connections (reciprocal) of these nuclei to the cerebral cortex. The limbic aspects will be discussed in the next section (Section D).

As was noted, there are two ways of dividing up the nuclei of the thalamus, namely, functionally and topographically (review text with Figure 12). The functional aspects of the thalamus will be reviewed with color used to display the connections of the nuclei with the cortical areas (dorsolateral and medial aspects).

SPECIFIC RELAY NUCLEI

- Sensory:
  - **VPL**, ventral posterolateral nucleus: This nucleus receives input from the somatosensory systems of the body, mainly for discriminative touch and position sense, as well as the "fast'" pain system for localization. The fibers relay to the appropriate areas of the post-central gyrus, areas 1, 2, and 3, the sensory homunculus. The hand, particularly the thumb, is well represented (see Figure 33, Figure 34, and Figure 36).
  - **VPM**, ventral posteromedial nucleus: The fibers to this nucleus are from the trigeminal system (TG), i.e., the face, and the information is relayed to the facial area of the post-central gyrus. The tongue and lips are well represented (see Figure 35 and Figure 36).
  - **MGB**, medial geniculate body (nucleus): This is the nucleus for the auditory fibers from the inferior colliculus, which relay to the transverse gyri of Heschl on the superior temporal gyri in the lateral fissure (see Figure 38 and Figure 39).
  - **LGB**, lateral geniculate body (nucleus): This is the relay nucleus for the visual fibers from the ganglion cells of the retina to the calcarine cortex. This nucleus is laminated with different layers representing the visual fields of the ipsilateral and contralateral eyes (see Figure 41A and Figure 41C).

- Motor:
  - **VA and VL**, ventral anterior and ventral lateral: Fibers to these nuclei originate in the globus pallidus and substantia nigra (pars reticulata) as well as the cerebellum, and are relayed to the motor and premotor areas of the cerebral cortex, as well as the supplementary motor cortex (see Figure 53 and Figure 57).

ASSOCIATION NUCLEI

- **DM**, dorsomedial nucleus: This most important nucleus relays information from many of the thalamic nuclei as well as from parts of the limbic system (hypothalamus and amygdala) to the prefrontal cortex (see Figure 77B).
- **AN**, anterior nuclei: These nuclei are part of the limbic system and relay information to the cingulate gyrus; they are part of the Papez circuit (see Figure 77A).
- **LD**, lateral dorsal nucleus: The function of this nucleus is not well established.
- **LP**, lateral posterior nucleus: This nucleus relays to the parietal association areas of the cortex; again it is not a well-known nucleus.
- **Pul**, pulvinar: This nucleus is part of the visual relay, but relays to visual association areas of the cortex, areas 18 and 19 (see Figure 41B).

NONSPECIFIC NUCLEI

- **IL, Mid, Ret**, intralaminar, midline, and reticular nuclei (not shown here, see Figure 12): These nuclei receive from other thalamic nuclei and from the ascending reticular activating system, as well as receiving fibers from the "slow" pain system; they relay to widespread areas.
- **CM**, centromedian nucleus: This nucleus is part of an internal loop receiving from the globus pallidus and relaying to the neostriatum, the caudate and putamen (see Figure 52).

There is definitely a processing of information in these nuclei of the thalamus, not simply a relay. On the sensory side, some aspects of a "crude" touch and particularly pain are located in the thalamus (see Figure 34). The nonspecific thalamic nuclei are part of the ascending reticular activating system (ARAS), which is required for consciousness (see Figure 42A and Figure 42B). The connection between the dorsomedial nucleus (DM) and the prefrontal cortex is known to be extremely important for the processing of limbic (emotional) aspects of behavior (discussed in Section D).

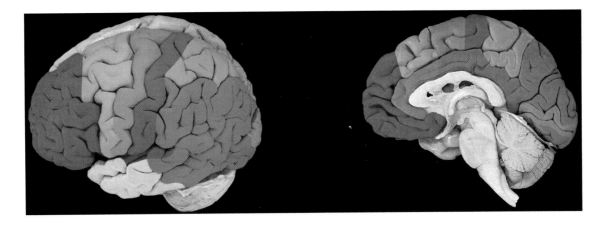

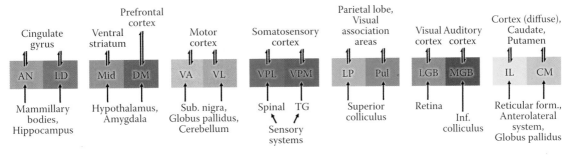

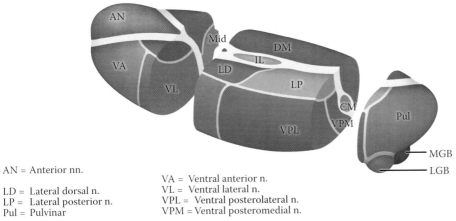

AN = Anterior nn.

LD = Lateral dorsal n.
LP = Lateral posterior n.
Pul = Pulvinar

DM = Dorsomedial n.
Mid = Midline nn.

VA = Ventral anterior n.
VL = Ventral lateral n.
VPL = Ventral posterolateral n.
VPM = Ventral posteromedial n.

LGB = Lateral geniculate body
MGB = Medial geniculate body

IL = Intralaminar nn.
CM = Centromedian n.

FIGURE 63: Thalamus: Nuclei and Connections

FIGURE 64A
BRAINSTEM HISTOLOGY

VENTRAL VIEW — SCHEMATIC

Study of the brainstem will be continued by examining its histological neuroanatomy through a series of cross-sections. Since it is well beyond the scope of the nonspecialist to know all the details, certain salient points have been selected, namely:

- The cranial nerve nuclei
- The ascending and descending tracts
- Certain brainstem nuclei that belong to the reticular formation
- Other select special nuclei

As has been indicated, the attachment of the cranial nerves to the brainstem is one of the keys to being able to understand this part of the brain (see Figure 6 and Figure 7). Wherever one sees a cranial nerve attached to the brainstem, one knows that its nucleus (or one of its nuclei) will be located at that level (see Figure 8A and Figure 8B). Therefore, if one visually recalls or "memorizes" the attachment of the cranial nerves, one has a key to understanding the brainstem. In the clinical setting, knowledge of which cranial nerve is involved is usually the main clue to localize a lesion in the brainstem.

Since the focus is on the cranial nerves, only a limited number of cross-sections will be studied. This diagram shows the ventral view of the brainstem, with the attached cranial nerves; the motor nuclei are shown on the right side (see Figure 8A), and the sensory cranial nerve nuclei are shown on the left side (see Figure 8B). The lines indicate the sections that will be depicted in the series to follow.

There are *eight* cross-sections that will be studied through the three parts of the brainstem; each is preceded by a photographic view of that part of the brainstem.

- Two through the midbrain
  - CN III, upper midbrain (superior colliculus level)
  - CN IV, lower midbrain (inferior colliculus level)
- Three through the pons
  - Upper pons (level for a special nucleus at this level)
  - CN V mid-pons (through the principal sensory and motor nuclei)
  - CN VI, VII, and part of VIII, the lower pons
- Three through the medulla
  - CN VIII (some parts), the upper medulla

- CN IX, X, and XII, the mid-medullary level
- Lower medulla, with some special nuclei

Two important points should be noted for the student-user of this atlas:

1. A small image of this view of the brainstem, both the ventral view and the sagittal view (see, for example, Figure 65A) will be shown with each cross-sectional level with the plane of the cross-section indicated.
2. These cross-sectional levels are the ones shown alongside the pathways in Section B (Functional Systems) of this atlas (see Figure 31).

HISTOLOGICAL STAINING

A variety of histological stains are available that can feature different normal and abnormal components of tissue. For the nervous system, there are many older stains and an ever-increasing number of newer stains using specific antibody markers, often tagged with fluorescent dyes. In general, the stains include those for:

- Cellular components, the cell bodies of neurons and glia (and cells lining blood vessels); these are general stains such as Hematoxylin & Eosin (H & E).
- The neurons, particularly the dendritic tree (including dendritic spines) and often the axons; the best known of these is the Golgi stain.
- Axonal fibers, either normal or degenerating.
- Glial elements (normal or reactive astrocytes).
- Myelin (normal or degenerating myelin).

The stain used for the histological sections in this atlas combines a cellular stain with a myelin stain; the combined stain is officially known as the Kluver-Barrera stain. Since the myelinated fibers are often compacted in certain areas, these tend to stand out clearly. The cellular neuronal areas are usually lightly stained as the cells are more dispersed, but the cell bodies can be visualized at higher magnification.

BLOOD SUPPLY

The vertebro-basilar system supplies the brainstem in the following pattern (see Figure 58 and Figure 61). Penetrating branches from the **basilar** artery supply nuclei and tracts that are adjacent to the midline; these are called the paramedian branches. The lateral territory of the brainstem, both tracts and nuclei, is supplied by one of the cerebellar **circumferential** arteries, posterior inferior, anterior inferior, and superior (see Figure 58).

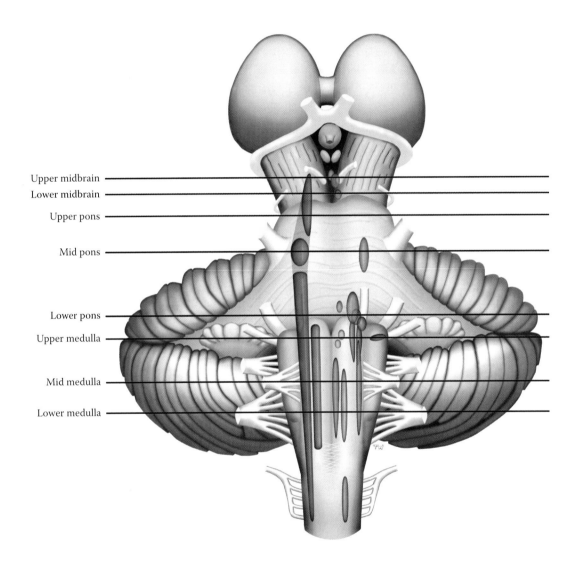

Upper midbrain
Lower midbrain
Upper pons
Mid pons
Lower pons
Upper medulla
Mid medulla
Lower medulla

FIGURE 64A: Brainstem Histology: Ventral View

FIGURE 64B
BRAINSTEM HISTOLOGY

SAGITTAL VIEW — SCHEMATIC

This is a schematic drawing of the brainstem seen in a midsagittal view (see Figure 17 and Figure 18). This view is being presented because it is one that is commonly used to portray the brainstem. The learner should try to correlate this view with the ventral view shown in the previous diagram. This schematic also will be shown in each of the cross-section diagrams, with the exact level indicated, in order to orient the learner to the plane of section through the brainstem.

The location of some nuclei of the brainstem can be visualized using this sagittal view, including the red nucleus in the upper midbrain, the pontine nuclei that form the "bulging" of the pons, and the inferior olivary nucleus of the medulla (not illustrated). Some of the cranial nerve attachments are shown as well but are not labeled.

Using this orientation, one can approach the description of the eight cross-sections in a systematic manner. This is sometimes referred to as the floor plan of the brainstem:

- **Ventral or basal**: The most anterior portion of each area of the brainstem contains some representation of the descending cortical fibers, specifically the cortico-bulbar, cortico-pontine, and cortico-spinal pathways (see Figure 45 and Figure 46). In the midbrain, the cerebral peduncles include all these axon systems. The cortico-bulbar fibers are given off to the various brainstem and cranial nerve nuclei. In the pons, the cortico-pontine fibers terminate in the pontine nuclei, which form the bulge known as the pons proper; the cortico-spinal fibers are dispersed among the pontine nuclei. In the medulla, the cortico-spinal fibers regroup to form the pyramids. The medulla ends at the point where these fibers decussate (see Figure 7).
- **Central**: The central portion of the brainstem is called the **tegmentum**. The reticular formation occupies the core region of the tegmentum (see Figure 42A and Figure 42B). This area contains virtually all the cranial nerve nuclei, and other nuclei including the red nucleus and the inferior olive, as well as the remaining tracts.
- **CSF**: The ventricular system is found throughout the brainstem (see Figure 20A, Figure 20B, and Figure 21). The brainstem level can often be identified according to the ventricular system

that passes through this region, namely, the aqueduct in the midbrain region and the fourth ventricle lower down.
- **Dorsal or roof**: The four colliculi, which collectively form the tectum, are located behind (dorsal to) the aqueduct of the midbrain. The fourth ventricle separates the pons and medulla from the cerebellum. The upper part of the roof of the fourth ventricle is called the superior medullary velum (see Figure 10 and Figure 41B).

CLINICAL ASPECT

The information that is being presented in this series should be sufficient to allow a student to recognize the clinical signs that would accompany a lesion at a particular level, particularly as it involves the cranial nerves. Such lesions would also interrupt the ascending or descending tracts, and this information would assist in localizing the lesion. Specific lesions will be discussed with the cross-sectional levels.

PLAN OF STUDY:

- A schematic of each section is presented in the upper figure, and the corresponding histological section of the human brainstem is presented below.
- The various nuclei of the brainstem have been colored differently, consistent with the color used in the tracts (see Section B of this atlas). This visual cataloging is maintained uniformly throughout the brainstem cross-sections (see page xviii).

The brainstem is being described starting from the midbrain downward through to the medulla for two reasons:

1. This order follows the numbering of the cranial nerves, from midbrain downward
2. This is the sequence that has been described for the fibers descending from the cortex

Others may prefer to start the description of the cross-sections from the medulla upward.

Note to the Learner: The presentation of the histology is the same on the accompanying CD-ROM, with the added feature that the structure to be identified is highlighted in both the schematic and histological section, at the same time. It is suggested that the learner review these cross-sections using the text together with the CD-ROM. The histological images of the brainstem will be more understandable after this combined approach.

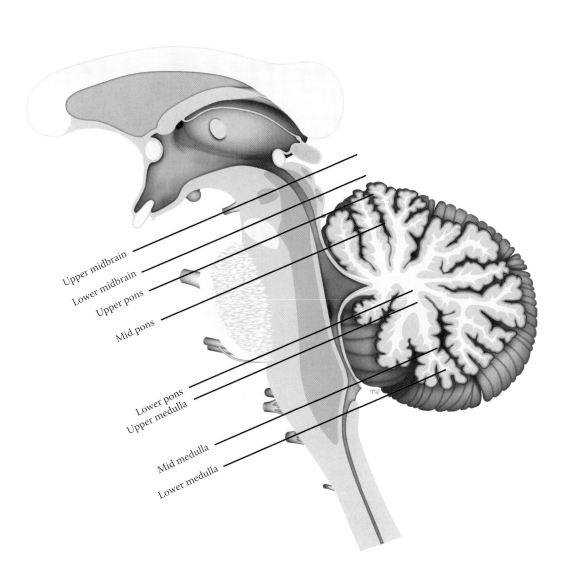

Upper midbrain

Lower midbrain

Upper pons

Mid pons

Lower pons
Upper medulla

Mid medulla

Lower medulla

FIGURE 64B: Brainstem Histology: Sagittal View

THE MIDBRAIN
FIGURE 65, FIGURE 65A, AND FIGURE 65B

The midbrain is the smallest of the three parts of the brainstem. The temporal lobes of the hemispheres usually obscure its presence on an inferior view of the brain (see Figure 15A).

The midbrain area is easily recognizable from the anterior view in a dissected specimen of the isolated brainstem (see Figure 7). The massive cerebral peduncles are located most anteriorly. These peduncles contain axons that are a direct continuation of the fiber systems of the internal capsule (see Figure 26). Within them are found the pathways descending from the cerebral cortex to the brainstem (cortico-bulbar, see Figure 46 and Figure 48), to the cerebellum via the pons (cortico-pontine, see Figure 48 and Figure 55), and to the spinal cord (cortico-spinal tracts, see Figure 45 and Figure 48).

The tegmentum contains two special nuclei in the midbrain region — the substantia nigra and the red nucleus, both involved in motor control.

- The **substantia nigra** is found throughout the midbrain and is located behind the cerebral peduncles. It derives its name from the dark melanin-like pigment found (not in all species) within its neurons in a freshly dissected specimen, as seen in the present illustration (see also Figure 15B). The pigment is not retained when the tissue is processed for sectioning. Therefore, this nuclear area is clear (appearing white) in most photographs in atlases, despite its name. With myelin-type stains, the area will appear "empty"; with cell stains, the neuronal cell bodies will be visible. Its function is related to the basal ganglia (see Figure 52 and Figure 53).
- The **red nucleus** derives its name from the fact that this nucleus has a reddish color in a freshly dissected specimen, presumably due to its marked vascularity. The red nucleus is found at the superior collicular level. Its function is discussed with the motor systems (see Figure 47).

The reticular formation is found in the core area of the tegmentum, and is particularly important for the maintenance of consciousness (see Figure 42A and Figure 42B). The **periaqueductal gray**, surrounding the aqueduct, has been included as part of the reticular formation (see Figure 42B); this area participates as part of the descending control system for pain modulation (see Figure 43).

The aqueduct of the midbrain helps to identify this cross-section as the midbrain area (see Figure 21). Posterior to the aqueduct are the two pairs of colliculi, which can also be seen on the dorsal view of the isolated brainstem (see Figure 9A and Figure 10). The four nuclei together form the tectal plate, or tectum, also called the quadrigeminal plate.

The **pretectal region**, located in front of and somewhat above the superior colliculus, is the nuclear area for the pupillary light reflex (see Figure 41C).

FIGURE 65: UPPER MIDBRAIN (PHOTOGRAPHIC VIEW)

This is a photographic image, enlarged, of the sectioned midbrain. As shown in the upper left image, the brainstem was sectioned at the level of the cerebral peduncles; the corresponding level is shown on a medial view of the brain, indicating that the section is through the superior colliculus. Many of the structures visible on this "gross" specimen will be seen in more detail on the histological sections.

The distinctive features identifying this section as midbrain are:

- Anteriorly, the outline of the cerebral peduncles with the fossa in between.
- Immediately behind is a dark band, the substantia nigra, pars compacta, with pigment present in the cell bodies.
- A faint outline of the red nucleus can be seen in the tegmentum, which identifies this section as the superior collicular level.
- In the middle toward the back of the specimen is a narrow channel, which is the aqueduct of the midbrain, surrounded by the periaqueductal gray.
- The gray matter behind the ventricle is the superior colliculus at this level.

There are two levels presented for a study of the midbrain:

- Figure 65A: Upper midbrain, which includes CN III nucleus and the superior colliculus.
- Figure 65B: Lower midbrain, at the level of the CN IV nucleus and the inferior colliculus, and the decussation of the superior cerebellar peduncles.

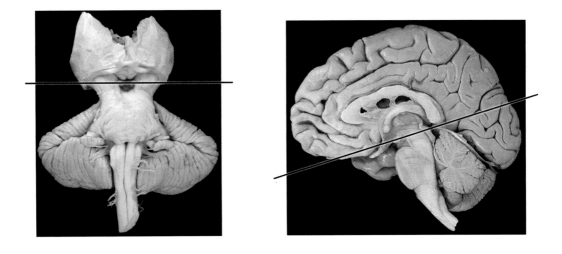

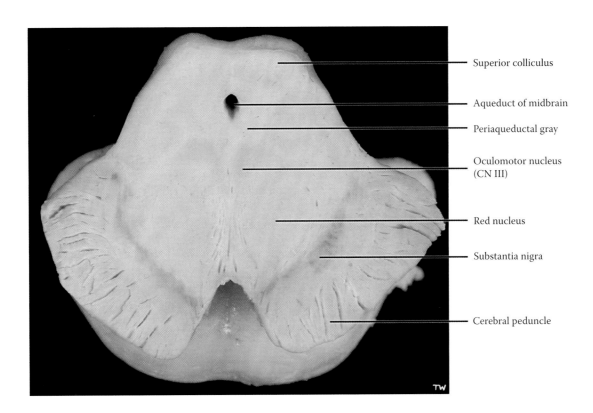

Superior colliculus

Aqueduct of midbrain

Periaqueductal gray

Oculomotor nucleus
(CN III)

Red nucleus

Substantia nigra

Cerebral peduncle

FIGURE 65: Brainstem Histology — Midbrain (upper — photograph)

FIGURE 65A
UPPER MIDBRAIN:
CROSS-SECTION

The identifying features of this cross-section of the midbrain include the cerebral peduncle ventrally, with the substantia nigra posterior to it. The aqueduct is surrounded by the periaqueductal gray. The remainder of the midbrain is the tegmentum, with nuclei and tracts. Dorsally, behind the aqueduct, is a colliculus.

The descending fiber systems are segregated within the cerebral peduncles (see Figure 45, Figure 46, and Figure 48). The substantia nigra consists, in fact, of two functionally distinct parts — the pars compacta and the pars reticulata. The **pars reticulata** lies adjacent to the cerebral peduncle and contains some widely dispersed neurons; these neurons connect the basal ganglia to the thalamus as one of the output nuclei of the basal ganglia (similar to the globus pallidus internal segment, see Figure 53). The **pars compacta** is a cell-rich region, located more dorsally, whose neurons contain the melanin-like pigment. These are the dopaminergic neurons that project to the neostriatum (discussed with Figure 52). Loss of these neurons results in the clinical entity Parkinson's disease (discussed with Figure 52).

The red nucleus is located within the tegmentum; large neurons are typical of the ventral part of the nucleus. With a section that has been stained for myelin, the nucleus is seen as a clear zone. The red nucleus gives origin to a descending pathway, the rubro-spinal tract, which is involved in motor control (see Figure 47 and Figure 48).

The oculomotor nucleus (CN III) is quite large and occupies the region in front of the periaqueductal gray, near the midline; this identifies the level as upper midbrain with the superior colliculus. These motor neurons are large in size and easily recognizable. The parasympathetic portion of this nucleus is incorporated within it and is known as the Edinger-Westphal (EW) nucleus (see Figure 8A). The fibers of CN III pass anteriorly through the medial portion of the red nucleus and exit between the cerebral peduncles, in the interpeduncular fossa (see Figure 6 and Figure 7).

The ascending (sensory) tracts present in the midbrain are a continuation of those present throughout the brainstem. The medial lemniscus, the ascending trigeminal pathway, and the fibers of the anterolateral system incorporated with them (see Figure 36 and Figure 40) are located in the outer part of the tegmentum, on their way to the nuclei of the thalamus (see Figure 63).

The nuclei of the reticular formation are found in the central region of the brainstem (the tegmentum); they are functionally part of the ascending reticular activating system and play a significant role in consciousness (discussed with Figure 42A and Figure 42B). The periaqueductal gray surrounding the cerebral aqueduct is involved with the descending pathway for the modulation of pain (see Figure 43).

The superior colliculus is a subcortical center for certain visual movements (see Figure 41B). These nuclei give rise to a fiber tract, the tecto-spinal tract, a descending pathway that is involved in the control of eye and neck movements; it descends to the cervical spinal cord as part of the medial longitudinal fasciculus (MLF) (see Figure 51B).

The MLF stains heavily with a myelin-type stain and is found anterior to the cranial nerve motor nucleus, next to the midline, at this level as well as other levels of the brainstem. Also to be noted at this level is the brachium of the inferior colliculus, a part of the auditory pathway (see Figure 10, Figure 37, and Figure 38).

CLINICAL ASPECT

A specific lesion involving a thrombosis of the basilar artery may destroy much of the brainstem yet leave the inner part of the midbrain intact. Few people actually survive this cerebrovascular damage, but those that do are left in a suspended (rather tragic) state of living, known by the name **"locked-in" syndrome**. The patient retains consciousness, with intellectual functions generally intact, meaning that they can think and feel as before. However, usually, all voluntary movements are gone, except perhaps for some eye movements, or occasionally some small movements in the hands and fingers. This means that they require a respirator to breathe and 24-hour total care. There may also be a loss of all sensations, or some sensation from the body may be retained.

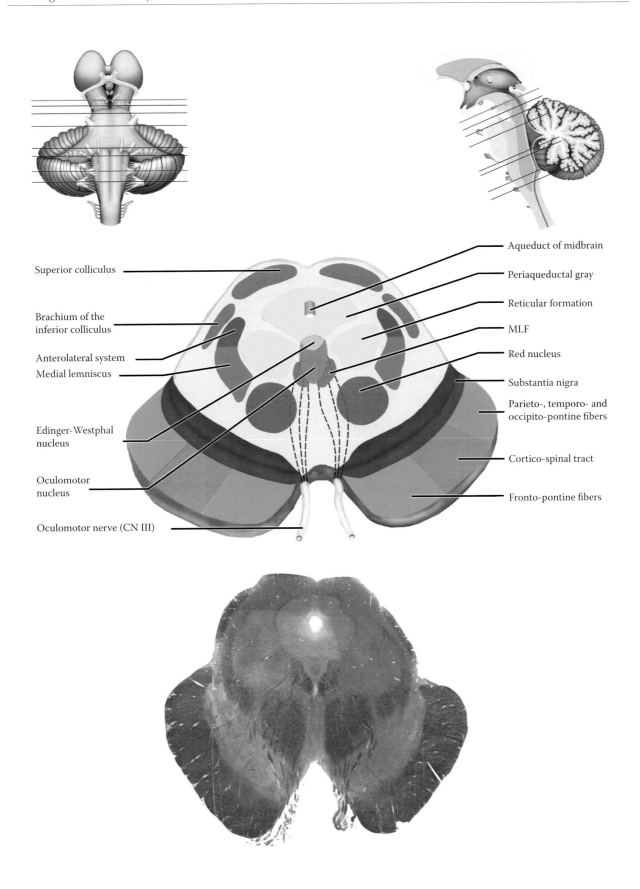

Superior colliculus

Brachium of the
inferior colliculus

Anterolateral system

Medial lemniscus

Edinger-Westphal
nucleus

Oculomotor
nucleus

Oculomotor nerve (CN III)

Aqueduct of midbrain

Periaqueductal gray

Reticular formation

MLF

Red nucleus

Substantia nigra

Parieto-, temporo- and
occipito-pontine fibers

Cortico-spinal tract

Fronto-pontine fibers

FIGURE 65A: Brainstem Histology — Upper Midbrain

FIGURE 65B
LOWER MIDBRAIN:
CROSS-SECTION

This cross-section includes the cerebral peduncles, still located anteriorly and the substantia nigra located immediately behind these fibers. The unique feature in the lower midbrain is the decussation (crossing) of the superior cerebellar peduncles, which occupies the central area of the section; this identifies the section as the inferior collicular level. Posteriorly the aqueduct is surrounded by the periaqueductal gray, and behind the aqueduct is the inferior colliculus. Often, the cross-section at this level includes some of the pontine nuclei. (as is seen in the histological section below). Therefore, one may see a somewhat confusing mixture of structures.

The arrangement of the fibers in the cerebral peduncle is the same as found in the upper midbrain. The tegmentum contains the ascending tracts, the medial lemniscus, the trigeminal pathway, and the anterolateral fibers (system), which are situated together at the outer edge of the lower midbrain (see Figure 40).

In sections through the lower levels of the midbrain, there is a brief appearance of a massive fiber system (as seen with a myelin-type stain) occupying the central region of the lower midbrain. These fibers are the continuation of the superior cerebellar peduncles, which are crossing (decussating) at this level (see Figure 10 and Figure 40). The fibers are coming from the deep cerebellar nuclei (the intracerebellar nuclei), mainly the dentate nucleus, and are headed for the ventral lateral nucleus of the thalamus, and then on to the motor cortex (discussed with Figure 57). Some of the fibers that come from the intermediate deep cerebellar nucleus will synapse in the red nucleus.

The nuclei of the reticular formation found in the central region (the tegmentum) at this level are function-ally part of the ARAS and play a significant role in consciousness (see Figure 42A and Figure 42B). Between the cerebral peduncles is a small nucleus, the interpeduncular nucleus, which belongs with the limbic system. The periaqueductal gray surrounding the aqueduct of the midbrain is involved with pain and also with the descending pathway for the modulation of pain (see Figure 43).

The nucleus of CN IV, the trochlear nucleus, is located in front of the periaqueductal gray, next to the midline. Because it supplies only one extra-ocular muscle, it is a smaller nucleus than the oculomotor nucleus. CN IV heads dorsally and will exit from the brainstem below the inferior colliculus (see Figure 48), on the posterior aspect of the brainstem. The MLF lies just anterior to the trochlear nucleus. Some unusually large round cells are often seen at the edges of the periaqueductal gray; these cells are part of the mesencephalic nucleus of the trigeminal nerve, CN V (see Figure 8B).

The lateral lemniscus, the ascending auditory pathway, is still present at this level, and its fibers are terminating in the inferior colliculus, a relay nucleus in the auditory pathway (see Figure 37 and Figure 38). After synapsing here, the fibers are relayed to the medial geniculate nucleus via the brachium of the inferior colliculus, seen at the upper midbrain level (previous illustration).

CLINICAL ASPECT

The presence of the pain and temperature fibers that are found at this level at the outer edge of the midbrain has prompted the possibility, in very select cases, to surgically sever the sensory ascending pathways at this level. This highly dangerous neurosurgical procedure would be done particularly for cancer patients who are suffering from intractable pain. Nowadays it would only be considered as a measure of last resort. Pain control is currently managed through the use of drugs, either as part of palliative care or in the setting of a pain "clinic," accompanied by other measures.

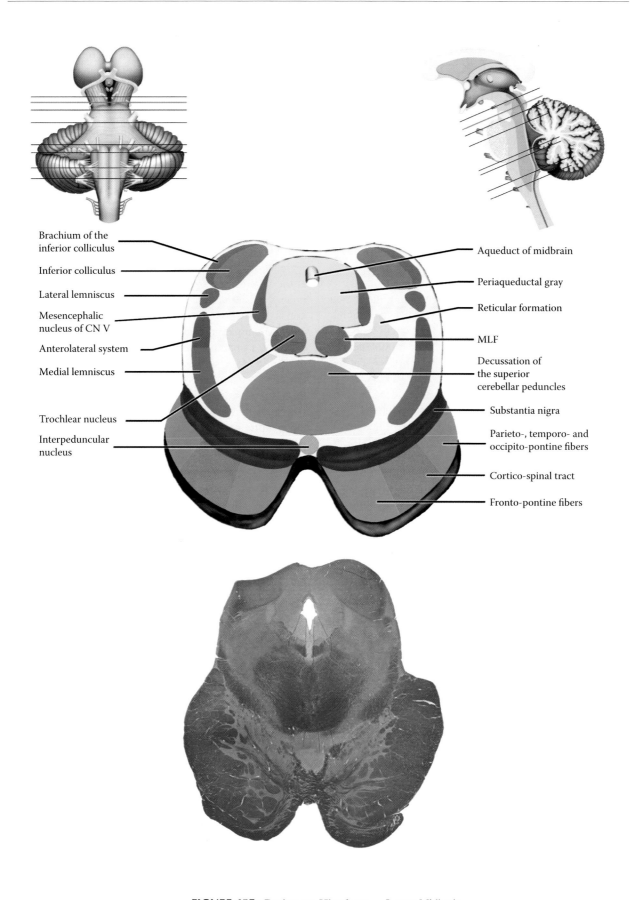

Brachium of the
inferior colliculus

Inferior colliculus

Lateral lemniscus

Mesencephalic
nucleus of CN V

Anterolateral system

Medial lemniscus

Trochlear nucleus

Interpeduncular
nucleus

Aqueduct of midbrain

Periaqueductal gray

Reticular formation

MLF

Decussation of
the superior
cerebellar peduncles

Substantia nigra

Parieto-, temporo- and
occipito-pontine fibers

Cortico-spinal tract

Fronto-pontine fibers

FIGURE 65B: Brainstem Histology — Lower Midbrain

THE PONS
FIGURE 66, FIGURE 66A,
FIGURE 66B, AND FIGURE 66C

The pons is characterized by its protruding anterior (ventral) portion, the pons proper, also called the basilar portion of the pons, with the basilar artery lying on its surface (see Figure 15A and Figure 58). This area contains the pontine nuclei, the site of relay of the cortico-pontine fibers (see Figure 48); the ponto-cerebellar fibers then cross and enter the cerebellum via the middle cerebellar peduncle (see Figure 55). Intermingled with the pontine nuclei are the dispersed fibers, which belong to the cortico-spinal system (see Figure 45 and Figure 48).

Behind the pons proper is the tegmentum, the region of the brainstem that contains the cranial nerve nuclei, most of the ascending and descending tracts, and the nuclei of the reticular formation. The cranial nerves attached to the pons include the trigeminal (CN V) at the mid-pontine level, and the abducens (CN VI), the facial (CN VII), and part of CN VIII (the vestibulocochlear) at the lowermost pons; the fibers of VII form an internal loop over the abducens nucleus in the pons (see Figure 48). The fibers of CN VII and CN VIII are located adjacent to each other at the cerebello-pontine angle (see Figure 6, Figure 7, and Figure 8A).

The ascending tracts present in the tegmentum are those conveying sensory information from the body and face. These include the medial lemniscus and the antero-lateral fibers (system). The medial lemniscus shifts its position in its course through the brainstem (see Figure 40), moving from a central to a lateral position. The ascending trigeminal pathways join with the medial lemniscus in the upper pons. The lateral lemniscus (auditory) is also located in the tegmentum.

One of the distinctive nuclei of the pons is the locus ceruleus, a pigment-containing nucleus located in the upper pontine region (to be discussed with Figure 66A). The nuclei of the reticular formation of the pons have their typical location in the tegmentum (see Figure 42A and Figure 42B). Their role in the motor systems has been described with the reticular formation, as well as giving rise to descending tracts (see Figure 49A and Figure 49B).

The fourth ventricle begins in the pontine region as a widening of the aqueduct and then continues to enlarge so it is widest at about the level of the junction between the pons and medulla (see Figure OA, Figure 20A, Figure 20B, and Figure 21). This ventricle separates the pons and medulla anteriorly from the cerebellum posteriorly. There is no pontine nucleus dorsal to the fourth ventricle; the cerebellum is located above (posterior to) the roof of the ventricle.

FIGURE 66: UPPER PONS (PHOTOGRAPHIC VIEW)

This is a photographic image, enlarged, of the pontine region, with the cerebellum attached. The section is done at the level of the upper pons, as indicated in the upper images of the ventral view of the brainstem and in the midsagittal view.

The unique nucleus present at this level is the locus ceruleus, a small nucleus whose cells have pigment, much like those of the substantia nigra, pars compacta (see Figure 65). As with that nucleus, the pigment is lost during histological processing.

The ventral region has the distinctive appearance of the pontine nuclei, with the cortico-spinal and cortico-pontine fibers dispersed among them. The pontine tegmentum seems quite compressed. The space in the middle of the tissue section is the fourth ventricle, as it begins to widen. Behind the ventricle is a small area of white matter, called the superior medullary velum (see Figure 10 and Figure 41B). The thin folia of the cerebellum are easily recognized, with an inner strip of white matter bounded on either side by the thin gray matter of the cerebellar cortex.

The pons is to be represented by three sections:

- Figure 66A: The upper pons, at the level of the locus ceruleus.
- Figure 66B: The mid (middle) pons, at the level of the attachment of the trigeminal nerve. It includes the massive middle cerebellar peduncles.
- Figure 66C: The lower pons, just above the junction with the medulla. This lowermost level has the nuclei of cranial nerves VI, VII, and parts of both divisions of CN VIII.

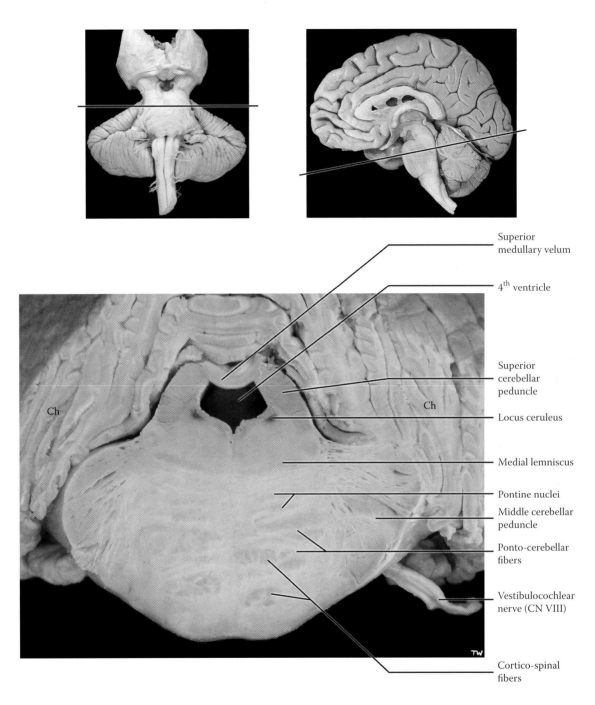

Ch = Cerebellar hemisphere

FIGURE 66: Brainstem Histology — Pons (upper — photograph)

FIGURE 66A
UPPER PONS:
CROSS-SECTION

This level is presented mainly to allow an understanding of the transition of midbrain to pons. This particular section is taken at the uppermost pontine level, where the trochlear nerve, CN IV, exits (below the inferior colliculus, see Figure 7). This is the only cranial nerve that exits posteriorly; its fibers cross (decussate) before exiting (see Figure 48).

Anteriorly, the pontine nuclei are beginning to be found. Cortico-pontine fibers will be terminating in the pontine nuclei. From these cells, a new tract is formed that crosses and projects to the cerebellum forming the middle cerebellar peduncle. The cortico-spinal fibers become dispersed between these nuclei and course in bundles between them (see Figure 45 and Figure 48).

The ascending tracts include the medial lemniscus and anterolateral system (somatosensory from the body, see Figure 33, Figure 34, and Figure 40), the ascending trigeminal pathway (see Figure 35 and Figure 40) and the lateral lemniscus (auditory, see Figure 37). The fibers of the trigeminal system that have crossed in the pons (discriminative touch from the principal nucleus of V), and those of pain and temperature (from the descending nucleus of V) that crossed in the medulla join together in the upper pons with the medial lemniscus (see Figure 35, Figure 36, and Figure 40). The medial lemniscus is located midway between its more central position inferiorly, and the lateral position found in the midbrain (see Figure 40). In sections stained for myelin, it has a somewhat "comma-shaped" configuration. The auditory fibers are located dorsally, just before terminating in the inferior colliculus in the lower midbrain (see Figure 38 and Figure 40). Centrally, the cerebral aqueduct is beginning to enlarge, becoming the fourth ventricle. The MLF is found in its typical location ventral to the fourth ventricle, next to the midline.

The nuclei of the reticular formation are located in the tegmentum (see Figure 42A and Figure 42B). The special nucleus at this level, the locus ceruleus, is located in the dorsal part of the tegmentum not too far from the edges of the fourth ventricle. The nucleus derives its name from its bluish color in fresh specimens, as seen in the photographic image in the previous illustration. As explained, the pigment is lost when the tissue is processed for histology. The locus ceruleus is usually considered part of the reticular formation (as discussed with Figure 42B) because of its widespread connections with virtually all parts of the brain. It is also unique because noradrenaline is its catecholamine neurotransmitter substance.

The superior cerebellar peduncle is found within the tegmentum of the pons. These fibers carry information from the cerebellum to the thalamus and the red nucleus. The fibers, which are the axons from the deep cerebellar nuclei, leave the cerebellum and course in the roof of the fourth ventricle (the superior medullary velum, see Figure 10 and Figure 40). They then enter the pontine region and move toward the midline, finally decussating in the lower midbrain (see Figure 57 and Figure 65B).

The uppermost part of the cerebellum is found at this level. One of the parts of the vermis, the midline portion of the cerebellum, the lingula, is identified. This particular lobule is a useful landmark in the study of the cerebellum and was identified when the anatomy of the cerebellum was explained (see Figure 54).

ADDITIONAL DETAIL

Several very large neurons belonging to the mesencephalic nucleus of the trigeminal may be found near the edges of the fourth ventricle (see Figure 8B). This small cluster of cells may not be found in each and every cross-section of this particular region.

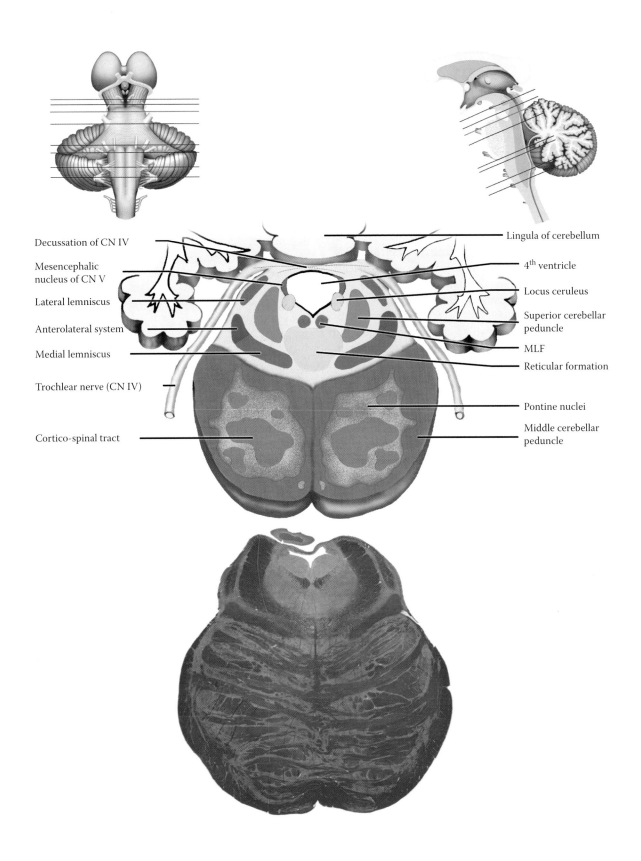

Decussation of CN IV

Mesencephalic
nucleus of CN V

Lateral lemniscus

Anterolateral system

Medial lemniscus

Trochlear nerve (CN IV)

Cortico-spinal tract

Lingula of cerebellum

4th ventricle

Locus ceruleus

Superior cerebellar
peduncle

MLF

Reticular formation

Pontine nuclei

Middle cerebellar
peduncle

FIGURE 66A: Brainstem Histology — Upper Pons

FIGURE 66B
MID-PONS: CROSS-SECTION

This section is taken through the level of the attachment of the trigeminal nerve. Anteriorly, the pontine nuclei and the bundles of cortico-spinal fibers are easily recognized. The pontine cells (nuclei) and their axons, which cross and then become the middle cerebellar peduncle, are particularly numerous at this level (see Figure 55). The cortico-spinal fibers are seen as distinct bundles that are widely dispersed among the pontine nuclei at this level (see Figure 45 and Figure 48).

The trigeminal nerve enters and exits the brainstem along the course of the middle cerebellar peduncle. CN V has several nuclei with different functions (see Figure 8B and Figure 35). This level contains only two of its four nuclei: the principal (or main) sensory nucleus and the motor nucleus. The principal (main) sensory nucleus subserves discriminative (i.e., two-point) touch sensation and accounts for the majority of fibers; the face area is extensively innervated, particularly the lips, and also the surface of the tongue. The motor nucleus supplies the muscles of mastication and usually is found as a separable nerve as it exits alongside the large sensory root. Within the pons, these nuclei are separated by the fibers of CN V; the sensory nucleus (with smaller cells) is found more laterally, and the motor nucleus (with larger cells) more medially.

The ascending fiber systems are easily located at this cross-sectional level. The medial lemniscus has moved away from the midline, as it ascends (see Figure 40). The anterolateral fiber system has become associated with it by this level. In addition, the ascending trigeminal pathway joins with the medial lemniscus. The lateral lemniscus is seen as a distinct tract, lying just lateral to the medial lemniscus. The MLF is found in its typical location anterior to the fourth ventricle.

The core area of the tegmentum is occupied by the nuclei of the reticular formation. Some of the nuclei here are called the oral portion of the pontine reticular formation (see Figure 42B). This "nucleus" contributes fibers to a descending medial reticulo-spinal tract, which is involved in the indirect voluntary pathway for motor control and plays a major role in the regulation of muscle tone (discussed with Figure 49B).

The fourth ventricle has become quite wide at this level. The superior cerebellar peduncles are found at its edges, exiting from the cerebellum and heading toward the midbrain (red nucleus) and thalamus. The thin sheet of white matter that connects these peduncles is called the superior medullary velum (see Figure 10). The cerebellum, which is quite large at this level, is situated behind the ventricle. The lingula of the cerebellum is again labeled and is sometimes seen actually intruding into the ventricular space.

ADDITIONAL DETAIL

The superior cerebellar peduncles and the superior medullary velum can be located in a specimen (such as the one shown in Figure 9A), a dorsal view of the isolated brainstem. These structures would be found below the inferior colliculi, just below the exiting fibers of CN IV dorsally.

Note on the cerebellum: The cerebellum is usually not included in the histological sections of the pons because of the technical difficulty of sectioning such a large fragment of tissue, transferring the section through the various staining solutions, and mounting the section on large slides.

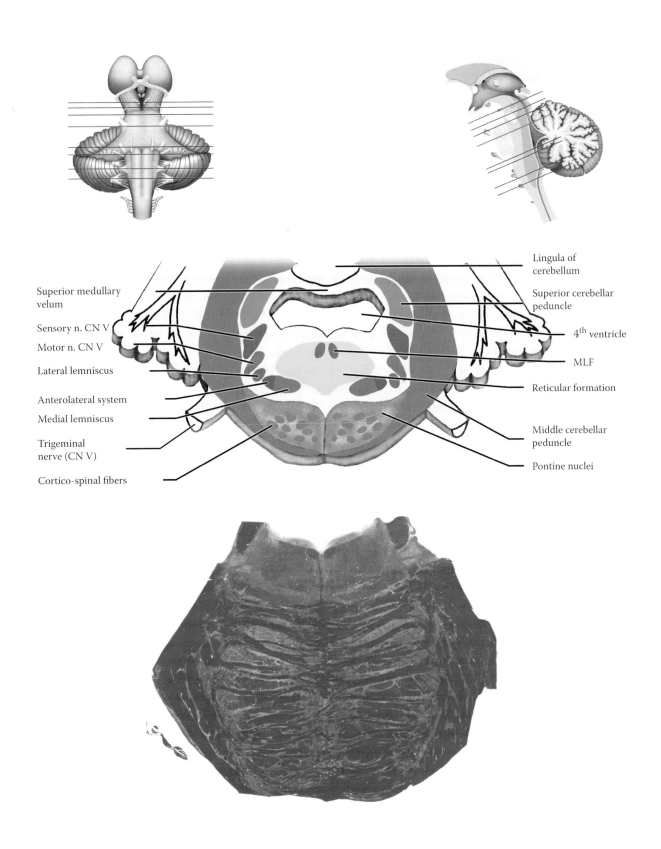

Superior medullary velum

Sensory n. CN V

Motor n. CN V

Lateral lemniscus

Anterolateral system

Medial lemniscus

Trigeminal nerve (CN V)

Cortico-spinal fibers

Lingula of cerebellum

Superior cerebellar peduncle

4th ventricle

MLF

Reticular formation

Middle cerebellar peduncle

Pontine nuclei

FIGURE 66B: Brainstem Histology — Mid-Pons

FIGURE 66C
LOWER PONS:
CROSS-SECTION

This section is very complex because of the number of nuclei related to the cranial nerves located in the tegmental portion, including CN V, VI, VII, and VIII. Some of the tracts are shifting in position or forming. Anteriorly, the pontine nuclei have all but disappeared, and the fibers of the cortico-spinal tract are regrouping into a more compact bundle, which will become the pyramids in the medulla (below).

CN V: The fibers of the trigeminal nerve carrying pain and temperature, that entered at the mid-pontine level, form the descending trigeminal tract, also called the spinal tract of V; medial to it is the corresponding nucleus (see Figure 8B). The descending fibers synapse in this nucleus as this pathway continues through the medulla, cross, and then ascend (see Figure 35), eventually joining the medial lemniscus in the upper pons (see Figure 36).

CN VI: The abducens nucleus, motor to the lateral rectus muscle of the eye (see Figure 8A), is located in front of the ventricular system. The MLF is found just anterior to these nuclei, near the midline. Some of the exiting fibers of CN VI may be seen as the nerve emerges anteriorly, at the junction of the pons and medulla.

CN VII: The motor neurons of the facial nerve nucleus, supplying the muscles of facial expression, are located in the ventrolateral portion of the tegmentum. As explained, the fibers of CN VII form an internal loop over the abducens nucleus (see Figure 48). The diagram is drawn as if the whole course of this nerve is present in a single section, but only part of this nerve is found on an actual section through this level of the pons.

CN VIII — Cochlear division: CN VIII enters the brainstem slightly lower, at the ponto-cerebellar angle (see Figure 6 and Figure 7). The auditory fibers synapse in the dorsal and ventral cochlear nuclei, which will be seen in the medulla in a section just below this level (see also Figure 8B). The two distinctive parts of this nerve at this histological level are the crossing fibers, which form the trapezoid body, and the superior olivary complex (see Figure 37 and Figure 40). After one or more synapses, the fibers then ascend and form the lateral lemniscus, which actually commences at this level.

CN VIII — Vestibular division: Of the four vestibular nuclei (see Figure 51A and Figure 51B), three are found at this level. The lateral vestibular nucleus, with its giant-size cells, is located at the lateral edge of the fourth ventricle; this nucleus gives rise to the lateral vestibulo-spinal tract (see Figure 50). The medial vestibular nucleus is also present at this level, an extension from the medullary region. There is also a small superior vestibular nucleus in this region. The latter two nuclei contribute fibers to the MLF, relating the vestibular sensory information to eye movements (discussed with Figure 51B).

The tegmentum of the pons also includes the ascending sensory tracts and the reticular formation. The medial lemniscus, often somewhat obscured by the fibers of the trapezoid body, is situated close to the midline but has changed its orientation from that seen in the medullary region (see Figure 40; see also cross-sections of the medulla, Figure 67B and Figure 67C). The anterolateral system is too small to be identified. The nuclei of the reticular formation include the caudal portion of the pontine reticular formation, which also contributes to the pontine reticulo-spinal tract (see Figure 49B).

The fourth ventricle is very large but often seems smaller because the lobule of the cerebellar vermis, called the nodulus (part of the flocculonodular lobe, refer to Figure 54), impinges upon its space. The MLF is found anterior to it, near the midline.

The lowermost part of the middle cerebellar peduncle can still be identified at this level. Also present is the inferior cerebellar peduncle, which entered the cerebellum at a lower level (see Figure 7); it is found more internally within the cerebellum. The intracerebellar (deep cerebellar) nuclei are also found at this cross-sectional level and are located within the white matter of the cerebellum (discussed with Figure 56A and Figure 56B).

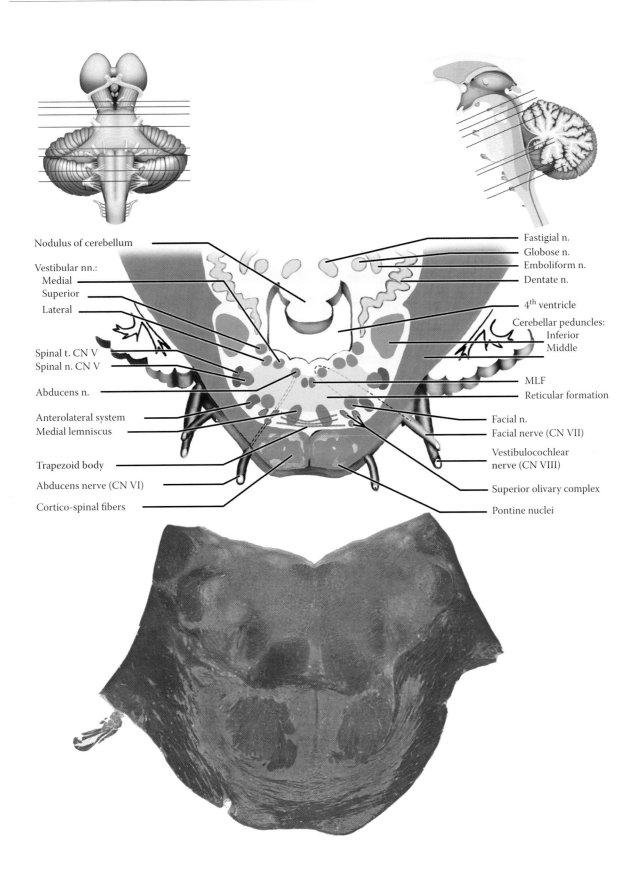

Nodulus of cerebellum

Vestibular nn.:
  Medial
  Superior
  Lateral

Spinal t. CN V
Spinal n. CN V

Abducens n.

Anterolateral system
Medial lemniscus

Trapezoid body

Abducens nerve (CN VI)

Cortico-spinal fibers

Fastigial n.
Globose n.
Emboliform n.
Dentate n.

4th ventricle

Cerebellar peduncles:
  Inferior
  Middle

MLF
Reticular formation

Facial n.
Facial nerve (CN VII)

Vestibulocochlear
nerve (CN VIII)

Superior olivary complex

Pontine nuclei

FIGURE 66C: Brainstem Histology — Lower Pons

THE MEDULLA
FIGURE 67, FIGURE 67A,
FIGURE 67B, AND FIGURE 67C

This part of the brainstem has a different appearance from the midbrain and pons because of the presence of two distinct structures: the pyramids and the inferior olivary nucleus.

The pyramids, located ventrally, are an elevated pair of structures located on either side of the midline (see Figure 6 and Figure 7). They contain the cortico-spinal fibers that have descended from the motor areas of the cortex and now emerge as a distinct bundle (see Figure 45 and Figure 48). Most of its fibers cross (decussate) at the lowermost part of the medulla. The inferior olive (nucleus) is a prominent structure that has a distinct scalloped profile when seen in cross-section. It is so large that it forms a prominent bulge on the lateral surface of the medulla (see Figure 6 and Figure 7). Its fibers relay to the cerebellum (see Figure 55).

The tegmentum is the area of the medulla that contains the cranial nerve nuclei, the nuclei of the reticular formation, the ascending tracts, and two special nuclei, the inferior olivary nucleus (discussed above) and the dorsal column nuclei (dorsally).

Cranial nerves IX, X, and XII are attached to the medulla and have their nuclei here; part of CN VIII is also represented in the uppermost medulla. The most prominent nucleus of the reticular formation in this region is the nucleus gigantocellularis (see Figure 42A and Figure 42B); the descending fibers form the lateral reticulo-spinal tract (see Figure 49B).

Included in the tegmentum are the two ascending tracts, the large medial lemniscus and the small anterolateral system, both conveying the sensory modalities from the opposite side of the body. The spinal trigeminal tract and nucleus, conveying the modalities of pain and temperature from the ipsilateral face and oral structures, is also found throughout the medulla. The solitary nucleus and tract, which subserve both taste and visceral afferents, are likewise found in the medulla. The MLF is still a distinct tract in its usual location (see Figure 51B).

The nuclei gracilis and cuneatus, the relay nuclei for the dorsal column tracts, are found in the lower part of the medulla, on its dorsal aspect (discussed with Figure 67C). The fourth ventricle lies behind the tegmentum, separating the medulla from the cerebellum (see Figure 20B). The roof of this (lower) part of the ventricle has choroid plexus (see Figure 21). CSF escapes from the fourth ventricle via the various foramina located here, and then flows into the subarachnoid space, the cisterna magna (see Figure 18 and Figure 21).

FIGURE 67: MID-MEDULLA
(PHOTOGRAPHIC VIEW)

This is a photographic image, enlarged, at the middle level of the medulla, with the cerebellum attached. This specimen shows the principal identifying features of the medulla, the pyramids ventrally on either side of the midline and the more laterally placed inferior olivary nucleus, with its scalloped borders.

Between the olivary nuclei, on either side of the midline, are two dense structures, the medial lemniscus. The other dense tract that is recognizable in this specimen is the inferior cerebellar peduncle located at the outer posterior edge of the medulla. Other tracts and cranial nerve nuclei, including the reticular formation, are found in the central region of the medulla, the tegmentum.

The space behind is the fourth ventricle, narrowing in its lower portion (see Figure 20B). There is no "roof" to the ventricle in this section, and it is likely that the plane of the section has passed through the median aperture, the foramen of Magendie (see Figure 21).

The cerebellum remains attached to the medulla, with the prominent vermis and the large cerebellar hemispheres. The cerebellar lobe adjacent to the medulla is the tonsil (see Figure 18; discussed with Figure 9B). The extensive white matter of the cerebellum is seen, as well as the thin outer layer of the cerebellar cortex.

The medulla is to be represented by three sections:

- Figure 67A: The upper medullary level typically includes CN VIII (both parts) and its nuclei.
- Figure 67B: This section through the mid-medulla includes the nuclei of cranial nerves IX, X, and XII.
- Figure 67C: The lower medullary section is at the level of the dorsal column nuclei, the nuclei gracilis and cuneatus.

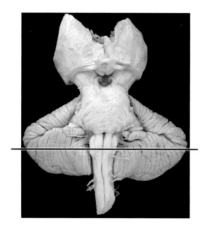
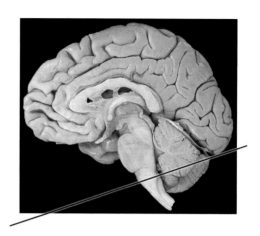

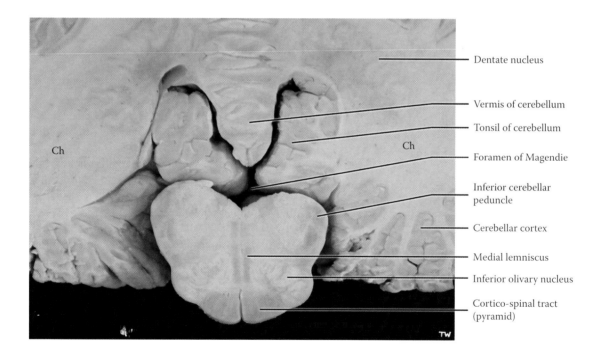

Dentate nucleus

Vermis of cerebellum

Tonsil of cerebellum

Foramen of Magendie

Inferior cerebellar
peduncle

Cerebellar cortex

Medial lemniscus

Inferior olivary nucleus

Cortico-spinal tract
(pyramid)

Ch = Cerebellar hemisphere

FIGURE 67: Brainstem Histology — Medulla (mid — photograph)

FIGURE 67A
UPPER MEDULLA:
CROSS-SECTION

This section has the characteristic features of the medullary region, namely the pyramids anteriorly with the inferior olivary nucleus situated just laterally and behind.

The cortico-spinal voluntary motor fibers from areas 4 and 6 go through the white matter of the hemispheres, funnel via the internal capsule (posterior limb), continue through the cerebral peduncles of the midbrain and the pontine region, and emerge as a distinct bundle in the medulla within the pyramids. The cortico-spinal tract is often called the pyramidal tract because its fibers form the pyramids (discussed with Figure 45).

The medial lemniscus is the most prominent ascending (sensory) tract throughout the medulla, carrying the modalities of discriminative touch, joint position, and vibration (see Figure 33 and Figure 40). The tracts are located next to the midline, oriented in the anteroposterior (ventrodorsal) direction (see Figure 40), just behind the pyramids; they will change orientation and shift more laterally in the pons. Dorsal to them, also along the midline, are the paired tracts of the MLF, situated in front of the fourth ventricle. The anterolateral tract, conveying pain and temperature, lies dorsal to the olive, although it is not of sufficient size to be clearly identified (see Figure 34 and Figure 40). Both the medial lemniscus and the anterolateral system are carrying fibers from the opposite side of the body at this level. The descending nucleus and tract of CN V are present more laterally, carrying fibers (pain and temperature) from the ipsilateral face and oral structures, before decussating (see Figure 35 and Figure 40).

The other prominent tract in the upper medullary region is the inferior cerebellar peduncle. This tract is conveying fibers to the cerebellum, both from the spinal cord and from the medulla, including the inferior olivary nucleus (discussed with Figure 55).

The VIIIth nerve enters the medulla at its uppermost level, at the cerebello-pontine angle, passing over the inferior cerebellar peduncle. The nerve has two nuclei along its course, the ventral and dorsal cochlear nuclei (see Figure 8B). The auditory fibers synapse in these nuclei and then go on to the superior olivary complex in the lower pons region. The crossing fibers are seen in the lowermost pontine region as the trapezoid body (see Figure 37 and Figure 40).

The vestibular part of the VIIIth nerve is represented at this level by two nuclei, the medial and inferior vestibular nuclei (see Figure 51A). Both these nuclei lie in the same position as the vestibular nuclei in the pontine section, adjacent to the lateral edge of the fourth ventricle. The inferior vestibular nucleus is distinct because of the many axon bundles that course through it. The vestibular nuclei contribute fibers to the MLF (discussed with Figure 51B).

The solitary nucleus is found at this level, surrounding a tract of the same name. This nucleus is the synaptic station for incoming taste fibers (mainly with CN VII, also with CN IX), and for visceral afferents entering with CN IX and X from the GI tract and other viscera. The solitary nucleus and tract are situated just beside (anterior to) the vestibular nuclei.

The core area is occupied by the cells of the reticular formation (see Figure 42A and Figure 42B). The most prominent of its nuclei at this level is the gigantocellular nucleus (noted for its large neurons), which gives rise to the lateral reticulo-spinal tract (see Figure 49B). The other functional aspects of the reticular formation should be reviewed at this point, including the descending pain system from the nucleus raphe magnus (discussed with Figure 43).

The fourth ventricle is still quite large at this level. The lower portion of its roof has choroid plexus (see Figure 20A and Figure 21); a fragment of this is present with the histological section, although the roof is torn. Behind the ventricle is the cerebellum, with the vermis (midline) portion and the cerebellar hemispheres. The dentate nucleus, the largest of the intracerebellar nuclei, is present at this level. Again, the cerebellum has not been processed with the histological specimen.

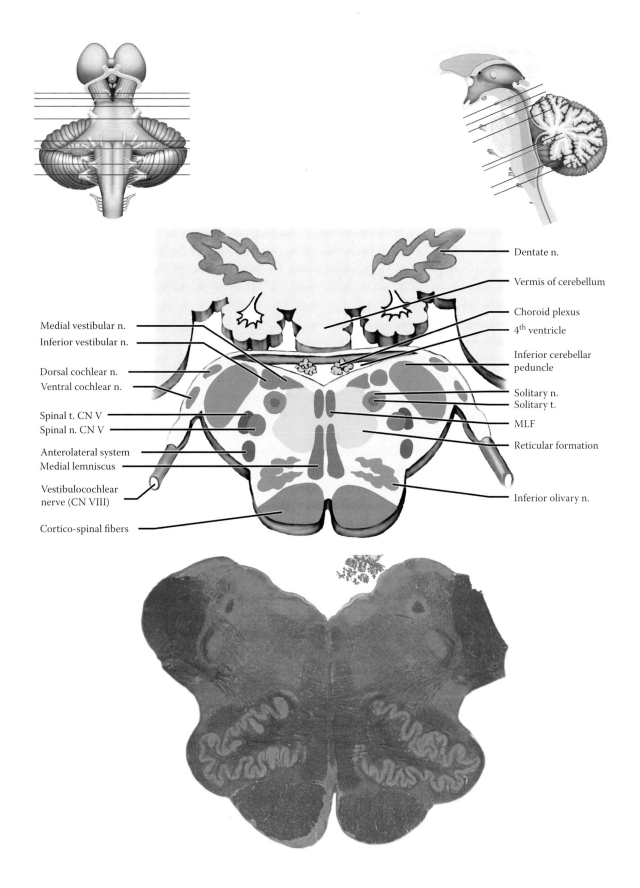

Dentate n.

Vermis of cerebellum

Choroid plexus

Medial vestibular n.

Inferior vestibular n.

4th ventricle

Inferior cerebellar
peduncle

Dorsal cochlear n.

Ventral cochlear n.

Solitary n.

Solitary t.

Spinal t. CN V

Spinal n. CN V

MLF

Anterolateral system

Reticular formation

Medial lemniscus

Vestibulocochlear
nerve (CN VIII)

Inferior olivary n.

Cortico-spinal fibers

FIGURE 67A: Brainstem Histology — Upper Medulla

FIGURE 67B
MID-MEDULLA:
CROSS-SECTION

This cross-sectional level is often presented as "typical" of the medulla. The pyramids and inferior olive are easily recognized anteriorly.

The medial lemniscus occupies the area between the olives, on either side of the midline (see Figure 40). The MLF lies behind (dorsal) the medial lemniscus, also situated adjacent to the midline. The fibers of the anterolateral system are situated dorsal to the olive. The descending nucleus and tract of the trigeminal system have the same location as seen previously in the lateral aspect of the tegmentum.

The hypoglossal nucleus (CN XII) is found near the midline and in front of the ventricle; its fibers exit anteriorly, between the pyramid and the olive (see Figure 6 and Figure 7). CN IX and CN X are attached at the lateral aspect of the medulla (see Figure 6 and Figure 7). Their efferent fibers are derived from two nuclei (indicated by the dashed lines): the dorsal motor nucleus, which is parasympathetic, and the nucleus ambiguus, which is motor to the muscles of the pharynx and larynx (see Figure 8A). The dorsal motor nucleus lies adjacent to the fourth ventricle just lateral to the nucleus of XII. The nucleus ambiguus lies dorsal to the olivary nucleus; in a single cross-section only a few cells of this nucleus are usually seen, making its identification difficult (i.e., "ambiguous") in actual sections. The taste and visceral afferents that are carried in these nerves synapse in the solitary nucleus, which is located in the posterior aspect of the tegmentum, surrounding the tract of the same name.

The reticular formation occupies the central core of the tegmentum; the nucleus gigantocellularis is located in this part of the reticular formation (see Figure 42B). These cells give rise to a descending tract, the lateral reticulospinal tract as part of the indirect voluntary motor system (see Figure 49B); there is also a strong influence on the excitability of the lower motor neuron, influencing the stretch reflex and muscle tone.

The inferior cerebellar peduncle is found at the lateral edge of this section, posteriorly, carrying fibers to the cerebellum (see Figure 55). The fourth ventricle is still a rather large space, behind the tegmentum, with the choroid plexus attached to its roof in this area; often the ventricle appears "open," likely because this thin tissue has been torn. There is no cerebellar tissue posteriorly since the section is below the level of the cerebellum (see the sagittal schematic accompanying this figure).

CLINICAL ASPECT

Vascular lesions in this area of the brainstem are not uncommon. The midline area is supplied by the paramedian branches from the vertebral artery (see Figure 58). The structures included in this territory are the corticospinal fibers, the medial lemniscus, and the hypoglossal nucleus.

The lateral portion is supplied by the posterior inferior cerebellar artery, a branch of the vertebral artery (see Figure 58, Figure 59A, and Figure 61), called **PICA** by neuroradiologists. This artery is prone to infarction for some unknown reason. Included in its territory are the cranial nerve nuclei and fibers of CN IX and X, the descending trigeminal nucleus and tract, fibers of the anterolateral system, and the solitary nucleus and tract, as well as descending autonomic fibers. The inferior cerebellar peduncle or vestibular nuclei may also be involved. The whole clinical picture is called the **lateral medullary syndrome** (of Wallenberg).

Interruption of the descending autonomic fibers gives rise to a clinical condition called **Horner's syndrome**. In this syndrome, there is loss of the autonomic sympathetic supply to one side of the face, ipsilaterally. This leads to ptosis (drooping of the upper eyelid), a dry skin, and constriction of the pupil. The pupillary change is due to the competing influences of the parasympathetic fibers, which are still intact. Other lesions elsewhere that interrupt the sympathetic fibers in their long course can also give rise to Horner's syndrome.

Note to the Learner: It is instructive to work out the clinical symptomatology of both of these vascular lesions, using a drawing, indicating which function is lost with each of the tracts or nucleus involved in the lesion, and which side of the body would be affected.

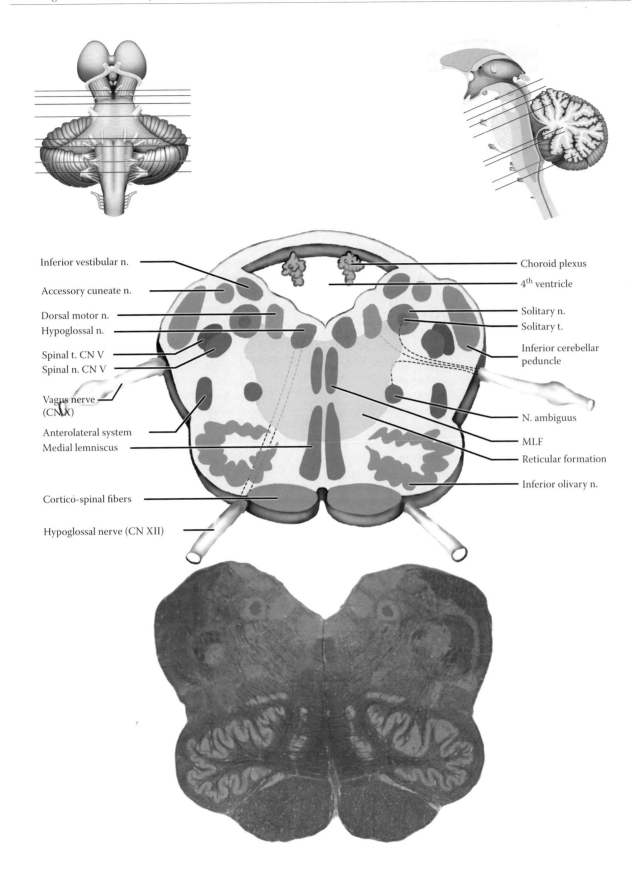

Inferior vestibular n.

Accessory cuneate n.

Dorsal motor n.

Hypoglossal n.

Spinal t. CN V

Spinal n. CN V

Vagus nerve
(CN X)

Anterolateral system

Medial lemniscus

Cortico-spinal fibers

Hypoglossal nerve (CN XII)

Choroid plexus

4th ventricle

Solitary n.

Solitary t.

Inferior cerebellar
peduncle

N. ambiguus

MLF

Reticular formation

Inferior olivary n.

FIGURE 67B: Brainstem Histology — Mid-Medulla

FIGURE 67C
LOWER MEDULLA:
CROSS-SECTION

The medulla seems significantly smaller in size at this level, approaching the size of the spinal cord below. The section is still easily recognized as medullary because of the presence of the pyramids anteriorly (the cortico-spinal tract) and the adjacent inferior olivary nucleus.

The tegmentum contains the cranial nerve nuclei, the reticular formation and the other tracts. The nuclei of CN X and CN XII, as well as the descending nucleus and tract of V, are present as before (as in the mid-medullary section, see Figure 67B). The MLF and anterolateral fibers are also in the same position. The solitary tract and nucleus are still found in the same location. The internal arcuate fibers are present at this level; these are the fibers from the nuclei gracilis and cuneatus, which cross (decussate) to form the medial lemniscus (see below). These fibers usually obscure visualization of the nucleus ambiguus. Finally, the reticular formation is still present.

The dorsal aspect of the medullary tegmentum is occupied by two large nuclei: the nucleus cuneatus (cuneate nucleus) laterally, and the nucleus gracilis (gracile nucleus) more medially. These are found on the dorsal aspect of the medulla (see Figure 9B and Figure 40). These nuclei are the synaptic stations of the tracts of the same name that have ascended the spinal cord in the dorsal column (see Figure 33, Figure 68, and Figure 69). The gracilis is mainly for the upper limb and upper body; the cuneatus carries information from the lower body and lower limb. The fibers relay in these nuclei and then move through the medulla anteriorly as the internal arcuate fibers, cross (decussate), and form the medial lemniscus on the opposite side (see Figure 40). At this level, the

medial lemniscus is situated between the olivary nuclei and dorsal to the pyramids, and is oriented anteroposteriorly.

Posteriorly, the fourth ventricle is tapering down in size, giving a "V-shaped" appearance to the dorsal aspect of the medulla (see Figure 20B). It is common for the ventricle roof to be absent at this level. This is likely accounted for by the presence of the foramen of Magendie, where the CSF escapes from the ventricular system into the subarachnoid space (see Figure 21). Posterior to this area is the cerebello-medullary cistern, otherwise known as the cisterna magna (see Figure 2, Figure 18, and Figure 21).

One special nucleus is found in the "floor" of the ventricle at this level, the **area postrema**. This forms a little bulge that can be appreciated on some sections. The nucleus is part of the system that controls vomiting, and it is often referred to as the vomiting "center." It is interesting to note that this region lacks a blood-brain barrier, allowing this particular nucleus to be "exposed" directly to whatever is circulating in the blood stream. It likely connects with the nuclei of the vagus nerve, which are involved in the act of vomiting.

ADDITIONAL DETAIL

The accessory cuneate nucleus is found at this level, as well as at the mid-medullary level. This nucleus is a relay for some of the cerebellar afferents from the upper extremity (see Figure 55). The fibers then go to the cerebellum via the inferior cerebellar peduncle. The inferior cerebellar peduncle has not yet been formed at this level.

Cross-sections through the lowermost part of the medulla may include the decussating cortico-spinal fibers, i.e., the pyramidal decussation (see Figure 40); this would therefore alter significantly the appearance of the structures in the actual section.

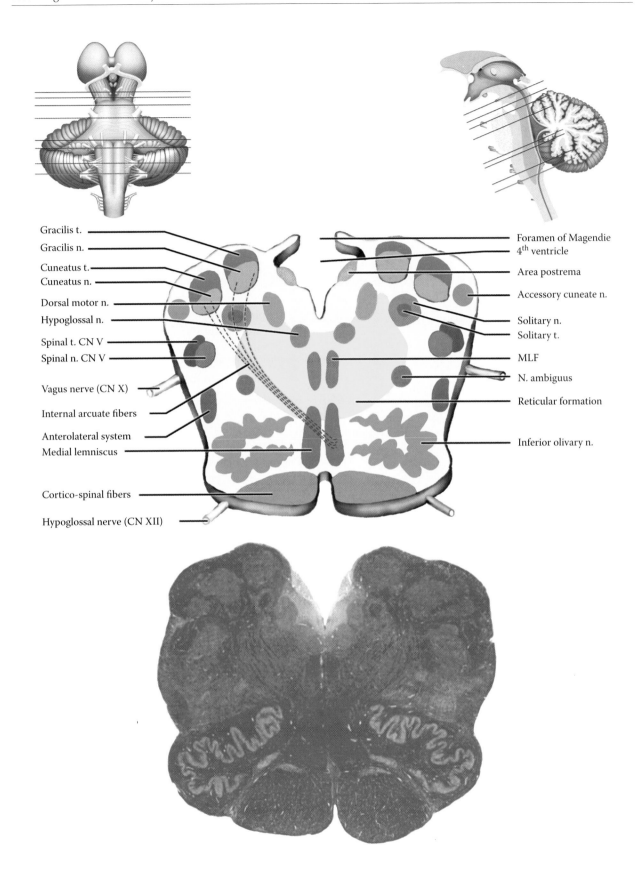

Gracilis t.
Gracilis n.
Cuneatus t.
Cuneatus n.
Dorsal motor n.
Hypoglossal n.
Spinal t. CN V
Spinal n. CN V
Vagus nerve (CN X)
Internal arcuate fibers
Anterolateral system
Medial lemniscus
Cortico-spinal fibers
Hypoglossal nerve (CN XII)

Foramen of Magendie
4th ventricle
Area postrema
Accessory cuneate n.
Solitary n.
Solitary t.
MLF
N. ambiguus
Reticular formation
Inferior olivary n.

FIGURE 67C: Brainstem Histology – Lower Medulla

FIGURE 68
SPINAL CORD:
CROSS-SECTIONS

UPPER ILLUSTRATION: NUCLEI

This diagram shows all the nuclei of the gray matter of the spinal cord — both sensory and motor (see Figure 4, Figure 32, and Figure 44).

LOWER ILLUSTRATION: TRACTS: C8 LEVEL

The major tracts of the spinal cord are shown on this diagram, with the descending tracts on the left side and the ascending ones on the right side. In fact, both sets of pathways are present on both sides. Some salient features of each will be presented.

DESCENDING TRACTS

- **Lateral cortico-spinal**, from the cerebral (motor) cortex (see Figure 45 and Figure 48): These fibers for direct voluntary control supply mainly the lower motor neurons in the lateral ventral horn to control fine motor movements of the hand and fingers. This pathway crosses in the lowermost medulla.
- **Anterior (ventral) cortico-spinal**, also from the motor cortex (see Figure 45): These fibers, which do not cross in the pyramidal decussation, go to the motor neurons that supply the proximal and axial musculature.
- **Rubro-spinal**, from the red nucleus (see Figure 47 and Figure 48): This tract crosses at the level of the midbrain. Its role in human motor function is not certain.
- **Medial and lateral reticulo-spinal tracts**, from the pontine and medullary reticular formation, respectively (see Figure 49A and Figure 49B): These pathways are the additional ones for indirect voluntary control of the proximal joints and for posture, as well as being important for the control of muscle tone.
- **Lateral vestibulo-spinal,** from the lateral vestibular nucleus (see Figure 50): Its important function is participating in the response of the axial muscles to changes in gravity. This pathway remains ipsilateral.
- **Medial longitudinal fasciculus** (MLF, see Figure 51B): This mixed pathway is involved in the response of the muscles of the eyes and of the neck to vestibular and visual input. It likely descends only to the cervical spinal cord level.

ASCENDING TRACTS

- **Dorsal column tracts**, consisting at this level of both the fasciculus cuneatus and fasciculus gracilis (see Figure 33 and Figure 40): These are the pathways for discriminative touch sensation, joint position and "vibration" from the same side of the body, with the lower limb fibers medially (gracile) and the upper limb pathway laterally (cuneate).
- **Anterolateral system**, consisting of the anterior (ventral) spino-thalamic and lateral spino-thalamic tracts (see Figure 34): These pathways carry pain and temperature, as well as crude touch information from the opposite side of the body, with the lower limb fibers more lateral and the upper limb fibers medial.
- **Spino-cerebellar tracts**, anterior (ventral) and posterior (dorsal) (reviewed with the cerebellum, see Figure 54 and Figure 55): These convey information from the muscle spindles and other sources to the cerebellum.

SPECIAL TRACT

The dorsolateral fasciculus, better known as the tract of Lissauer (see Figure 32), carries intersegmental information, particularly relating to pain afferents.

CLINICAL ASPECT

The functional aspects of each of these tracts should be reviewed at this time by noting the loss of function that would be found following a lesion of the various pathways.

An acute injury to the cord, such as severing of the cord following an accident, will usually result in a complete shutdown of all spinal cord functions, called **spinal shock** (discussed with Figure 5). After a period of about 3–4 weeks, the spinal cord reflexes will return. In a matter of weeks, due to the loss of all the descending influences on the spinal cord, there is an increase in the reflex responsiveness (hyperreflexia) and a marked increase in tone (spasticity), along with the Babinski response (discussed with Figure 49B).

A classic lesion of the spinal cord is the **Brown-Sequard syndrome**, which is a lesion of one-half of the spinal cord on one side. Although rare, this is a useful lesion for the learner to review the various deficits, sensory and motor, that would be found after such a lesion. In particular, it helps the learner understand which side of the body would be affected because of the various crossing of the pathways (sensory and motor) at different levels.

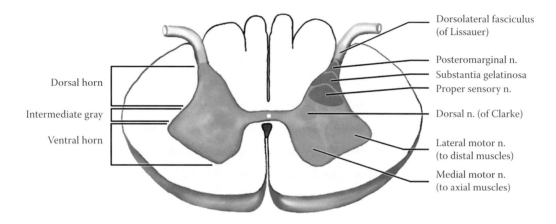

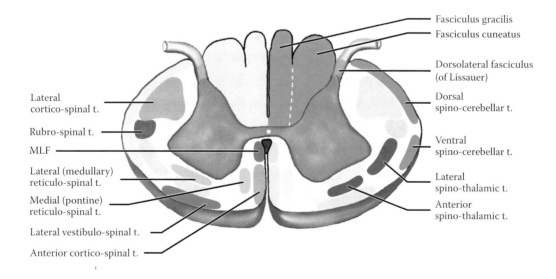

FIGURE 68: Spinal Cord — Nuclei and Tracts

FIGURE 69
SPINAL CORD:
CROSS-SECTIONAL VIEWS —
HISTOLOGICAL

The spinal cord was introduced in the orientation section of this atlas (Section A, see Figure 1–Figure 5). The organization of the nervous tissue in the cord has the gray matter inside, in a typical "butterfly" or "H-shaped" configuration, with the white matter surrounding (see Figure 1, Figure 4, and Figure 5). The functional aspects of the spinal cord were presented in Section B, including the nuclei and connections for the afferent fibers (sensory, see Figure 32), and the efferent circuits with some reflexes (motor, see Figure 44).

The white matter surrounding the gray matter is divided by it into three areas: the dorsal, the lateral, and the anterior areas. These zones are sometimes referred to as funiculi (singular funiculus). Various tracts are located in each of these three zones, some ascending and some descending, which were reviewed (see previous illustration).

The following are cross-sectional views of various levels of the spinal cord, stained with a myelin and cell stain.

CERVICAL LEVEL — C8

This is a cross-section of the spinal cord through the cervical enlargement. This level has been used in many of the illustrations of the various pathways (in Section B). Since the cervical enlargement contributes to the formation of the brachial plexus to the upper limb, the gray matter ventrally is very large because of the number of neurons involved in the innervation of the upper limb, particularly the muscles of the hand. The dorsal horn is likewise large, because of the amount of afferents coming from the skin of the fingers and hand.

The white matter is comparatively larger at this level because:

- All the ascending tracts are present and are carrying information from the lower parts of the body as well as the upper limb.
- All the descending tracts are fully represented, as many of the fibers will terminate in the cervical region of the spinal cord. In fact, some of them do not descend to lower levels.

Various layers of meninges are seen in these cross-sections, as well as dorsal and ventral roots in the subarachnoid (CSF) space.

THORACIC LEVEL — T6

The thoracic region of the spinal cord presents an altered morphology because of the decrease in the amount of gray matter. There are fewer muscles and less dense innervation of the skin in the thoracic region. The gray matter has, in addition, a lateral horn, which represents the sympathetic preganglionic neurons. The lateral horn is present from T1 to L2.

LUMBAR LEVEL — L3

This cross-sectional level of the spinal cord has been used in the various illustrations of the pathways in Section B of this atlas. This cross-section is similar in appearance to the cervical section, because both are innervating the limbs. There is, however, proportionately less white matter at the lumbar level. The descending tracts are smaller because many of the fibers have terminated at higher levels. The ascending tracts are smaller because they are conveying information only from the lower regions of the body.

The **sacral region** of the spinal cord is the smallest in size and is therefore easy to recognize (not shown). The white matter is quite reduced in size. There is still a fair amount of gray matter because of the innervation of the pelvic musculature.

This region of the spinal cord, roughly the conus medullaris (see Figure 2A), also contains the preganglionic parasympathetic neurons of the autonomic nervous system. These neurons innervate the bowel and the bladder.

BLOOD SUPPLY

The anterior spinal artery, the main blood supply to the spinal cord, comes from branches from each of the vertebral arteries that join (see Figure 58); it descends in the midline (see Figure 2B) and supplies the ventral horn and the anterior and lateral group of tracts, including the lateral cortico-spinal pathway. The posterior spinal arteries supply the dorsal horn and the dorsal columns.

CLINICAL ASPECT

The blood supply to the spinal cord was reviewed with Figure 2B; it is known that this blood supply is marginal, particularly in the mid-thoracic region. The learner is encouraged to work out the clinical symptomatology following lesions of the spinal cord at various levels.

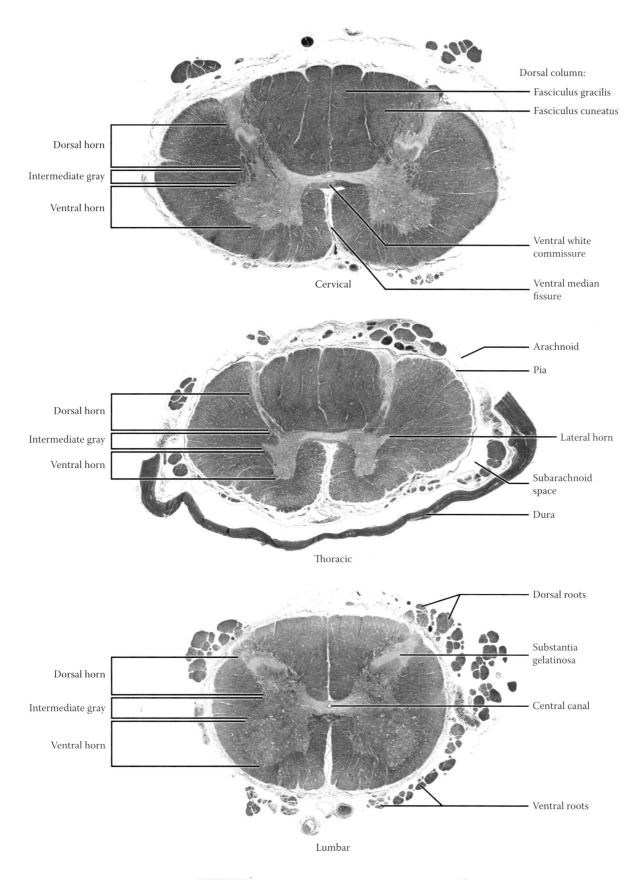

Dorsal column:
Fasciculus gracilis
Fasciculus cuneatus

Dorsal horn

Intermediate gray

Ventral horn

Ventral white commissure

Cervical

Ventral median fissure

Arachnoid
Pia

Dorsal horn

Intermediate gray

Ventral horn

Lateral horn

Subarachnoid space

Dura

Thoracic

Dorsal roots

Substantia gelatinosa

Dorsal horn

Intermediate gray

Central canal

Ventral horn

Ventral roots

Lumbar

FIGURE 69: Spinal Cord Histology — Cross-Sections

THE LIMBIC SYSTEM

INTRODUCTION

The term limbic is almost synonymous with the term emotional brain — the parts of the brain involved with our emotional state. In 1937, Dr. James Papez initiated the limbic era by proposing that a number of limbic structures in our brain formed the anatomical substratum for emotion.

EVOLUTIONARY PERSPECTIVE

Dr. Paul MacLean has postulated that there are in fact three separable "brains" that have evolved. The pre-mammalian (reptilian) brain has the capacity to look after the basic life functions and has organized ritualistic stylized patterns of behavior. In higher species, including mammals, forebrain structures have evolved that relate to the external world (e.g., visual input). These are adaptive, allowing for a modification of behavior depending upon the situation.

MacLean has suggested that the limbic system arises in early mammals to link these two brain functions; according to this scheme, the limbic system relates the reptilian brain, which monitors the internal milieu, with the newer forebrain areas of mammals responsible for analyzing the external environment. Many now view the limbic system from this perspective.

DEFINITION

Most of us are quite aware or have a general sense of what we mean when we use the term "emotion" or feelings, yet it is somewhat difficult to explain or define precisely. Stedman's Medical Dictionary defines emotion as "a strong feeling, aroused mental state, or intense state of drive or unrest directed toward a definite object and evidenced in both behavior and in psychological changes." Thus, emotions involve:

- Physiological changes: This includes basic drives such as thirst, sexual behavior, and appetite. These changes are often manifested as alterations of the autonomic nervous system or the endocrine system.
- Behavior: The animal or human does something, that is, performs some type of motor activity, for example fighting, fleeing, displaying anger, mating; in humans, this may include facial expression.

- Alterations in the mental state: This can be understood as a subjective change in the way the organism "feels" or reacts to the state of being or to events occurring in the outside world. In humans, we use the term psychological reaction.

It is clear, at least in humans, that some of these psychological functions and behaviors must engage the cerebral cortex. In addition, many of these alterations are conscious and involve association areas. In fact, humans are sometimes able to describe and verbalize their reactions or the way they feel. Both cortical and subcortical areas (e.g., basal ganglia) may be involved in the behavioral reactions associated with emotional responses. The hypothalamus controls the autonomic changes, along with brainstem nuclei, and also the activity of the pituitary gland underlying the endocrine responses.

Therefore, we can finally arrive at a *definition* of the limbic system as an interrelated group of cortical and subcortical (noncortical) structures that are involved in the regulation of the internal or emotional state, with the accompanying physiological, behavioral, and psychological responses.

NEURAL STRUCTURES

In neuroanatomical terms, the limbic system is thought to include cortical and noncortical (subcortical, diencephalic, and brainstem) structures. The following is a listing of the structures:

- *Core* structures are those definitely associated with the limbic system.
- *Extended* structures are those closely connected with limbic functions.

a. Cortical:
   - **Core**: The hippocampal formation, which consists of three subparts (which are "buried" in the medial temporal lobe in humans), parahippocampal gyrus, cingulate gyrus
   - **Extended**: Parts of the prefrontal and orbitofrontal cortex (the limbic forebrain)

b. Noncortical:
   - Forebrain:
     - Core: Amygdala, septal region, ventral portions of the basal ganglia, including the nucleus accumbens

- Extended: the basal forebrain
- Diencephalic and Brainstem:
  - Core: Certain nuclei of the thalamus, the hypothalamus
  - Extended: Parts of the midbrain (the limbic midbrain), and medulla

All of these structures are collectively called **the limbic system**. The particular role of the olfactory system and its connections will be discussed in the context of the limbic system.

OVERVIEW OF "KEY" LIMBIC STRUCTURES

There are key structures of the limbic system that integrate information and relate the external and internal worlds — the hippocampal formation, the parahippocampal gyrus, the amygdala, and the hypothalamus.

- The hippocampal formation is an older cortical region that is involved with integrating information; its role in the formation of memory for facts and events will be discussed below.
- The parahippocampal gyrus has widespread connections with many cortical (particularly sensory) areas and is probably the source of the most significant afferents to the hippocampal information.
- The amygdala is in part a subcortical nucleus involved with internal (visceral afferent) information, as well as receiving sensory input about olfaction (our sense of smell).
- The hypothalamus oversees autonomic physiological and hormonal regulation.

Both the amygdala and the hypothalamus are involved with the motor (i.e., behavioral) responses of the organism (the amygdala, in part, via the hypothalamus), and both are involved along with other structures in generating "emotional" reactions.

LIMBIC CONNECTIONS

The limbic system has internal circuits connecting the key structures; these link the hippocampal formation, the parahippocampal gyrus, the amygdala, and the hypothalamus, as well as other structures of the limbic system. There are multiple interconnections within and between these structures, and knowledge of the circuits of the limbic system (which are quite complex) allows one to trace pathways within the limbic system. Only some of these pathways will be presented. The best known of these functionally (and for historical reasons) is the Papez circuit (discussed with Figure 77A). Additional pathways that connect the limbic structures to the remainder of the nervous system and through which the limbic system influences the activity of the nervous system will be discussed.

MEMORY

Unfortunately, the definition and description of the limbic system does not include one aspect of brain function that seems to have evolved in conjunction with the limbic system — memory. Memory systems are usually grouped into two types:

- Memory for skills and procedures called **procedural** memory
- Memory for facts and events called **declarative** or **episodic** memory

A part of the hippocampal formation is specifically necessary for the initial formation of episodic memories. It is critical to understand that this initial step is an absolute prerequisite to the formation of any *new* memory trace. Once encoded by the hippocampal formation, the memory trace is then transferred to other parts of the brain for short- and long-term storage. The limbic system seems not to be involved in the storage and retrieval of long term memories.

It is interesting to speculate that forgetting may be theoretically more appropriate for this unique aspect of limbic function. This idea proposes that to undo or unlock the fixed behavioral patterns of the old reptilian brain, some part of the brain must be assigned the function of "recording" that something has happened. In order to change a response, the organism needs to "remember" what happened the last time when faced with a similar situation, hence the development of memory functions of the brain in association with the evolution of the limbic system. The availability of stored memories makes it possible for mammals to override or overrule the stereotypical behaviors of the reptile, allowing for more flexibility and adaptiveness when faced with a changing environment or altered circumstances. Therefore, we have suggested that the "F" mnemonic — forgetting — may be applicable for this "memory" function.

OTHER "LIMBIC" FUNCTIONS

In summary, the limbic system — both cortical and noncortical components — includes a set of "F" functions: feeding (and other basic drives), fornication (reproduction), fighting and fleeing (behavioral), feeling (psychological), and "forgetting" (memory).

It has also been suggested that some mammalian behavior associated with caring for its young is associated with the limbic structures, such as recognizing and responding to the vocalizations of the "pups" in rodents, cats, and other animals; a mother responds to the unique tone of her own baby's crying. The cingulate gyrus seems to be the area of the brain involved in this activity. This notion of rearing and "family" would add another "F" to our list of limbic functions.

It is also interesting to speculate that the elaboration of limbic functions is closely associated with the development of self-awareness, consciousness of the self (not an "F" word).

These functions will be reviewed and discussed at the end of this section of the atlas.

FIGURE 70A
THE LIMBIC LOBE 1

CORTICAL STRUCTURES

The limbic lobe refers to cortical areas of the limbic system. These cortical areas, which were given the name "limbic," form a border (limbus) around the inner structures of the diencephalon and midbrain (see Figure 17 and Figure 70B). The core cortical areas include the hippocampal formation, the parahippocampal gyrus, and the cingulate gyrus.

There are a number of cortical areas located in the most medial (also called mesial) aspects of the temporal lobe in humans that form part of this "limbus." These areas are collectively called the **hippocampal formation**; it is made up of three portions — the hippocampus proper, the dentate gyrus, and the subicular region (see Figure 72A and Figure 72B). The **hippocampus proper** is, in fact, no longer found at the surface of the brain as would be expected for any cortical area. The **dentate gyrus** is a very small band of cortex, part of which can be found at the surface, and the **subicular** region is located at the surface but far within the temporal area. These structures are the central structures of the limbic lobe.

The typical cortex of the various lobes of the brain consists of six layers (and sometimes sublayers), called the neocortex. One of the distinguishing features of the limbic cortical areas is the fact that, for the most part, these are older cortical areas consisting of three to five layers, termed the allocortex. The hippocampus proper and the dentate cortex are three-layered cortical areas, while the subicular region has four to five layers.

Note to the Learner: At this stage, it is very challenging to understand where these structures are located. The component parts of the hippocampal formation are "buried" in the temporal lobe and remain somewhat obscure. It is suggested that the learner preview some of the illustrations of the "hippocampus" (see Figure 73), as well as sections through the hippocampal formation (see Figure 38 and Figure 72B) in order to better understand the configuration of the three component parts and the relationship to the parahippocampal gyrus. The details of these various limbic structures, including their important connections and the functional aspects, will be discussed with the appropriate diagram.

The **parahippocampal gyrus,** which is situated on the inferior aspect of the brain (see Figure 15A and Figure 15B) is a foremost structure of the limbic lobe. It is also composed of a five- and six-layered cortex. It is heavily connected (reciprocally) with the hippocampal formation. This gyrus also has widespread connections with many areas of the cerebral cortex, including all the sensory cortical regions, as well as the cingulate gyrus. It is thought to play a key role in memory function.

The **cingulate gyrus**, which is situated above the corpus callosum (see Figure 17), consists of a five-layered cortex, as well as neocortex. The cingulate gyrus is connected reciprocally with the parahippocampal gyrus via a bundle of fibers in the white matter, known as the cingulum bundle (see next illustration). This connection unites the various portions of the limbic "lobe." It also has widespread connections with the frontal lobe.

Of the many tracts of the limbic system, two major tracts have been included in this diagram, the fornix and the anterior commissure.

- The **fornix** is one of the more visible tracts and is often encountered during dissections of the brain (e.g., see Figure 17). This fiber bundle connects the hippocampal formation with other areas (to be discussed with Figure 72A and Figure 72B).
- The **anterior commissure** is an older commissure than the corpus callosum and connects several structures of the limbic system on the two sides of the brain; these include the amygdala, the hippocampal formation, and parts of the parahippocampal gyrus, as well as the anterior portions of the temporal lobe. The anterior commissure will be seen on many of the limbic diagrams and can also be a useful reference point for orientation (e.g., see Figure 75B).

The other structures shown in this diagram include the diencephalon (the thalamus) and the brainstem. The corpus callosum "area" is indicated as a reference point in these illustrations (see next illustration).

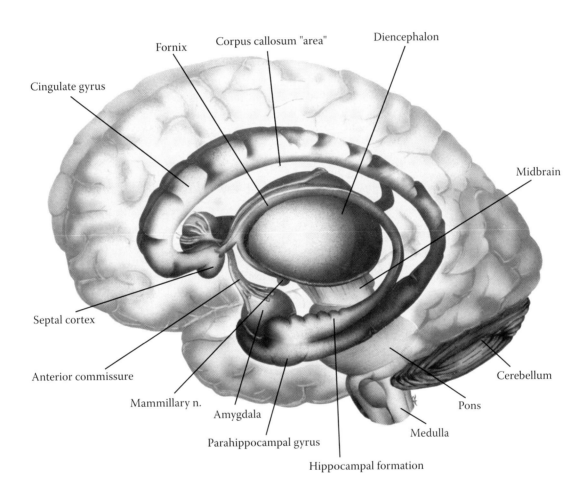

FIGURE 70A: Limbic Lobe 1 — Cortical

FIGURE 70B
LIMBIC LOBE 2

CINGULUM BUNDLE
(PHOTOGRAPHIC VIEW)

This is a dissection of the brain, from the medial perspective, as depicted in the previous illustration (see also Figure 17). The brainstem and cerebellum have been removed from this specimen. The specimen has been tilted slightly to show more of the inferior aspect of the temporal lobe. The thalamus (diencephalon) has been excised, revealing the fibers of the internal capsule (see Figure 26).

The cortex of the cingulate gyrus has been scraped away (with a blunt instrument), revealing a bundle of fibers just below the surface. The dissection is continued to the parahippocampal gyrus, as demarcated by the collateral sulcus/fissure (see Figure 15A and Figure 15B). This fiber bundle, called the cingulum bundle, is seen to course between these two gyri of the limbic system. This association tract will be discussed as part of a limbic circuit known as the Papez circuit (discussed with Figure 77A).

The brain is dissected in such a way to reveal the fornix (of both sides) as this fiber tract courses from the hippocampal formation in the temporal lobe, passes over the diencephalon, and heads toward its connections (see Figure 72A and Figure 72B).

CINGULATE GYRUS

MacLean's studies have indicated that the development of this gyrus is correlated with the evolution of the mammalian species. He has postulated that this gyrus is important for nursing and play behavior, characteristics that are associated with the rearing of the young in mammals. It is this cluster of behavioral patterns that forms the basis for the other "F" in the list of functions of the limbic system — family (see Introduction to this section). The cingulate gyrus also seems to have an important role in attention, an important aspect of behavior, with connections to the prefrontal cortex.

A small cortical region under the anterior part (the rostrum) of the corpus callosum is also included with the limbic system. These small gyri (not labeled; located just in front of the anterior commissure in Figure 41B) are named the septal cortex (see previous illustration); this area along with the septal nuclei (to be shown in the next illustration) are collectively called the septal region (see Figure 78B).

EXTENDED LIMBIC LOBE

Other areas of the brain are now known to be involved in limbic functions and are included in the functional aspects of the limbic system. This includes large parts of the "prefrontal cortex," particularly cortical areas lying above the orbit, the orbitofrontal cortex (not labeled), as well as the cortex on the medial aspect of the frontal lobe (to be discussed with Figure 77B).

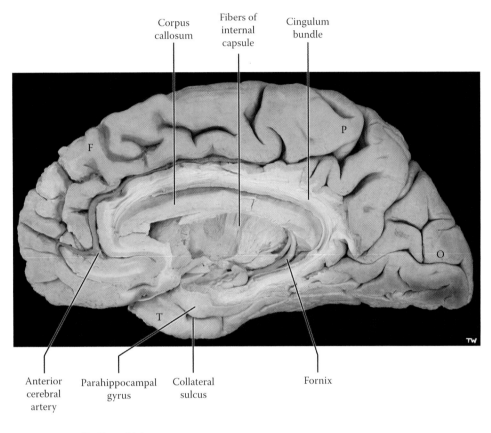

Corpus
callosum

Fibers of
internal
capsule

Cingulum
bundle

F

P

O

T

Anterior
cerebral
artery

Parahippocampal
gyrus

Collateral
sulcus

Fornix

F = Frontal lobe
P = Parietal lobe
T = Temporal lobe
O = Occipital lobe

FIGURE 70B: Limbic Lobe 2 — Cingulum Bundle (photograph)

FIGURE 71
LIMBIC SYSTEM

NONCORTICAL STRUCTURES

The term limbic system is the concept now used to include those parts of the brain that are associated with the functional definition of the limbic system.

This is an overall diagram focusing on the noncortical components of the limbic system, both core and extended. These structures are found in the forebrain, the diencephalon, and also in the midbrain. Each of the structures, including the connections, will be discussed in greater detail in subsequent illustrations when this diagram, indicated appropriately, will be used showing only the structures of the limbic system that are being described.

The noncortical areas include:

- Amygdala
- Septal nuclei (region)
- Basal forebrain
- Basal ganglia
- Thalamus
- Hypothalamas
- Limbic midbrain
- Olfactory system

FOREBRAIN

The **amygdala**, also called the amygdaloid nucleus, a core limbic structure, is anatomically one of the basal ganglia (as discussed earlier with Figure 22; see Figure OL and Figure 25). Functionally, and through its connections, it is part of the limbic system. Therefore, it will be considered in this section of the atlas (see Figure 75A and Figure 75B).

The **septal region** includes two components, the cortical gyri below the rostrum of the corpus callosum, the septal cortex (see Figure 70A), and some nuclei deep to them, the septal nuclei; these nuclei are not located within the septum pellucidum in humans. The term septal region includes both the cortical gyri and the nuclei (see Figure 78B).

Not represented in this diagram is the area known as the **basal forebrain**. This subcortical region is composed of several cell groups located beside the hypothalamus and below the anterior commissure (see Figure 80A and Figure 80B). This somewhat obscure region has connections with several limbic areas and the prefrontal cortex.

BASAL GANGLIA

The ventral portions of the putamen and globus pallidus are now known to be connected with limbic functions and are part of the extended limbic system (see Figure 80B). These functional parts are being identified as the ventral striatum and ventral pallidum.

The **nucleus accumbens** is a specific nuclear area adjacent to the septal nuclei and the neostriatum (see Figure 24). It has recently been found to have a critically important function in activities where there is an aspect of reward and punishment; this is now thought to be the critical area of the brain involved in addiction.

DIENCEPHALON

Two of the nuclei of the **thalamus**, the anterior group of nuclei and the dorsomedial nucleus (see Figure 12 and Figure 63), are part of the pathways of the limbic system, relaying information from subcortical nuclei to limbic parts of the cortex (the cingulate gyrus and areas of the prefrontal cortex).

The **hypothalamus** lies below and somewhat anterior to the thalamus (see Figure 17). Many nuclei of the hypothalamus function as part of the core limbic system. Only a few of these nuclei are shown, and among these is the prominent **mammillary nucleus**, which is visible on the inferior view of the brain (see Figure 15B). The connection of the hypothalamus to the pituitary gland is not shown.

MIDBRAIN

The extended limbic system also includes nuclei of the midbrain, the "**limbic midbrain**." Some of the descending limbic pathways terminate in this region, and it is important to consider the role of this area in limbic functions. An important limbic pathway, the medial forebrain bundle interconnects the septal region, the hypothalamus, and the limbic midbrain (see Figure 78B).

OLFACTORY

The **olfactory system** is described with the limbic system, as many of its connections are directly with limbic areas. Years ago it was commonplace to think of various limbic structures as part of the "smell brain," the rhinencephalon. We now know that this is only partially correct. The olfactory input connects directly into the limbic system (and not via the thalamus, see Figure 79), but the limbic system is now known to have many other functional capabilities.

TRACTS

The various tracts that interconnect the limbic structures — fornix, stria terminalis, ventral amygdalofugal pathway — will be discussed at the appropriate time with the relevant structure(s).

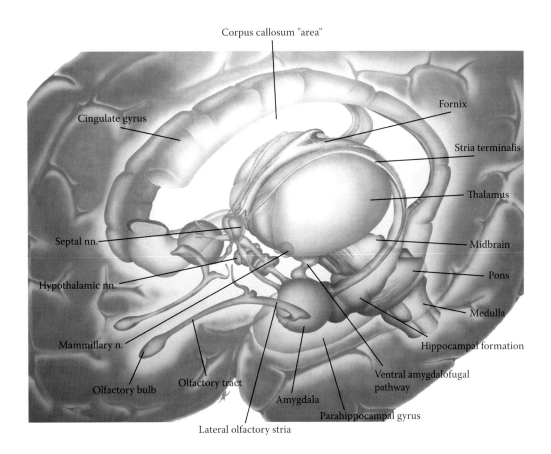

Corpus callosum "area"

Cingulate gyrus

Fornix

Stria terminalis

Thalamus

Septal nn.

Midbrain

Pons

Hypothalamic nn.

Medulla

Mammillary n.

Hippocampal formation

Ventral amygdalofugal
pathway

Olfactory bulb

Olfactory tract

Amygdala

Parahippocampal gyrus

Lateral olfactory stria

FIGURE 71: Limbic System — Noncortical

FIGURE 72A
"HIPPOCAMPUS" 1

HIPPOCAMPAL FORMATION

This diagram, which is the same as Figure 71, highlights the functional portion of the limbic lobe to be discussed — the "hippocampus" (i.e., the hippocampal formation) and the pathway known as the fornix.

The hippocampal formation includes older cortical regions, all consisting of less than six layers, which are located deep within the most medial aspect of the temporal lobe in humans. The location and complex arrangement of the structures are illustrated and explained in this series of diagrams.

In the rat, the hippocampal formation is located dorsally, above the thalamus. During the evolution of the temporal lobe, these structures have migrated into the temporal lobe, leaving behind a fiber pathway, the fornix, which is located above the thalamus.

The term hippocampal formation includes (see Figure 72B):

- The **hippocampus proper**, a three-layered cortical area that, during development, becomes "rolled-up" and is no longer found at the surface of the hemispheres (as is the case for all other cortical regions)
- The **dentate gyrus**, a three-layered cortical area that is partly found on the surface of the brain, although its location is so deep that it presents a challenge to nonexperts to locate and visualize this thin ridge of cortex
- The **subicular region**, a transitional cortical area of three to five layers that becomes continuous with the parahippocampal gyrus located on the inferior aspect of the brain (see Figure 15B)

The fornix is a fiber bundle that is visible on medial views of the brain (see Figure 17 and Figure 41B). These fibers emerge from the hippocampal formation (shown in Figure 73; see also Figure 70B) and course over the thalamus, where they are found just below the corpus callosum (see coronal sections, Figure 29 and Figure 74). The fibers end in the septal region and in the mammillary nucleus of the hypothalamus (shown in the next illustration). Some fibers in the fornix are conveying information from these regions to the hippocampal formation. It is perhaps best to regard the fornix as an association bundle, part of the limbic pathways. It has attracted much attention because of its connections and because of its visibility and accessibility for research into the function of the hippocampal formation, particularly with regard to memory.

MEMORY

Recent studies in humans have indicated that the neurons located in one portion of the hippocampus proper, called the CA3 region, are critical for the formation of new memories — declarative or episodic types of memories (not procedural). This means that in order to "remember" some new fact or event, the new information must be registered within the hippocampal formation. This information is "processed" through some complex circuitry in these structures and is retained for a brief period of seconds. In order for it to be remembered for longer periods, some partially understood process occurs so that the transient memory trace is transferred to other parts of the brain, and this is now stored in working memory or as a long-term memory. The process of memory storage consolidation may require a period of hours, if not days.

In the study of the function of the hippocampus in animals, there is considerable evidence that the hippocampal formation is involved in constructing a "spatial map." According to this literature, this part of the brain is needed to orient in a complex environment (such as a maze). It is not quite clear whether this is a memory function or whether this spatial representation depends upon the connections of the hippocampal formation and parahippocampal gyrus with other parts of the brain.

CLINICAL ASPECT

The clinical implications of the functional involvement of the hippocampal formation in memory will be further elaborated with Figure 73.

It is now possible to view the hippocampal area in detail on MRI and to assess the volume of tissue. Bilateral damage here apparently correlates with the loss of memory function in humans with Alzheimer's dementia, particularly for the formation of memories for new events or for new information (further discussed with Figure 73).

ADDITIONAL DETAIL

A vestigial part of the hippocampal formation is still found above the corpus callosum, as shown in this illustration — not labeled.

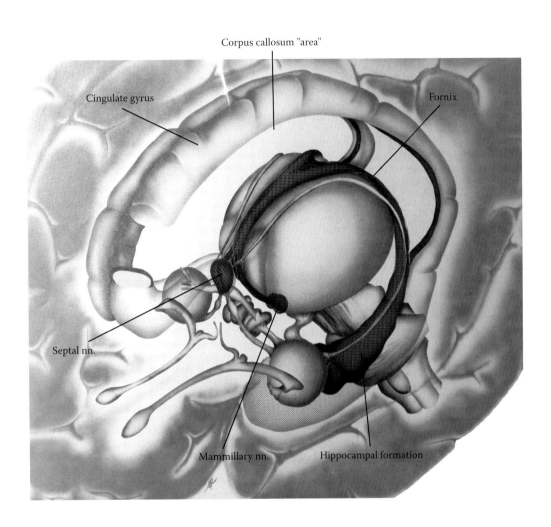

FIGURE 72A: Hippocampus 1 — Hippocampal Formation

FIGURE 72B
"HIPPOCAMPUS" 2

HIPPOCAMPAL FORMATION: THREE PARTS

The hippocampal formation is one of the most important structures of the limbic system in humans. It is certainly the most complex. This diagram isolates the component parts of the hippocampal formation, on both sides.

One expects a cortical area to be found at the surface of the brain, even if this surface is located deep within a fissure. During the evolution and development of the hippocampal formation, these areas became "rolled up" within the brain. The photographic view of the hippocampal formation is shown in Figure 74.

Note to the Learner: The learner is advised to consult Williams and Warwick, one of the reference books, for a detailed visualization and understanding of this developmental phenomenon.

THE HIPPOCAMPUS PROPER

The hippocampus proper consists of a three-layered cortical area. This forms a large mass, which actually intrudes into the ventricular space of the inferior horn of the lateral ventricle (see Figure 73 and Figure 74). In a coronal section through this region, there is a certain resemblance of the hippocampal structures to the shape of a seahorse (see Figure 38 and Figure 74). It is from this shape that the name "hippocampus" is derived, from the French word for seahorse. The other name for this area is **Ammon's horn** or *cornu ammonis* (**CA**), named after an Egyptian deity with ram's horns because of the curvature of the hippocampus in the brain. (This cortical region has been divided into a number of subportions, CA 1–4, usually studied in more advanced courses.)

THE DENTATE GYRUS

The dentate gyrus is also a phylogenetically older cortical area consisting of only three layers. During the formation discussed above, the leading edge of the cortex detaches itself and becomes the dentate gyrus. Parts of it remain visible at the surface of the brain. Since this small surface is buried on the most medial aspect of the temporal lobe and is located deep within a fissure, it is rarely located in studies of the gross brain. Its cortical surface has serrations, which led to its name, dentate (referring to teeth).

The appearance of the dentate gyrus is shown on the view of the medial aspect of the temporal lobe (on the far side of the illustration; see also Figure 76). A "cut" section through the temporal lobe (as seen in the lower portion of this illustration) indicates that the dentate gyrus is more extensive than its exposed surface portion.

THE SUBICULAR REGION

The next part of the cortically rolled-in structures that make up the hippocampal formation is the subicular region (see also Figure 29 and Figure 74). The cortical thickness is transitional, starting from the three-layered hippocampal formation to the six-layered parahippocampal gyrus. (Again, there are a number of subparts of this area, which are rarely studied in an introductory course.)

CONNECTIONS AND FUNCTION

In the temporal lobe, the six-layered parahippocampal gyrus provides extensive input to the adjacent hippocampal formation. The hippocampal formation also receives input from the amygdala. There are extensive interconnections within the component parts of the hippocampal formation itself.

Part of the output of the hippocampal formation is directed back to the parahippocampal gyrus, establishing a strong reciprocal connection. This is analogous to the cortical association pathways described earlier. The parahippocampal gyrus has widespread connections with other cortical areas of the brain, particularly sensory areas.

The other major output of the hippocampal formation is through the **fornix**. Only the hippocampus proper and the subicular region project fibers into the fornix. This tract can be regarded as a subcortical pathway that terminates in the septal region (via the precommissural fibers, discussed with Figure 78B) and in the mammillary nucleus of the hypothalamus (via the post-commissural fibers, discussed with Figure 78A). There are also connections in the fornix from the septal region back to the hippocampal formation. The dentate gyrus only connects with other parts of the hippocampal formation and does not project beyond.

CLINICAL ASPECT

The term medial or mesial temporal sclerosis is a general term for damage to the hippocampal region and adjacent structures located in this part of the brain. Lesions in this area are known to be associated with epilepsy (particularly psychomotor seizures), classified as a partial complex seizure disorder.

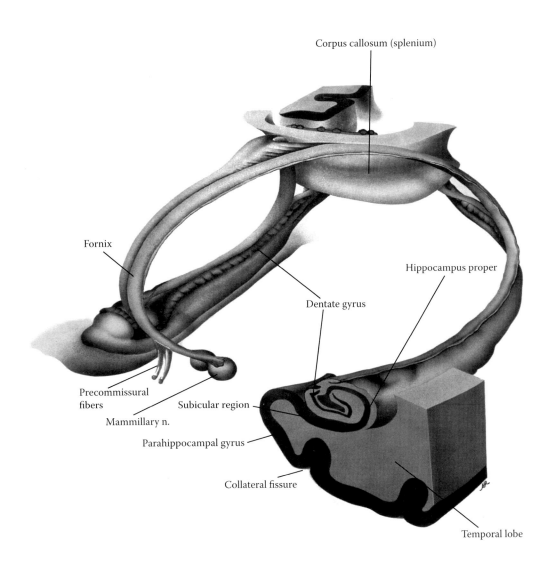

FIGURE 72B: **Hippocampus 2** — Hippocampal Formation (3 parts)

FIGURE 73
"HIPPOCAMPUS" 3

THE HIPPOCAMPAL FORMATION (PHOTOGRAPHIC VIEW)

The brain is being shown from the dorsolateral aspect (as in Figure 14A). The left hemisphere has been dissected by removing the cortex and white matter above the corpus callosum: the lateral ventricle has been exposed from this perspective. The choroid plexus tissue has been removed from the ventricle in order to improve visualization of the structures (see Figure 20A). This dissection also shows the lateral aspect of the lenticular nucleus, the putamen, and the fibers of the internal capsule emerging between it and the thalamus (see Figure OA, Figure OL, Figure 7, Figure 25, and Figure 27).

A similar dissection has been performed in the temporal lobe, thereby exposing the inferior horn of the lateral ventricle (see Figure 20A). A large mass of tissue is found protruding into the inferior horn of this ventricle — named the hippocampus, a visible gross brain structure. In fact, the correct term now used is the hippocampal formation. In a coronal section through this region the protrusion of the hippocampus into the inferior horn of the lateral ventricle also can be seen, almost obliterating the ventricular space (shown in the next illustration; see also Figure 29, Figure 30, Figure 38, and Figure 76).

The hippocampal formation is composed of three distinct regions — the hippocampus proper (Ammon's horn), the dentate gyrus, and the subicular region, as explained in the previous diagram. The fiber bundle that arises from the visible "hippocampus," the fornix, can be seen adjacent to the hippocampus in the temporal lobe (see Figure 70A and Figure 70B), and it continues over the top of the thalamus to the septal region and mammillary nucleus (discussed with the previous illustration).

CLINICAL ASPECT — MEMORY

We now know that the hippocampal formation is one of the critical structures for memory. This function of the hippocampal formation became understood because of an individual known in the literature as H.M., who has been extensively studied by neuropsychologists. H.M. had surgery several decades ago for a valid therapeutic reason — the removal of an epileptic area in the temporal lobe of one side, which was the source of intractable seizures. Most importantly, the surgeons did not know, and could not know according to the methods available at that time, that the contralateral hippocampal area was also severely damaged. This surgery occurred, unfortunately, before the functional contribution of this area to memory formation was known. Since the surgery, H.M. has not been able to form any new memory for events or facts, although he has been taught new motor skills (called procedural memory). (The full story of H.M. and his deficits is found in Kolb and Whishaw — see the Annotated Bibliography.)

We now know that bilateral damage or removal of the anterior temporal lobe structures, including the amygdala and the hippocampal formation, leads to a unique condition in which the person can no longer form new declarative or episodic memories, although older memories are intact. The individual cannot remember what occurred moments before. Therefore, the individual is unable to learn (i.e., to acquire new information) and is not able to function independently. If surgery is to be performed in this region nowadays, special testing is done to ascertain that the side contralateral to the surgery is intact and functioning.

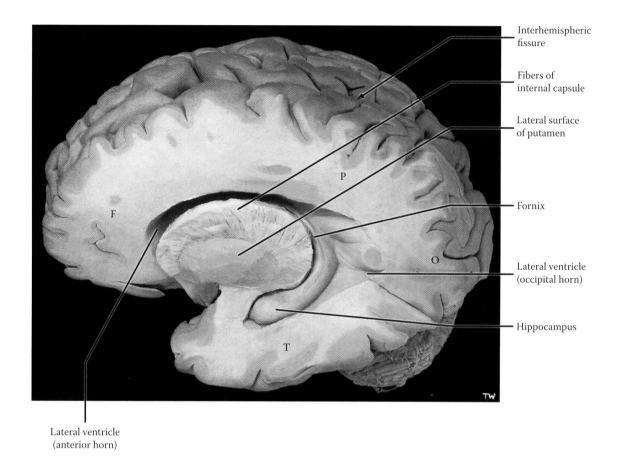

Interhemispheric
fissure

Fibers of
internal capsule

Lateral surface
of putamen

Fornix

Lateral ventricle
(occipital horn)

Hippocampus

Lateral ventricle
(anterior horn)

F = Frontal lobe
P = Parietal lobe
T = Temporal lobe
O = Occipital lobe

FIGURE 73: Hippocampus 3 — The Hippocampus (photograph)

FIGURE 74
"HIPPOCAMPUS" 4

CORONAL BRAIN SECTION (PHOTOGRAPHIC VIEW)

This section is taken posterior to the one shown in Figure 29 and includes the inferior horn of the lateral ventricle (see Figure 20A and Figure 73). The basal ganglia, putamen and globus pallidus, are no longer present (see Figure 22 and Figure 25). The corpus callosum is seen in the depth of the interhemispheric fissure, and at this plane of section the fornix is found just below the corpus calluosum. The lateral ventricles are present, as the body of the ventricle, and choroid plexus is seen on its medial corner (see Figure 20A). The section passes through the midbrain (with the red nucleus and the substantia nigra) and the pons, as shown in the upper right image.

The inferior horn of the lateral ventricles is found in the temporal lobes on both sides and is seen as only a small crescent-shaped cavity (shown also in Figure 38). The inferior horn of the lateral ventricle is reduced to a narrow slit because a mass of tissue protrudes into this part of the ventricle from its medial-inferior aspect. Closer inspection of this tissue reveals that it is gray matter; this gray matter is in fact the hippocampal formation.

LOWER INSERT

This higher magnification of the hippocampal area allows one to follow the gray matter from the hippocampus proper medially and through an intermediate zone, known as the subicular region (as in Figure 72B), until it becomes continuous with the gray matter of the parahippocampal gyrus. The hippocampus proper has only three cortical layers. The subicular region consists of four to five layers; the parahippocampal gyrus is mostly a six-layered cortex. The configuration of the dentate gyrus also can be seen. This view also allows us to understand that the parahippocampal gyrus is named because it lies beside the "hippocampus."

CLINICAL ASPECT

The neurons of the hippocampal area are prone to damage for a variety of reasons, including vascular conditions. The key neurons for the memory function are located in area CA 3 of the hippocampus proper, and these neurons are extremely sensitive to anoxic states. An acute hypoxic event, such as occurs in a cardiac arrest, is thought to trigger a delayed death of these neurons, several days later, termed *apoptosis*, programmed cell death. Much research is now in progress to try to understand this cellular phenomenon and to devise methods to stop this reaction of these neurons.

Currently, studies indicate that in certain forms of dementia, particularly Alzheimer's, there is a loss of neurons in this same region of the hippocampus proper. This loss is due to involvement of these neurons in the disease process. Again, this correlates with the type of memory deficit seen in this condition — loss of short-term memory — although the disease clearly involves other neocortical areas, which goes along with the other cognitive deficits typical for this disease.

ADDITIONAL DETAIL

The relationship of the caudate nucleus with the lateral ventricle is shown in two locations, the body with the body of the ventricle, and the tail in the "roof" of the inferior horn (see Figure 25).

The space between the thalamic areas in this section is not the third ventricle; it is a cistern of the subarachnoid space, outside the brain, because this coronal section has been taken at the posterior tip of the diencephalic region (see Figure 29 and Figure 30). It is located posterior to the pineal and the colliculi, named the quadrigeminal cistern (the four colliculi are also called the quadrigeminal plate, see Figure 10, Figure 21, and Figure 28A). It has extensions or wings laterally called the cisterna ambiens (see Figure 28A). The posterior commissure also is seen.

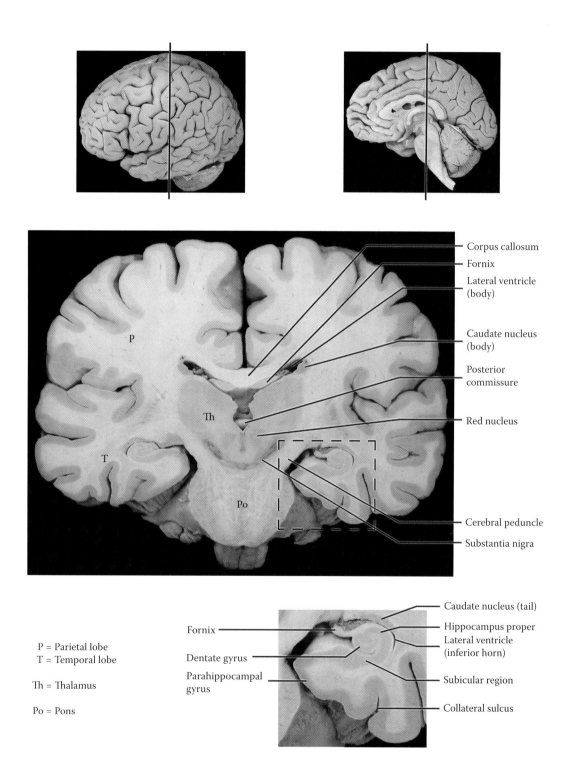

P = Parietal lobe
T = Temporal lobe

Th = Thalamus

Po = Pons

Corpus callosum
Fornix
Lateral ventricle
(body)

Caudate nucleus
(body)

Posterior
commissure

Red nucleus

Cerebral peduncle

Substantia nigra

Caudate nucleus (tail)
Hippocampus proper
Lateral ventricle
(inferior horn)

Subicular region

Collateral sulcus

Fornix

Dentate gyrus

Parahippocampal
gyrus

FIGURE 74: Hippocampus 4 — Coronal View (photograph)

FIGURE 75A
AMYGDALA 1

AMYGDALA — LOCATION AND FUNCTION

This diagram, which is the same as Figure 71, highlights a functional portion of the limbic system — the amygdala and its pathways, the stria terminalis and the ventral amygdalofugal pathway. The septal region and functionally connected portions of the midbrain and medulla are also marked.

The amygdala (amygdaloid nucleus) is a subcortical nuclear structure located in the temporal lobe in humans (see Figure 25 and Figure 29). As a subcortical nucleus of the forebrain, it belongs by definition with the basal ganglia, but its connections are with limbic structures and it is now almost always described with the limbic system.

The amygdala is located between the temporal pole (the most anterior tip of the temporal lobe) and the end of the inferior horn of the lateral ventricle (in the temporal lobe, see Figure OL and Figure 25). The nucleus is located "inside" the uncus, which is seen on the inferior aspect of the brain as a large medial protrusion of the anterior aspect of the temporal lobe (see Figure 15A and Figure 15B).

The amygdala receives input from the olfactory system, as well as from visceral structures. Two fiber tracts are shown connecting the amygdala to other limbic structures, a dorsal one (the stria terminalis) and a ventral one (the ventral amygdalofugal pathway, consisting of two parts). These will be described in detail with the following diagram.

The amygdala in humans is now being shown, using functional MRI imaging, to be the area of the brain that is best correlated with emotional reactions. The emotional aspect of the response of the individual is passed on to the frontal cortex (discussed with the connections in the next illustration), where "decisions" are made regarding possible responses. In this way, the response of the individual incorporates the emotional aspect of the situation.

Stimulation of the amygdaloid nucleus produces a variety of vegetative responses, including licking and chewing movements. Functionally, in animal experimentation, stimulation of the amygdala may produce a rage response, whereas removal of the amygdala (bilaterally) results in docility. Similar responses are also seen with stimulation or lesions in the hypothalamus. Some of these responses may occur through nuclei in the midbrain and medulla.

In monkeys, bilateral removal of the anterior parts of the temporal lobe (including the amygdala) produces a number of behavioral effects which are collectively called the **Kluver-Bucy syndrome**. The monkeys evidently become tamer after the surgery, put everything into their mouths, and display inappropriate sexual behavior.

The amygdala is also known to contain a high amount of enkephalins. It is not clear why this is so and what may be the functional significance.

CLINICAL ASPECT

The amygdala is known to have a low threshold for electrical discharges, which may make it prone to be the focus for the development of seizures. This seems to occur in *kindling*, an experimental model of epilepsy. In humans, epilepsy from this part of the brain (anterior and medial temporal regions) usually gives rise to complex partial seizures, sometimes called temporal lobe seizures, in which oral and licking movements are often seen, along with a loss of conscious activity (see also Figure 72B).

In very rare circumstances, bilateral destruction of the amygdala is recommended in humans for individuals whose violent behavior cannot be controlled by other means. This type of treatment is called psychosurgery.

The role of the amygdala in the formation of memory is not clear. Bilateral removal of the anterior portions of the temporal lobe in humans, for the treatment of severe cases of epilepsy, results in a memory disorder, which has been described with the hippocampal formation (discussed with Figure 73). It is possible that the role of the amygdala in the formation of memories is mediated either through the connections of this nuclear complex with the hippocampus, or with the dorsomedial nucleus of the thalamus.

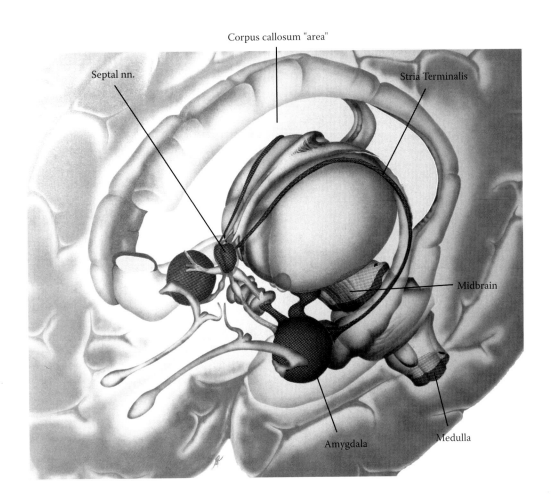

FIGURE 75A: Amygdala 1 — Location

FIGURE 75B
AMYGDALA 2

THE AMYGDALA — CONNECTIONS

One of the major differences between the amygdala and the other parts of the basal ganglia is that the amygdala is not a homogeneous nuclear structure but is composed of different component parts. These are not usually studied in an introductory course.

The amygdala receives a variety of inputs from other parts of the brain, including the adjacent parahippocampal gyrus (not illustrated). It receives olfactory input directly (via the lateral olfactory stria, see Figure 79) and indirectly from the cortex of the uncal region (as shown on the left side of the diagram).

The amygdaloid nuclei are connected to the hypothalamus, thalamus (mainly the dorsomedial nucleus), and the septal region. The connections, which are reciprocal, travel through two routes:

- A dorsal route, known as the **stria terminalis**, which follows the ventricular curve and is found on the upper aspect of the thalamus (see previous illustration). The stria terminalis lies adjacent to the body of the caudate nucleus in this location (see Figure 76). This connects the amygdala with the hypothalamus and the septal region.
- A ventral route, known as the ventral pathway or the **ventral amygdalofugal** pathway. This pathway, which goes through the basal forebrain region (see Figure 80B), connects the amygdala to the hypothalamus (as shown) and to the thalamus (the fibers are shown "en route"), particularly the dorsomedial nucleus (see Figure 63 and Figure 77B).

The connection with the hypothalamus is likely the basis for the similarity of responses seen in animals with stimulation of the amygdala and the hypothalamus (see previous illustration and Figure 78A). This pathway to the hypothalamus may result in hormonal responses, and the connections with the midbrain and medulla may lead to autonomic responses (see Figure 78A).

Further possible connections of the amygdala with other limbic structures and other parts of the brain can occur via the septal region (see Figure 78B), and via the dorsomedial nucleus of the thalamus to the prefrontal cortex (see Figure 77B).

The anterior commissure conveys connections between the nuclei of the two sides.

CLINICAL ASPECT

Seizure activity in the anterior temporal region may spread to the orbitofrontal region, via a particular group of fibers called the uncinate bundle.

ADDITIONAL DETAIL

The association pathway, called the uncinate fasciculus, is a "U-shaped" bundle of fibers between the anterior temporal region and the inferior portion of the frontal lobe. (It is suggested that the learner consult Carpenter — see the Annotated Bibliography — for an illustration of this structure).

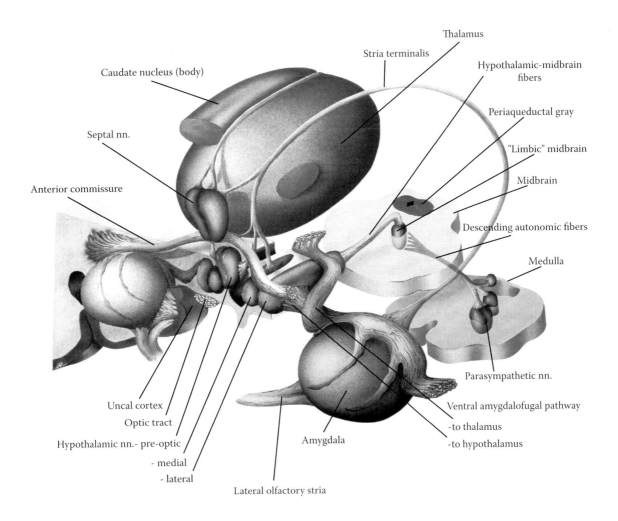

Caudate nucleus (body)

Septal nn.

Anterior commissure

Thalamus

Stria terminalis

Hypothalamic-midbrain
fibers

Periaqueductal gray

"Limbic" midbrain

Midbrain

Descending autonomic fibers

Medulla

Parasympathetic nn.

Ventral amygdalofugal pathway
-to thalamus
-to hypothalamus

Amygdala

Lateral olfactory stria

Uncal cortex
Optic tract
Hypothalamic nn.- pre-optic
- medial
- lateral

FIGURE 75B: Amygdala 2 — Connections

FIGURE 76
LIMBIC "CRESCENT"

LIMBIC STRUCTURES AND THE LATERAL VENTRICLE

The temporal lobe is a more recent addition in the evolution of the hemispheres and develops later in the formation of the brain. During the development of the temporal lobe, a number of structures migrate into it — the lateral ventricle, the hippocampal formation, the caudate nucleus, as well as various tracts, the fornix and stria terminalis.

The lateral ventricle and associated structures form a crescent in the shape of a reverse letter C (see Figure OL and Figure 20A). These relationships are shown in this diagram by showing detailed "cuts" at various points along the lateral ventricle:

- The first section is through the anterior horn of the ventricle, in front of the interventricular foramen (of Monro).
- The following section is through the body of the ventricle, over the dorsal aspect of the thalamus.
- The next section shows the ventricle at its curvature into the temporal lobe, the area called the atrium or the trigone.
- The last section is through the inferior horn of the ventricle, in the temporal lobe, including the hippocampal formation.

Note to the Learner: The initials used in these sections to identify structures are found in brackets after the labeled structure in the main part of the diagram.

- **Caudate Nucleus** (see Figure OL, Figure 23, Figure 24, and Figure 25):
  The various parts of the caudate nucleus, the head, the body, and the tail, follow the inner curvature of the lateral ventricle. The large head is found in relation to the anterior horn of the lateral ventricle, where it bulges into the space of the ventricle (see Figure 27 and Figure 28A). The body of the caudate nucleus is coincident with the body of the lateral ventricle, on its lateral aspect (see Figure 29, Figure 30, and Figure 74). As the caudate follows the ventricle into the temporal lobe, it becomes the tail of the caudate nucleus, where it is found on the upper aspect of the inferior horn, its roof (see Figure 38).

- **Hippocampal formation** (see Figure 72A, Figure 72B, Figure 73, and Figure 74):
  The hippocampal formation is found in the temporal lobe situated medial and inferior to the ventricle. It bulges into the ventricle, almost obliterating the space; it is often difficult to visualize the small crevice of the ventricle in specimens and radiograms. The dentate gyrus is again seen (on the far side) with its indented surface (see also Figure 72B). The configuration of the three parts of the hippocampal formation is shown in the lower inset (see Figure 74).

- **Fornix**:
  The fornix is easily found in studies of the gross brain (e.g., see Figure 17 and Figure 41B). Its fibers can be seen as a continuation of the hippocampal formation (see Figure 72B and Figure 73), and these fibers course on the inner aspect of the ventricle as they sweep forward above the thalamus. In the area above the thalamus and below the corpus callosum (see coronal section, Figure 29 and Figure 30), the fornix is found at the lower edge of the septum pellucidum. In this location, the fornix of one side is adjacent to that of the other side (see also Figure 71); there are some interconnections between the two sides in this area.

  The fibers of the fornix pass in front of the interventricular foramen (see medial view of brain in Figure 17). It then divides into pre-commissural fibers to the septal region (see Figure 78B), and post-commissural fibers, through the hypothalamus, to the mammillary nucleus (which is not portrayed in this diagram, see Figure 72B and Figure 78B).

- **Amygdala** (see Figure 25 and Figure 75A):
  The amygdala is clearly situated anterior to the inferior horn of the lateral ventricle and in front of the hippocampal formation.

- **Stria Terminalis**:
  The stria terminalis follows essentially the same course as the fornix (see Figure 71), connecting the amygdala with the septal region and hypothalamus (see Figure 78B).

ADDITIONAL DETAIL

In the temporal lobe, the stria is found in the roof of the inferior horn of the lateral ventricle. The stria terminalis is found slightly more medially than the fornix on the dorsal aspect of the thalamus, in the floor of the body of the lateral ventricle.

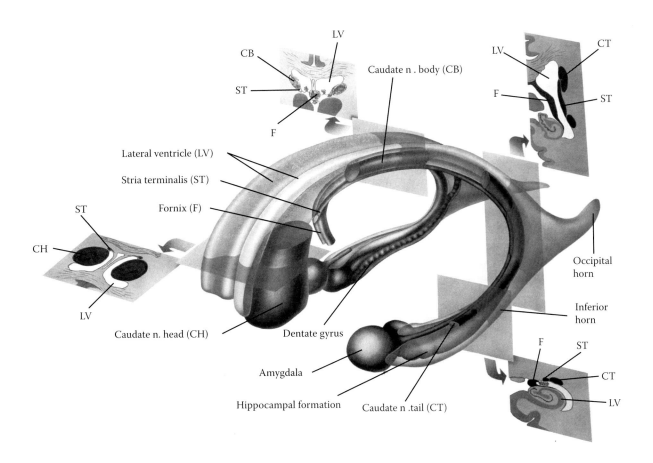

FIGURE 76: Limbic Structures and Lateral Ventricle

FIGURE 77A
LIMBIC DIENCEPHALON 1

ANTERIOR NUCLEUS

This detailed diagram shows one of the major connections of the limbic system via the thalamus. This diagram shows an enlarged view of the thalamus of one side (see Figure 11 and Figure 12), the head of the caudate nucleus, as well as a small portion of the cingulate gyrus (see Figure 17). Immediately below is the hypothalamus, with only the two mammillary nuclei being shown (see Figure 71).

ANTERIOR NUCLEUS — CINGULATE
GYRUS

The fibers of the fornix (carrying information from the hippocampal formation) have been followed to the mammillary nuclei (as the post-commissural fibers, see Figure 72B). A major tract leaves the mammillary nuclei, the **mammillo-thalamic tract**, and its fibers are headed for a group of association nuclei of the thalamus called the **anterior nuclei** (see Figure 12 and Figure 63). (**Note to the Learner**: The learner is advised to refer to the classification of the thalamic nuclei, see Figure 12 and also Figure 63.)

Axons leave the anterior nuclei of the thalamus and course through the anterior limb of the internal capsule (see Figure 26). These fibers course between the caudate nucleus (head and body) and the lentiform nucleus (which is just visible in the background). The axons terminate in the cortex of the cingulate gyrus after passing through the corpus callosum (see Figure 17). The continuation of this circuit is discussed below.

PAPEZ CIRCUIT

About 60 years ago, James Papez described a pathway involving some limbic and cortical structures and associated pathways. These, he postulated, formed the anatomical substrate for emotional experiences. The pathway forms a series of connections, which has since been called the Papez circuit. We have continued to learn about many other pathways and structures involved in processing "emotion," but this marked the beginning of the unfolding of our understanding.

To review, fibers leave the hippocampal formation and proceed through the fornix, and some of these fibers have been shown to terminate in the mammillary nuclei of the hypothalamus. From here, a new pathway, the mammillo-thalamic tract, ascends to the anterior group of thalamic nuclei. This group of nuclei projects to the cingulate gyrus (see Figure 63).

From the cingulate gyrus, there is an association bundle, the cingulum, which connects the cingulate gyrus (reciprocally) with the parahippocampal gyrus as part of the limbic lobe (refer to Figure 70A and Figure 70B). The parahippocampal gyrus projects to the hippocampal formation, which processes the information and sends it via the fornix to the mammillary nuclei of the hypothalamus (and the septal region). Hence, the circuit is formed.

We now have a broader view of the limbic system, and the precise functional role of the Papez circuit is not completely understood. It should be realized that although there is a circuitry that forms a loop, the various structures have connections with other parts of the limbic system and other areas of the brain, and thus can influence other neuronal functions (to be discussed with the limbic system synthesis at the end of this section).

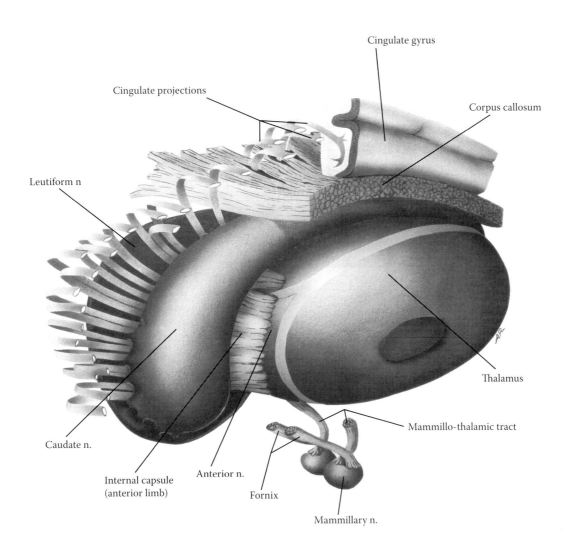

FIGURE 77A: Limbic Diencephalon 1 — Anterior Nucleus

FIGURE 77B
LIMBIC DIENCEPHALON 2

DORSOMEDIAL NUCLEUS

The thalamus of both sides is shown in this diagram, focusing on the medial nuclear mass of the thalamus, the dorsomedial nucleus, one of the most important of the association nuclei of the thalamus (see Figure 11 and Figure 12).

Shown below is the amygdala with one if its pathways, the ventral amygdalofugal fibers, projecting to the dorsomedial nucleus (see Figure 75A and Figure 75B). This pathway brings "emotional" information to the thalamus. The dorsomedial nucleus collects information from a variety of sources, including other thalamic nuclei, as well as from various hypothalamic nuclei (see Figure 63).

The dorsomedial nucleus projects heavily to the frontal lobe, particularly to the cortical area that has been called the prefrontal cortex (see Figure 14A). The projection thus includes the emotional component of the experience. This pathway passes through the anterior limb of the internal capsule, between the head of the caudate nucleus and the lentiform nucleus (see Figure 26). The fibers course in the white matter of the frontal lobes.

Our expanded view of the limbic system now includes its extension to this prefrontal cortex, specifically the orbital and medial portions of the frontal lobe; this has been called the **limbic forebrain**. Widespread areas of the limbic system and association cortex of the frontal lobe, particularly the medial and orbital portions, are involved with human reactions to pain, particularly to chronic pain, as well as the human experiences of grief and reactions to the tragedies of life.

CLINICAL ASPECT — PSYCHOSURGERY

The projection of the dorsomedial nucleus to the prefrontal cortex has been implicated as the key pathway that is interrupted in a now-banned surgical procedure. Before the era of medication for psychiatric disorders, when up to one-half of state institutions were filled with patients with mental illness, a psychosurgical procedure was attempted to help alleviate the distressing symptoms of these diseases.

The procedure involved the introduction of a blunt instrument into the frontal lobes, passing the instrument (bilaterally) through the orbital bone (which is a very thin plate of bone) above the eye. This interrupts the fibers projecting through the white matter, presumably including the projection from the dorsomedial nucleus. This operation became known as a *frontal lobotomy.*

Long-term studies of individuals who have had frontal lobotomies have shown profound personality changes in these individuals. These people become emotionally "flat" and lose some hard-to-define human quality in their interpersonal interactions. In addition, such an individual may perform socially inappropriate acts that are not in keeping with the personality of that individual prior to the surgery.

Once the long term effects of this surgery became clear, and since powerful and selective drugs became widely available for various psychiatric conditions, this surgery was abandoned in the 1960s and is not performed nowadays.

This same procedure had also been recommended for the treatment of pain in terminal cancer patients, as part of the palliative care of an individual. After the surgery, the individual is said to still have the pain but no longer "suffers" from it, that is, the psychic aspect of the pain has been removed. There may even be a reduced demand for pain medication such as morphine. Again, other approaches to pain management are now used.

PHINEAS GAGE

Phineas Gage has become a legendary figure in the annals of the history of the brain. In brief, Gage was working on the construction of a railway in the 1800s, when an untimely explosion drove a steel peg through his brain. The steel peg is said to have penetrated the orbit and the frontal lobes, much like the surgical procedure described above, emerging through the skull. He survived and lived on; his personality changes, which have been well documented, subsequent to this accident concur with those described following a frontal lobotomy. The story of Phineas showing a reconstruction of his injury and describing the changes in his personality can be found in Kolb and Whishaw (see the Annotateed Bibliography).

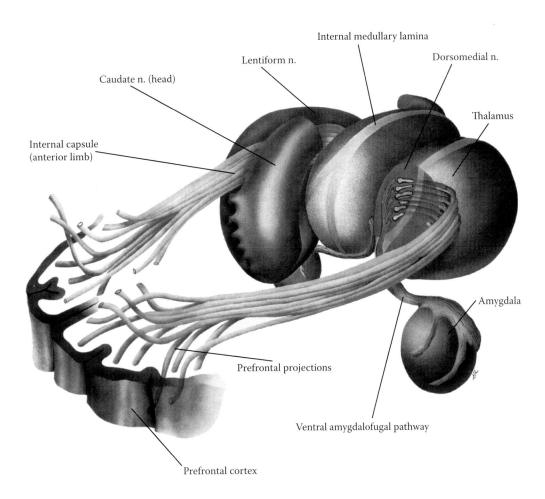

Internal medullary lamina

Lentiform n.

Dorsomedial n.

Caudate n. (head)

Thalamus

Internal capsule
(anterior limb)

Amygdala

Prefrontal projections

Ventral amygdalofugal pathway

Prefrontal cortex

FIGURE 77B: Limbic Diencephalon 2 — Dorsomedial Nucleus

FIGURE 78A
HYPOTHALAMUS

THE NEURAL HYPOTHALAMUS

This diagram, which is the same as Figure 71, highlights the hypothalamus, one of the core structures of the limbic system, with the prominent mammillary nuclei as part of the hypothalamus. The third ventricle is situated between the two diencephalic parts of the brain, (e.g. see Figure 9A and Figure 27) and the hypothalamic tissue of both sides joins together at its inferior portion as the median eminence (see next illustration and Figure 15A and Figure 15B).

The hypothalamus is usually divided into a medial and lateral group of nuclei (see next illustration), and pre-optic nuclei (see Figure 75B). A number of nuclei that control the anterior pituitary gland are located in the medial group. This occurs via the median eminence and the portal system of veins along the pituitary stalk; other nuclei in the supraoptic region (above the optic chiasm) connect directly with the posterior pituitary via the pituitary stalk (see Figure 15A and Figure 15B).

Some of the major inputs to the hypothalamus come from limbic structures, including the amygdala (via the stria terminalis and the ventral pathway, see Figure 75A and Figure 75B) and the hippocampal formation (via the fornix, see Figure 72B). Stimulation of particular small areas of the hypothalamus can lead to a variety of behaviors (e.g., sham rage), similar to that which occurs following stimulation of the amygdala.

Certain basic drives (as these are known in the field of psychology), such as hunger (feeding), thirst (drinking), sex (fornication), and body temperature, are regulated through limbic structures. Many of the receptor mechanisms for these functions are now known to be located in highly specialized hypothalamic neurons. The hypothalamus responds in two ways — as a neuroendocrine structure controlling the activities of the pituitary gland and as a neural structure linked to the limbic system.

In its neural role there are small areas of the hypothalamus that act as the "head ganglion" of the autonomic nervous system, influencing both sympathetic and parasympathetic activities. The response to hunger or thirst or a cold environment usually leads to a complex series of motor activities that are almost automatic, as well as autonomic adjustments and hormonal changes. In addition, in humans, there is an internal state of discomfort to being cold, or hungry, or thirsty, which we call an emotional response. Additional connections are required for the behavioral (motor) activities, and the accompanying psychological reaction requires the forebrain, as well as the limbic cortical areas (to be discussed with the limbic system synthesis at the end of this section).

The mammillary nuclei are of special importance as part of the limbic system. They receive a direct input from the hippocampal formation via the fornix (see Figure 72B) and give rise to the mammillo-thalamic tract to the thalamic anterior group of nuclei as part of the Papez circuit (discussed with Figure 77A). In addition, there are fibers that connect directly to the limbic midbrain (shown in the next illustration).

Running through the lateral mass of the hypothalamus is a prominent fiber tract, the medial forebrain bundle, which interconnects the hypothalamus with two areas, the septal region of the forebrain and certain midbrain nuclei associated with the limbic system, the "limbic midbrain" (both to be discussed with the next illustration). Other fiber bundles connect the hypothalamus with the "limbic midbrain." There are also some indirect connections to nuclei of the medulla via descending autonomic fibers. Both parts of the brainstem are therefore "highlighted" in this illustration.

ADDITIONAL DETAIL

The Habenula (not illustrated)

The habenular nuclei are a group of small nuclei situated at the posterior end of the thalamus on its upper surface (see Figure 11). The pineal gland is attached in this region (see Figure 9A).

There is another circuit whereby septal influences are conveyed to the midbrain. The first part of the pathway is the stria medullaris (note the possible confusion of terminology), which connects the septal nuclei (region) with the habenular nuclei. The stria medullaris is found on the medial surface of the thalamus. From the habenular nuclei, the habenulo-interpeduncular tract descends to the midbrain reticular formation, mainly to the interpeduncular nucleus located between the cerebral peduncles (see midbrain cross-section, Figure 65B). (This tract is also called the fasciculus retroflexus.)

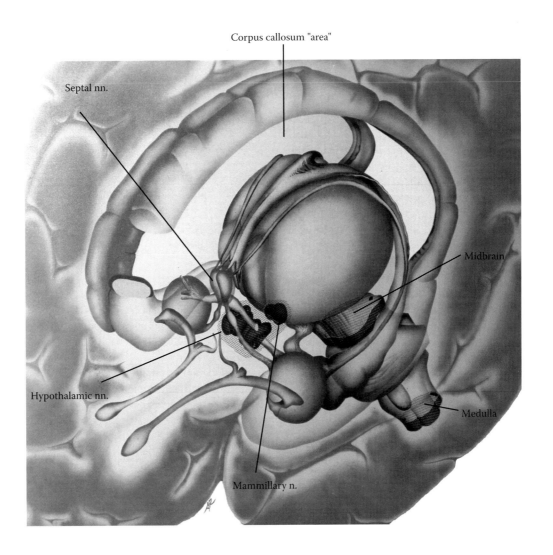

FIGURE 78A: Hypothalamus

FIGURE 78B
MEDIAL FOREBRAIN BUNDLE

SEPTAL REGION AND LIMBIC MIDBRAIN

This illustration provides detailed information about other important parts of the limbic system, the septal region and the limbic midbrain. The pathway that interconnects the hypothalamus and these areas is the **medial forebrain bundle**.

THE SEPTAL REGION

The septal region includes both cortical and subcortical areas that belong to the forebrain. The cortical areas, named the septal cortex, are found under the rostrum of the corpus callosum (the thin "inferior" portion of the corpus callosum, see Figure 17 and Figure 70A). Nuclei lying deep to this region are called the septal nuclei and in some species (not humans) are located within the septum pellucidum (the septum that separates the anterior horns of the lateral ventricles, see Figure 17 and Figure 30). In this atlas, both areas are included in the term septal region.

The septal region receives input from the hippocampal formation (via the precommissural fibers of the fornix, see Figure 72B) and from the amygdala (via the stria terminalis, see Figure 75B). The major connection of the septal region with the hypothalamus and the limbic midbrain occurs via the medial forebrain bundle. (Refer also to the Additional Detail with the previous illustration.)

Several decades ago, experiments were done in rats with a small electrode implanted in the septal region; pressing of the bar completed an electrical circuit that resulted in a tiny (harmless) electric current going through this area of brain tissue. It was shown that rats will quickly learn to press a bar to deliver a small electric current to the septal region. In fact, the animals will continue pressing the bar virtually nonstop, even in preference to food. From this result it has been inferred that the animals derive some type of "pleasant sensation" from stimulation of this region, and it was named the "pleasure center"; it has since been shown that there are other areas where a similar behavior can be produced. However, this type of positive

effect is not seen in all parts of the brain, and in some areas an opposite (negative) reaction may be seen.

THE LIMBIC MIDBRAIN

A number of limbic pathways terminate within the reticular formation of the midbrain, including the periaqueductal gray, leading to the notion that these areas are to be incorporated in the structures that comprise the extended limbic system (discussed in the Introduction to this section). This has led to the use of the term limbic midbrain.

The two major limbic pathways, the medial forebrain bundle and a descending tract from the mammillary nuclei (the mammillo-tegmental tract), terminate in the midbrain reticular formation. From here, there are apparently descending pathways that convey the "commands" to the parasympathetic and other nuclei of the pons and medulla (e.g., the dorsal motor nucleus of the vagus, the facial nucleus for emotional facial responses), and areas of the reticular formation of the medulla concerned with cardiovascular and respiratory control mechanisms (discussed with Figure 42A and Figure 42B). Other connections are certainly made with autonomic neurons in the spinal cord (i.e., for sympathetic-type responses).

MEDIAL FOREBRAIN BUNDLE

Knowledge of this bundle of fibers is necessary if one is to understand the circuitry of the limbic system and how the limbic system influences the activity of the nervous system.

The medial forebrain bundle (MFB) connects the septal region with the hypothalamus and extends into the limbic midbrain; it is a two-way pathway. Part of its course is through the lateral part of the hypothalamus where the fibers become somewhat dispersed (as illustrated). There are further connections to nuclei in the medulla. It is relatively easy to understand how the septal region and the hypothalamus can influence autonomic activity and the behavior of the animal.

ADDITIONAL DETAIL

There are other pathways from the hypothalamus to the limbic midbrain, such as the dorsal longitudinal bundle.

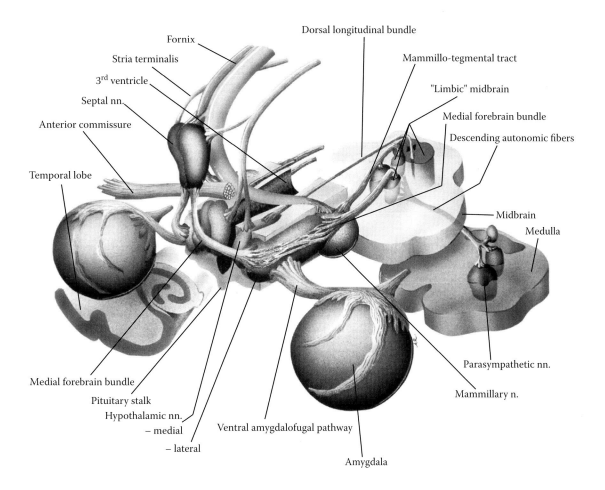

Fornix

Stria terminalis

3rd ventricle

Septal nn.

Anterior commissure

Temporal lobe

Dorsal longitudinal bundle

Mammillo-tegmental tract

"Limbic" midbrain

Medial forebrain bundle

Descending autonomic fibers

Midbrain

Medulla

Medial forebrain bundle

Pituitary stalk

Hypothalamic nn.

– medial

– lateral

Ventral amygdalofugal pathway

Amygdala

Parasympathetic nn.

Mammillary n.

FIGURE 78B: **Medial Forebrain Bundle** — Septal Region and Limbic Midbrain

FIGURE 79
OLFACTORY SYSTEM

SENSE OF SMELL

The olfactory system, our sense of smell, is a sensory system that inputs directly into the limbic system and does not have a thalamic nucleus (see Figure 12 and Figure 63).

The olfactory system is a phylogenetically older sensory system. Its size depends somewhat on the species, being larger in animals that have a highly developed sense of smell; this is not the case in humans in whom the olfactory system is small. Its component parts are the olfactory nerve, bulb, and tract, and various areas where the primary olfactory fibers terminate, including the amygdala and the cortex over the uncal region.

OLFACTORY NERVE, BULB, AND TRACT

The sensory cells in the nasal mucosa project their axons into the CNS. These tiny fibers, which constitute the actual peripheral olfactory nerve (CN I), pierce the bony (cribriform) plate in the roof of the nose and terminate in the olfactory bulb, which is a part of the CNS. There is a complex series of interactions in the olfactory bulb, and one cell type then projects its axon into the olfactory tract, a CNS pathway.

The olfactory tract runs posteriorly along the inferior surface of the frontal lobe (see Figure 15A and Figure 15B) and divides into lateral and medial tracts, called stria. At this dividing point there are a number of small holes for the entry of several blood vessels to the interior of the brain, the striate arteries (see Figure 62 and Figure 80B); this triangular area is known as the anterior perforated space or area.

It is best to remember only the lateral tract as the principal tract of the olfactory system. It is said to have cortical tissue along its course for the termination of some olfactory fibers. The lateral tract ends in the cortex of the uncal area (see Figure 15A and Figure 15B), with some of the fibers terminating in an adjacent part of the amygdaloid nucleus (see Figure 75A and Figure 75B). It is important to note that the olfactory system terminates directly in primary olfactory areas of the cortex without a thalamic relay.

OLFACTORY CONNECTIONS

The connections of the olfactory system involve the limbic cortex, called the secondary olfactory areas. These include the cortex in the anterior portion of the parahippocampal gyrus, an area that has been referred to as the entorhinal cortex. (The term rhinencephalon refers to the olfactory parts of the CNS, the "smell brain.") This input of olfactory information into the limbic system makes sense if one remembers that one of the functions of the limbic system is procreation of the species. Smell is important in many species for mating behavior and for identification of the nest and territory.

Olfactory influences may spread to other parts of the limbic system, including the amygdala and the septal region. Through these various connections, information may reach the dorsomedial nucleus of the thalamus.

Smell is an interesting sensory system. We have all had the experience of a particular smell evoking a flood of memories, often associated with strong emotional overtones. This simply demonstrates the extensive connections that the olfactory system has with components of the limbic system and, therefore, with other parts of the brain.

CLINICAL ASPECT

One form of epilepsy often has a significant olfactory aura (which precedes the seizure itself). In such cases, the "trigger" area is often orbitofrontal cortex. This particular form of epilepsy has unfortunately been called "uncinate fits." The name is derived from a significant association bundle, the uncinate bundle (an association bundle), which interconnects this part of the frontal lobe and the anterior parts of the temporal lobe where olfactory connections are located (see also Figure 75B).

ADDITIONAL DETAIL

Diagonal Band

This obscure fiber bundle and nuclei associated with it are additional olfactory connections, some of which interconnect the amygdala with the septal region (see Figure 80B).

Corpus Callsum "area"

Diagonal
Band

Olfactory
bulb

Olfactory
tract

Lateral
Olfactory
stria

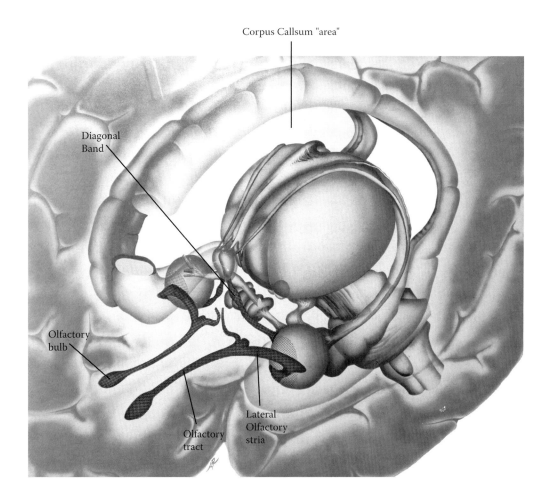

FIGURE 79: Olfactory System

FIGURE 80A
BASAL FOREBRAIN 1

BASAL FOREBRAIN REGION

The basal forebrain is shown using the same diagram of the limbic system (Figure 71). This area, previously called the substantia innominata, contains a variety of neurons.

This area is located below the anterior commissure and lateral to the hypothalamus. On the gross brain, this region can be found by viewing the inferior surface of the brain where the olfactory tract ends and divides into medial and lateral stria (see Figure 15A and Figure 15B). This particular spot is the location where a number of blood vessels, the striate arteries, penetrate the brain substance and is called the anterior perforated space (shown in the next illustration). The basal forebrain region is found "above" this area.

The basal forebrain contains a group of diverse structures:

- Clusters of large cells that are cholinergic, and which have been collectively called the basal nucleus (of Meynert)
- The ventral portions of the putamen and globus pallidus, namely, the ventral striatum and ventral pallidum
- The nucleus accumbens, which may include a number of diverse neurons within its boundaries
- Groups of cells that are continuous with the amygdala, now called the extended amygdala

CHOLINERGIC BASAL NUCLEUS

These rather large neurons are found in clusters throughout this region. These cells project to widespread areas of the prefrontal cortex, providing that cortical area with cholinergic innervation.

THE EXTENDED AMYGDALA

A group of cells extends medially from the amygdaloid nucleus and follows the ventral pathway (the ventral amygdalofugal pathway, Figure 75B and Figure 77B) through this basal forebrain region. These neurons receive a variety of inputs from the limbic cortical areas and from other parts of the amygdala. Its output projects to the hypothalamus and to autonomic-related areas of the brainstem, thereby influencing neuroendocrine, autonomic, and, perhaps, somatomotor activities.

CLINICAL ASPECT

Dementia is a general term for an acquired progressive decline of cognitive function whose hallmark is a loss of short-term memory. It is an age-related disease where the clinical manifestations become evident in older individuals; with the increase in lifespan in the industrialized world, there is an increase in the number of individuals afflicted with this disease. These people eventually require more and more care, often necessitating institutional placement. **Alzheimer's** dementia is the most prevalent clinical syndrome, accompanied by certain neuropathological changes in the brain.

Several years ago, it was reported that there was a depletion of acetylcholine in the frontal lobe areas in Alzheimer patients. Subsequent reports indicated that this was accompanied by a loss of these cholinergic cells in the basal forebrain. Many thought that the "cause" of Alzheimer's disease had been uncovered, that is, a cellular degeneration of a unique group of cells and a neurotransmitter deficit. (The model for this way of thinking is Parkinson's disease.) This was followed immediately by several therapeutic trials using medication to boost the acetylcholine levels of the brain.

It is currently thought that cortical degeneration is the primary event in Alzheimer's dementia, starting often in the parietal areas of the brain. We now know that several other neurotransmitters are depleted in the cortex in Alzheimer's disease. This information would lead us to postulate that the loss of the target neurons in the prefrontal cortex, the site of termination for the cholinergic neurons, would be followed, or accompanied, by the degeneration of the cholinergic cells of the basal forebrain. In addition, there is the hippocampal degeneration that goes along with the memory loss (discussed with Figure 74).

Notwithstanding this current state of our knowledge, therapeutic intervention to boost the cholinergic levels of the brain is currently considered a valid therapeutic approach, particularly in the early stages of this tragic human disease. New drugs that maintain or boost the level of acetylcholine in the brain are currently undergoing evaluation. The reports have shown some improvement, or at least a stabilization of the decline, in both memory and cognitive functions for a period of weeks or months.

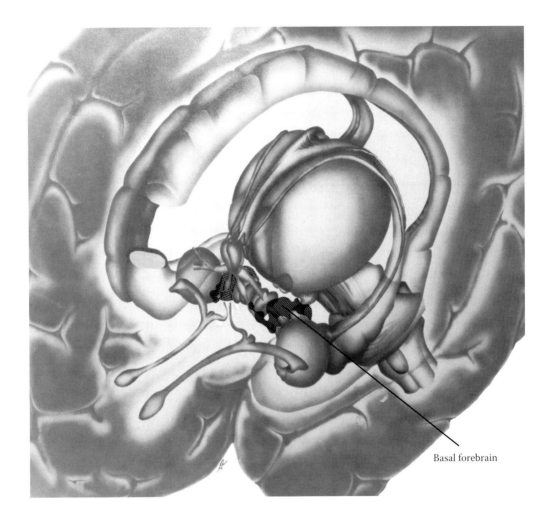

Basal forebrain

FIGURE 80A: Basal Forebrain 1 — Basal Nucleus

FIGURE 80B
BASAL FOREBRAIN 2

BASAL GANGLIA

This is a somewhat schematic view of the various "nuclei" located in the basal forebrain area. The hypothalamus is shown in the midline, with the third ventricle. The penetrating striate arteries are seen in the anterior perforated area (see Figure 62). This view shows the ventral pathway emerging from the amygdala and some of the fibers going to the hypothalamus, and the others on their way to the dorsomedial nucleus of the thalamus (see Figure 75B and Figure 77B). The anterior commissure demarcates the upper boundary of this area (see Figure 70A). The cell clusters that form the basal (cholinergic) nucleus are contained within this area but are not portrayed.

THE VENTRAL STRIATUM AND PALLIDUM

The lowermost portions of the putamen and globus pallidus are found in the basal forebrain area; here they are referred to as the ventral striatum and ventral pallidum (see Figure 29).

The ventral part of the striatum (the putamen) receives input from limbic cortical areas, as well as a dopaminergic pathway from a group of dopamine-containing cells in the midbrain. The information is then relayed to the ventral pallidum (both parts of the globus pallidus are seen on the left side of the diagram). This area has a significant projection to the dorsomedial nucleus of the thalamus (and, hence, to the prefrontal cortex).

The overall organization is therefore quite similar to that of the dorsal parts of the basal ganglia, although the sites of relay and termination are different. Just as the amygdala is now considered a limbic nucleus, many now argue that the ventral striatum and pallidum should be included with the limbic system.

THE NUCLEUS ACCUMBENS

This nucleus (see also Figure 24) is composed of various groups of neurons, some that are part of the basal ganglia and others, possibly limbic neurons. It has many of the connections of the ventral striatum as well as those of the extended amygdala. Functionally, this neural area becomes activated in situations that involve reward and punishment, integrating certain cognitive aspects of the situation with the emotional component. There is strong evidence that this area is involved in addiction behavior in animals and likely in humans.

In summary, the region of the basal forebrain has important links with other parts of the limbic system. There is a major output to the prefrontal cortex, via the dorsomedial nucleus of the thalamus, which is considered by some to be the forebrain component of the limbic system. The basal forebrain is thus thought to have a strong influence on "drives" and emotions, as well as higher cognitive functions that have an emotional component. The cholinergic neurons in this area may have a critical role in memory.

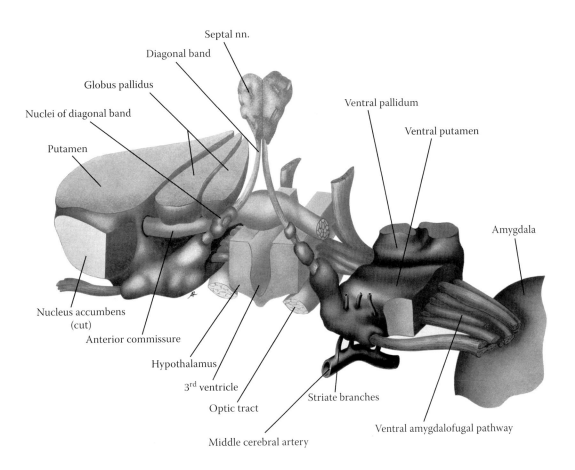

Septal nn.

Diagonal band

Globus pallidus

Nuclei of diagonal band

Putamen

Ventral pallidum

Ventral putamen

Amygdala

Nucleus accumbens
(cut)

Anterior commissure

Hypothalamus

3rd ventricle

Optic tract

Striate branches

Middle cerebral artery

Ventral amygdalofugal pathway

FIGURE 80B: Basal Forebrain 2 — Basal Ganglia

LIMBIC SYSTEM: SYNTHESIS

After studying the structures and connections of the limbic system in some detail a synthesis of the anatomical information with the notion of an "emotional" part of the brain seems appropriate. It is not easy to understand how the limbic system is responsible for the reactions required by the definition of "emotion" proposed in the Introduction to this section.

The "key" structures of the limbic system are the limbic lobe (the cortical regions, including the hippocampal formation and the parahippocampal and cingulate gyri), the amygdala, the hypothalamus, and the septal region. The limbic pathways interconnect these limbic areas (e.g., the Papez circuit). In many ways it seems that the limbic structures communicate only with each other. What is not clear is how activity in these structures influences the rest of the brain. How does the limbic system influence changes in the physiological systems (endocrine and autonomic), motor activity (behavior), and the mental state (psychological reactions)?

The following discussion is presented as a way of understanding the outcome or output of limbic function — the categories of responses are the same as those discussed in the Introduction to this section.

Physiological Responses:

- Hormonal and "homeostatic" responses: Hormonal changes, as regulated by the hypothalamus, are part of the physiological responses to emotional states, both acute and chronic. The work of Dr. Hans Selye, for example, has shown how chronic stress influences our body and may lead to structural damage to select areas of the brain (e.g., the hippocampal formation).

- Autonomic responses: A wide number of parasympathetic and sympathetic responses accompany emotional states, including the diameter of the pupil (in states of fear), salivation, respiration, blood pressure, pulse, and various gastrointestinal functions. These are controlled in part by the hypothalamus and by the limbic connections in the midbrain and medulla.

Behavioral Responses:

The physiologic adjustments often involve complex motor actions. Consider, for example, the motor activities associated with thirst, temperature regulation, and satisfying other basic drives. The amygdala and hypothalamus are likely involved in the motor patterns associated with these basic drives.

Limbic activity involves areas of the midbrain reticular formation and other brainstem nuclei in specific ways. The best examples are perhaps the facial expressions associated with emotions, the responses to pain that are generated in part in the brainstem, and the basic "fight or flight" response to emergency situations. All of these activate a considerable number of motor circuits. The ventral parts of the basal ganglia and various cortical areas are likely the areas of the CNS involved with the motor activities associated with emotional reactions.

Psychological Reactions:

Neocortical areas that are involved in limbic function include portions of the prefrontal cortex, the cingulate gyrus, and the parahippocampal gyrus. Activities in these limbic cortices (and the associated thalamic nuclei) are clearly candidates for the psychological (mental) reactions of emotion. These enter consciousness and become part of the substrate that is used by humans in decision making.

In summary, the limbic system has many connections outside itself through which it influences the hormonal, autonomic, motor, and psychological functions of the brain.

The older cortical regions of the hippocampal formation seem to have an additional function related to the formation of new episodic memories, specifically related to events and factual information. Why this is so and how this evolved is a matter of speculation.

The limbic system is intricate and intriguing, providing a window into human behavior beyond our sensory and motor activities. It is not always clear what each part contributes to the overall functional system. In addition, the pathways are obscure and, perhaps, confusing. Nevertheless, they are part of the neuroanatomical framework for a discussion of the contribution of the limbic system to the function of the human organism.

On a final note, one can only wish that the basic activities of the limbic system that are involved in preservation of the self and species can be controlled and tamed by higher-order cortical influences, leading humankind to a more human and hopefully a more humane future.

ANNOTATED BIBLIOGRAPHY

This is a select list of references with some commentary to help the learner choose additional learning resources about the structure, function, and diseases of the human brain.

The perspective is for medical students and practitioners not involved with neurology, as well as those in related fields in the allied health professions. The listing includes texts, atlases, and videotapes, as well as Web sites and CD-ROMs.

TEXTS AND ATLASES

This listing includes neuroanatomical textbooks and atlases, as well as clinical texts; recent publications (since 2000) have been preferentially selected.

NEUROANATOMICAL TEXTS

Afifi, A.K. and Bergman, R.A., *Functional Neuroanatomy Text and Atlas*, 2nd ed., Lange Medical Books, McGraw-Hill, New York, 2005.

This is a neuroanatomical text with the addition of functional information on clinical syndromes. A chapter on the normal is followed by a chapter on clinical syndromes (e.g., of the cerebellum). The book is richly illustrated (in two colors) using semi-anatomic diagrams and MRIs. Each chapter has key points at the beginning and terminology for that chapter at the end. It is a pleasant book visually and quite readable. There is an atlas of the CNS at the end, but it's not in color, and also several brain MRIs.

Arslan, O., *Neuroanatomical Basis of Clinical Neurology*, Parthenon Publishing, New York and London, 2001.

A traditional neuroanatomical textbook with many references to clinical disease entities (set in blue boxes). The text is nicely formatted, and there are many illustrations, photographs, histological sections, and diagrams (in two colors).

Carpenter, M.B., *Core Text of Neuroanatomy*, 4th ed., Williams and Wilkins, Baltimore, 1991.

This "classic" textbook by a highly respected author presents a detailed description of the nervous system, from the perspective of a neuroanatomist. A more complete version is also available as a reference text — Carpenter's Human Neuroanatomy, (1995), now with A. Parent as the author.

Fitzgerald, M.J.T. and Folan-Curran, J., *Clinical Neuroanatomy and Related Neuroscience*, 4th ed., Saunders, Philadelphia, 2002.

The authors have attempted to create an integrated text for medical and allied health professionals, combining the basic neuroscience with clinical entities. The book is richly illustrated, in full color, with large appealing explanatory diagrams and some MRIs, but there are few actual photographs. The clinical syndromes are in boxes accompanied by illustrations. A glossary has been added.

Haines, D.E., *Fundamental Neuroscience,* 2nd ed., Churchill Livingstone, Philadelphia, 2002.

This edited large text, with many color illustrations, is an excellent reference book, mainly for neuroanatomical detail.

Kandel, E.R., Schwartz, J.H., and Jessell, T.M., *Principles of Neural Science*, 4th ed., McGraw-Hill, New York, 2000.

This thorough textbook presents a physiological depiction of the nervous system, with experimental details and information from animal studies. It is suitable as a reference book and for graduate students.

Kiernan, J.A., *Barr's The Human Nervous System: An Anatomical Viewpoint*, 8th ed., Lippincott, Williams & Wilkins, Baltimore, 2005.

This new edition of Barr's book is a neuroanatomical textbook, now with added color, as well as clinical notes (in boxes) and MRIs. It is clearly written, clearly presented, and includes a glossary. There is an accompanying CD-ROM with questions and expanded versions of certain chapters.

Kolb, B. and Whishaw, I.Q., *Fundamentals of Human Neuropsychology*, 4th ed., W.H. Freeman and Co., New York, 1996.

A classic in the field and highly recommended for a good understanding of the human brain in action. Topics discussed include memory, attention, language, and the limbic system.

Martin, J.H., *Neuroanatomy: Text and Atlas*, 3rd ed., McGraw-Hill, New York. 2003.

A very complete text with a neuroanatomical perspective and accompanied by some fine (two-color) explanatory illustrations, written as the companion to Kandel et al. The material is clearly presented, with explanations of how systems function. A detailed atlas section is included at the end, as well as a glossary of terms.

Nolte, J., *The Human Brain*, 5th ed., Mosby, St. Louis, 2002.

This is a new edition of an excellent neuroscience text, with anatomical and functional (physiological) information on the nervous system, complemented with clinically relevant material. The textbook includes scores of illustrations in full color, stained brainstem and spinal cord cross-sections, along with three-dimensional brain reconstructions by John Sundsten. A glossary has been added.

Steward, O., *Functional Neuroscience*, Springer, Berlin, 2000.

According to the author, this is a book for medical students that blends the physiological systems approach with the structural aspects. The emphasis is on the "processing" of information, for example, in the visual system. Chapters at the end discuss arousal, attention, consciousness, and sleep. It is nicely formatted and readable.

Williams, P. and Warwick, R., *Functional Neuroanatomy of Man*, W.B. Saunders, Philadelphia, 1975.

This is the "neuro" section from *Gray's Anatomy*. Although somewhat dated, there is excellent reference material on the central nervous system, as well as the nerves and autonomic parts of the peripheral nervous system. The limbic system and its development are also well described.

Wilson-Pauwels, L., Akesson, E.J., and Stewart, P.A., *Cranial Nerves: Anatomy and Clinical Comments*, B.C. Decker, Toronto, 1988.

A handy resource on the cranial nerves, with some very nice illustrations. It is relatively complete and easy to follow.

NEUROANATOMICAL ATLASES

DeArmond, S.J., Fusco, M.M., and Dewey, M.M., *Structure of the Human Brain: A Photographic Atlas*, 3rd ed., Oxford University Press, Oxford, 1989.

An excellent and classic reference to the neuroanatomy of the human CNS. No explanatory text and no color.

England, M.A. and Wakely, J., *Color Atlas of the Brain and Spinal Cord*, Mosby, St. Louis, 1991.

A very well illustrated atlas, with most of the photographs and sections in color. Little in the way of explanatory text.

Felten, D.L. and Jozefowicz, R.F., *Netter's Atlas of Human Neuroscience*, Icon Learning Systems, Teterboro, NJ, 2003.

The familiar illustrations of Netter on the nervous system have been collected into a single atlas, each with limited commentary. Both peripheral and autonomic nervous systems are included. The diagrams are extensively labeled.

Haines, D., *Neuroanatomy: An Atlas of Structures, Sections and Systems*, 6th ed., Lippincott, Williams and Wilkins, Baltimore, 2004.

A popular atlas that has some excellent photographs of the brain, some color illustrations of the vascular supply, with additional radiologic material, all without explanatory text. The histological section of the brainstem is very detailed. There is a limited presentation of the pathways and functional systems, with text. This edition comes with a CD-ROM containing all the illustrations, with some accompanying text.

Netter, F.H., *The CIBA Collection of Medical Illustrations*, Volume 1, Part 1, CIBA, Summit, NJ, 1983.

A classic. Excellent illustrations of the nervous system, as well as of the skull, the autonomic and peripheral nervous systems, and embryology. The text is interesting but may be dated.

Nieuwenhuys, R., Voogd, J., and van Huijzen, C., *The Human Central Nervous System*, Springer Verlag, Berlin, 1981.

Unique three-dimensional drawings of the CNS and its pathways are presented, in tones of gray. These diagrams are extensively labeled, with no explanatory text.

Nolte, J. and Angevine, J.B., *The Human Brain in Photographs and Diagrams*, 2nd ed., Mosby, St. Louis, 2000.

A well illustrated (color) atlas, with text and illustrations, and neuroradiology. Functional systems are drawn onto the brain sections with the emphasis on the neuroanatomy; the accompanying text is quite detailed. Excellent three-dimensional brain reconstructions by J.W. Sundsten.

Woolsey, T.A., Hanaway, J., and Gado, M.H., *The Brain Atlas: A Visual Guide to the Human Central Nervous System*, 2nd ed., Wiley, Hoboken, NJ, 2003.

Part II of the book is a complete pictorial atlas of the human brain, with some color illustrations and radiographic material. Parts III and IV consist of histological sections of the hemispheres, brainstem, spinal cord, and limbic structures. Part V presents the pathways, accompanied by some explanatory text.

CLINICAL TEXTS

Aminoff, M.J., Greenberg, D.A., and Simon, R.P., *Clinical Neurology*, 6th ed., Lange Medical Books/McGraw-Hill, New York, 2005.

If a student wishes to consult a clinical book for a quick look at a disease or syndrome, then this is a suitable book of the survey type. Clinical findings are given, and investigative studies are included, as well as treatment. The illustrations are adequate (in two colors), and there are many tables with classifications and causes.

Asbury, A.K., McKhann, G.M., McDonald, W.I., Goodsby, P.J., and McArthur, J.C., *Diseases of the Nervous System: Clinical Neurobiology*, 3rd ed., Cambridge University Press, Cambridge, 2002.

A complete neurology text, in two volumes, on all aspects of basic and clinical neurology and the therapeutic approach to diseases of the nervous system.

Donaghy, M., *Brain's Disease of the Nervous System*, 11th ed., Oxford University Press, Oxford, 2001.

A very trusted source of information about clinical diseases and their treatments.

Fuller, G. and Manford, M., *Neurology: An Illustrated Colour Text*, Churchill Livingstone, London, 2000.

A concise explanation of select clinical entities is presented, with many illustrations (in full color); not a comprehensive textbook. The large format and presentation make this an appealing but limited book.

Harrison's Principles of Internal Medicine, 16th ed., Kasper, D.L., Braunwald, E., Fauci, A.S., Hauser, S.L., Longo, D.L., and Jameson, J.L., Eds, McGraw-Hill, New York, 2005.

Harrison's is a trusted, authoritative source of information, with few illustrations. Part 2 in Section 3 (Volume I) has chapters on the presentation of disease; Part 15 (Volume II) is on all neurologic disorders of the CNS, nerve and muscle diseases, as well as mental disorders. The online version of Harrison's has updates, search capability, practice guidelines, and online lectures and reviews, as well as illustrations.

Ropper, A.H. and Brown, R.H., *Adams and Victor's Principles of Neurology*, 8th ed., McGraw-Hill, New York, 2005.

A comprehensive neurology text — with part devoted to cardinal manifestations of neurologic diseases and part to major categories of diseases.

Rowland, J.P., *Merritt's Neurology*, 11th ed., Lippincott, Williams and Wilkins, Baltimore, 2005.

A well-known, complete, and trustworthy neurology textbook, now edited by L.P. Rowland.

Royden-Jones, H., *Netter's Neurology*, Icon Learning Systems, Teterboro, NJ, 2005.

Netter's neurological illustrations have been collected in one textbook, with the addition of Netter-style clinical pictures; these add an interesting dimension to the descriptive text. There is broad coverage of many disease states, though not in depth, with clinical scenarios in each chapter. It is now available with a CD-ROM.

PEDIATRIC NEUROLOGY

Fenichel, G.M., *Clinical Pediatric Neurology*, W. B. Saunders, Philadelphia, 2001.

This book is recommended for medical students and other novices by a highly experienced pediatric neurologist as a basic text with a clinical approach, using signs and symptoms.

NEUROPATHOLOGY

Robbin's Neuropathology

Robbins and Cotran Pathologic Basis of Disease, 7th ed., Kumar, V., Abbas, A.K., and Fausto, N., Eds, Elsevier Saunders, Philadelphia, 2005.

A complete source for information on all aspects of pathology for learners, including neuropathology. Purchase of the book includes a CD-ROM with interactive clinical cases, and access to the Web site.

Robbins Basic Pathology, 7th ed., Kumar, V., Cotran, R.S., and Robbins, S.L., Eds, Saunders, Philadelphia, 2003.

Not as complete as the other text (above).

WEB SITES

Web sites should only be recommended to students *after* they have been critically evaluated by the teaching faculty. If keeping up with various teaching texts is difficult, a critical evaluation of the various Web resources is an impossible task for any one person. This is indeed a task to be shared with colleagues, and perhaps by a consortium of teachers and students.

Additional sources of reliable information on diseases are usually available on the disease-specific Web site maintained by an organization, usually with clear explanatory text on the disease and often accompanied by excellent illustrations.

The following sites have been visited by the author, and several of them are gateways to other sites — clearly not every one of the links has been viewed. Although some are intended for the general public, they may contain good illustrations or other links.

The usual www precaution prevails — look carefully at who created the Web site and when.

One additional piece of advice — a high-speed connection is a must for this exploration.

SOCIETY FOR NEUROSCIENCE

http://web.sfn.org/
This is the official Web site for the Society for Neuroscience, a very large and vibrant organization with an annual meeting attended by more than 30,000 neuroscientists from all over the world.

The Society maintains an active educational branch, which is responsible for sponsoring a Brain Awareness Week aimed at the public at large and, particularly, at students in elementary and high schools. The following are examples of their publications.

Searching for Answers: Families and Brain Disorders

This four-part DVD shows the human face of degenerative brain diseases. Researchers tell how they are working to find treatments and cures for Huntington''s disease, Parkinson's disease, amyotrophic lateral sclerosis (ALS), and Alzheimer's disease. Patients and families describe the powerful physical, emotional, and financial impact of these devastating disorders.

Brain Facts

Brain Facts is a 52-page primer on the brain and nervous system, published by the Society for Neuroscience. It is a starting point for a general audience interested in neuroscience. This newly revised edition of *Brain Facts* is available in print and in pdf format. The new edition updates all sections and includes new information on brain development, addiction, neurological and psychiatric illnesses, and potential therapies.

DIGITAL ANATOMIST PROJECT

http://www9.biostr.washington.edu/da.html
Brain Atlas: The material includes two-dimensional and three-dimensional views of the brain from cadaver sections, MRI scans, and computer reconstructions. Authored by John W. Sundsten.

Neuroanatomy Interactive Syllabus: This syllabus uses the images in the Atlas (above) and many others. It is organized into functional chapters suitable as a laboratory guide, with an instructive caption accompanying each image. It contains three-dimensional computer graphic reconstructions of brain material; MRI scans; tissue sections, some enhanced with pathways; gross brain specimens and dissections; and summary drawings. Chapters include Topography and Development, Vessels and Ventricles, Spinal Cord, Brainstem and Cranial Nerves, Sensory and Motor Systems, Cerebellum and Basal Ganglia, Eye Movements, Hypothalamus and Limbic System, Cortical Connections, and Forebrain and MRI Scan Serial Sections. Authored by John W. Sundsten and Kathleen A. Mulligan.

Institution: Digital Anatomist Project, Department of Biological Structure, University of Washington, Seattle.

Atlas was formerly available on CD-ROM (JAVA program running on Mac and PC platform).

BRAINSOURCE

http://www.brainsource.com/
BrainSource is an informational Web site aimed at enriching professional, practical, and responsible applications of neuropsychological and neuroscientific knowledge. The Web site is presented by neuropsychologist Dennis P. Swiercinsky.

The site includes a broad and growing collection of information and resources about normal and injured brains, clinical and forensic neuropsychology, brain injury rehabilitation, creativity, memory and other brain processes, education, brain-body health, and other topics in brain science. BrainSource is also a guide to products, books, continuing education, and Internet resources in neuroscience.

This Web site originated in 1998 for promotion of clinical services and as a portal for dissemination of certain documents useful for attorneys, insurance professionals, students, families and persons with brain injury, rehabilitation specialists, and others working in the field of brain injury. The Web site is growing to expand content to broader areas of neuropsychological application.

DISEASES AND DISORDERS

http://www.mic.ki.se/Diseases/C10.html
This site was created by the Karolinska Institute University Library and contains links pertaining to Nervous System Diseases. It is a convenient starting point for all sources of information about the brain. Not all the sites are necessarily scientifically certified.

NEUROANATOMY AND NEUROPATHOLOGY ON THE INTERNET

http://www.neuropat.dote.hu
This site has been compiled and designed by Katalin Hegedus, Department of Neurology, University of Debrecen, Hungary. It is a source for other sites including neuroanatomy, neuropathology, and neuroradiology, and software (commercial and noncommercial) on the brain, and even includes quizzes.

HARDIN MD

http://www.lib.uiowa.edu/hardin/md/neuro.html
This site is a service of the Hardin Library for Health Sciences, University of Iowa. Hardin MD was first launched in 1996 as a source to find the best lists, or directories, of information in health and medicine. The name Hardin MD comes from *Hardin Meta Directory*, since the site was conceived as a "directory of directories." Providing links to high quality directory pages is still an important part of Hardin MD. In recent years, however, they have added other types of links: Just Plain Links pages have direct links to primary information in circumscribed subjects, and many of their pages have links to medical pictures.

SPECIFIC SITES

- Loyola University, Chicago, Stritch School of Medicine
  http://www.meddean.luc.edu/lumen/MedEd/Neuro/index.htm
- Harvard, The Whole Brain Imaging Atlas
  http://www.med.harvard.edu/AAN-LIB/home.html

THE BRAIN FROM TOP TO BOTTOM

http://www.thebrain.mcgill.ca/flash/index_d.html
This site is designed to let users choose the content that matches their level of knowledge. For every topic and subtopic covered on this site, you can choose from three different levels of explanation — beginner, intermediate, or advanced. The major topics include anatomy and function, memory, sensory and motor systems, pain and pleasure, emotion, evolution; other subject areas are under development.

This site focuses on five major levels of organization — social, psychological, neurological, cellular, and molecular. On each page of this site, you can click to move among these five levels and learn what role each plays in the subject under discussion.

THE NEUROLOGIC EXAM — ONLINE

http://medstat.med.utah.edu/neurologicexam/home_exam.html
This includes both an adult and pediatric neurological examination, with video and sound. In addition, there are four neurologic cases on this site, with possibly more to come.

THE DANA FOUNDATION

http://www.dana.org/
The Dana Foundation is a private philanthropic organization with a special interest in brain science, immunology, and arts education. It was founded in 1950.

The Dana Alliance is a nonprofit organization of more than 200 pre-eminent scientists dedicated to advancing education about the progress and promise of brain research.

The Brain Center of this site is a gateway to the latest research on the human brain. The Brain Information and Brain Web sections access links to validated sites related to more than 25 brain disorders.

NEUROSCIENCE FOR KIDS

http://faculty.washington.edu/chudler/neurok.html
Neuroscience for Kids was created for all students and teachers who would like to learn about the nervous system. The site contains a wide variety of resources, including images — not only for kids. Sections include exploring the brain, Internet neuroscience resources, neuroscience in the news, and reference to books, magazines articles, and newspaper articles about the brain.

Neuroscience for Kids is maintained by Eric H. Chudler and supported by a Science Education Partnership Award (R25 RR12312) from the National Center for Research Resources.

TELEVISION SERIES

http://www.pbs.org/wnet/brain/index.html
The Secret Life of the Brain, a David Grubin Production, reveals the fascinating processes involved in brain development across a lifetime. This five-part series, which was

shown nationally on PBS in the winter of 2002, informs viewers of exciting new information in the brain sciences, introduces the foremost researchers in the field, and utilizes dynamic visual imagery and compelling human stories to help a general audience understand otherwise difficult scientific concepts.

The material includes History of the Brain, 3-D Brain Anatomy, Mind Illusions, and Scanning the Brain. Episodes include: The Baby's Brain, The Child's Brain, The Teenage Brain, The Adult Brain, The Aging Brain.

The Secret Life of the Brain is a co-production of Thirteen/WNET New York and David Grubin Productions, © 2001 Educational Broadcasting Corporation and David Grubin Productions, Inc.

VIDEOTAPES (BY THE AUTHOR)

These edited videotape presentations are on the skull and the brain as the material would be shown to students in the gross anatomy laboratory. They have been prepared with the same teaching orientation as this atlas and are particularly useful for self-study or small groups. These videotapes of actual specimens are particularly useful for students who have limited or no access to brain specimens. The videotapes are fully narrated and each lasts for about 20–25 minutes.

The videotapes are handled by Health Sciences Consortium, a non-profit publishing cooperative for instructional media. They may now be requested in DVD format.

INTERIOR OF THE SKULL

This program includes a detailed look at the bones of the skull, the cranial fossa, and the various foramina for the cranial nerves and other structures. Included are views of the meninges and venous sinuses.

THE GROSS ANATOMY OF THE HUMAN BRAIN SERIES

Part I: The Hemispheres
    A presentation on the hemispheres, the functional areas of the cerebral cortex, including the basal ganglia.
Part II: Diencephalon, Brainstem, and Cerebellum
    A detailed look at the brainstem, with a focus on

the cranial nerves and a functional presentation of the cerebellum.
Part III: Cerebrovascular System and Cerebrospinal Fluid
    A presentation of these two subjects.
Part IV: The Limbic System
    A quite detailed presentation on the various aspects of the limbic system, with much explanation and special dissections.

NOTE: It is suggested that these videotapes be purchased by the library or by an institutional (or departmental) media or instructional resource center.

Information regarding the purchase of these and other videotapes may be obtained from: Health Sciences Consortium, 201 Silver Cedar Ct., Chapel Hill, NC, 27514-1517. Phone: (919) 942-8731. Fax: (919) 942-3689.

CD-ROMS

Numerous CDs are appearing on the market, and their evaluation by the teaching faculty is critical before recommending them to learners. In addition, several of the newer textbooks and atlases now have an accompanying CD-ROM. It is indeed a difficult task to obtain and review all the CDs now available and perhaps one that can be shared with students after they have completed their program of study on the nervous system.

A listing of the CD-ROMs available can be viewed on the Web site Neuroanatomy and Neuropathology on the Internet (above) — see http://www.neuro-pat.dote.hu/software.htm.

The following has been reviewed:

Brainstorm: Interactive Neuroanatomy
    By Gary Coppa and Elizabeth Tancred, Stanford University
    A highly interactive and well-integrated cross-linked presentation of the anatomy and some functional aspects of the nervous system.
    Published by Mosby, 11830 Westline Industrial Drive, P.O. BOX 46908, St. Louis, MO, 63146-9934.

GLOSSARY

Note to the Learner: This glossary contains neuronatomical terms, as well as terms commonly used clinically to describe neurological symptoms and physical findings of a neurological examination; few clinical syndromes are included.

Abducens nerve 6th cranial nerve (CN VI); to lateral rectus muscle for abduction of the eye

Accessory nerve 11th cranial nerve (CN XI) — see spinal accessory nerve

Afferent Conduction toward the central nervous system; usually means sensory

Agnosia Loss of ability to recognize the significance of sensory stimuli (tactile, auditory, visual), even though the primary sensory systems are intact

Agonist A muscle that performs a certain movement of the joint; the opposing muscle is called the antagonist

Agraphia Inability to write due to a lesion of higher brain centers, even though muscle strength and coordination are preserved

Akinesia Absence or loss of motor function; lack of spontaneous movement; difficulty in initiating movement (as in Parkinson's disease)

Alexia Loss of ability to grasp the meaning of written words; inability to read due to a central lesion; word blindness

Allocortex The phylogenetically older cerebral cortex, consisting of less than six layers; includes paleocortex (e.g., subicular region = three to five layers) and archicortex (e.g., hippocampus proper and dentate = three layers)

Alpha motor neuron Another name for the anterior (ventral) horn cell, also called the lower motor neuron

Ammon's horn The hippocampus proper, which has an outline in cross-section suggestive of a ram's horn; also called the Cornu Ammonis (CA)

Amygdala Amygdaloid nucleus or body in the temporal lobe of the cerebral hemisphere; a nucleus of the limbic system

Angiogram Display of blood vessels for diagnostic purposes, using, x-rays, MRI or CT, usually by using contrast medium injected into the vascular system

Anopia A defect in the visual field (e.g., hemianopia — loss of one-half of visual field; quadrantanopia — loss of one-quarter of visual field)

Antagonist A muscle that opposes or resists the action of another muscle, which is called the agonist

Antidromic Relating to the propagation of an impulse along an axon in a direction that is the reverse of the normal or usual direction

Aphasia An acquired disruption or disorder of language, specifically a deficit of expression using speech or of comprehending spoken or written language; global aphasia is a severe form affecting all language areas

Apopotosis Programmed cell death, either genetically determined or following an insult or injury to the cell

Apraxia Loss of ability to carry out purposeful or skilled movements despite the preservation of power, sensation, and coordination

Arachnoid The middle meningeal layer, forming the outer boundary of the subarachnoid space

Areflexia Loss of reflex as tested using the myotatic, stretch, deep tendon reflex

Archicerebellum A phylogenetically old part of the cerebellum, functioning in the maintenance of equilibrium; anatomically, the flocculonodular lobe

Archicortex Three-layered cortex included in the limbic system; located mainly in the hippocampus proper and dentate gyrus of the temporal lobe

Area postrema An area involved in vomiting; located in the caudal part of the floor of the fourth ventricle, with no blood-brain-barrier

Ascending tract Central sensory pathway, e.g., from spinal cord to brainstem, cerebellum, or thalamus

Association fibers Fibers connecting parts of the cerebral hemisphere, on the same side

Astereognosis Loss of ability to recognize the nature of objects or to appreciate their shape by touching or feeling them

Astrocyte A type of neuroglial cell with metabolic and structural functions; reacts to injury of the CNS by forming a gliotic "scar"

Asynergy Disturbance of the proper sequencing in the contraction of muscles, at the proper moment, and of the proper degree, so that an action is not executed smoothly or accurately

Ataxia A loss of coordination of voluntary movements; often associated with cerebellar dysfunction

Athetosis Slow writhing movements of the limbs, especially of the hands, not under voluntary control, caused by degenerative changes in the striatum

Autonomic Autonomic nervous system; usually taken to mean the efferent or motor innervation of viscera (smooth muscle and glands)

Autonomic nervous system (ANS) Visceral innervation; sympathetic and parasympathetic divisions system

Axon Efferent process of a neuron, conducting impulses to other neurons or to muscle fibers (striated and smooth) and gland cells

Babinski response Babinski reflex is not correct; stroking the outer border of the sole of the foot in an adult normally results in a plantar (downgoing) of the toes; the Babinski response consists of an upgoing of the first toe and a fanning of the other toes, indicating a lesion of the pyramidal (cortico-spinal) tract

Basal ganglia (nuclei) CNS nuclei involved in motor control, the caudate, putamen and globus pallidus (the lentiform nucleus); including, functionally, the subthalamus and the substantia nigra

Basilar artery The major artery supplying the brainstem and cerebellum, formed by the two vertebral arteries

Brachium A large bundle of fibers connecting one part with another (e.g., brachium associated with the inferior and superior colliculi of the midbrain)

Bradykinesia Abnormally slow initiation of voluntary movements (usually seen in Parkinson's disease)

Brainstem Includes the medulla, pons, and midbrain

Brodmann areas Numerical subdivisions of the cerebral cortex on the basis of histological differences between different functional areas (e.g. area 4 = motor cortex; area 17 = primary visual area)

Bulb Referred at one time to the medulla but in the context of "cortico-bulbar tract" refers to the whole brainstem in which the motor nuclei of cranial nerves and other nuclei are located

Carotid siphon Hairpin bend of the internal carotid artery within the skull

CAT or CT scan Computerized (Axial) Tomography; a diagnostic imaging technique that uses x-rays and computer reconstruction of the brain

Cauda equina "Horse's tail"; the lower lumbar, sacral, and coccygeal spinal nerve roots within the subarachnoid space of the lumbar (CSF) cistern

Caudal Toward the tail, or hindmost part of neuraxis

Caudate nucleus Part of the neostriatum, consists of a head, body, and tail (which extends into the temporal lobe)

Central nervous system (CNS) Brain (cerebral hemispheres), including diencephalon, cerebellum, brainstem, and spinal cord

Cerebellar peduncles Inferior, middle, and superior; fiber tracts linking the cerebellum and brainstem

Cerebellum The little brain; an older part of the brain with motor functions, dorsal to the brainstem, situated in the posterior cranial fossa

Cerebral aqueduct (of Sylvius) Aqueduct of the midbrain; passageway carrying CSF through the midbrain, as part of the ventricular system

Cerebral peduncle Descending cortical fibers in the "basal" (ventral) portion of the midbrain, sometimes includes the substantia nigra (located immediately behind)

Cerebrospinal fluid (CSF) Fluid in the ventricles, and in the subarachnoid space and cisterns

Cerebrum Includes the cerebral hemispheres and diencephalon but not the brainstem and cerebellum

Cervical Referring to the neck region; the part of the spinal cord that supplies the structures of the neck; C1–C7 vertebral; C1–C8 spinal segments

Chorda tympani Part of the 7th cranial nerve (CN VII) (see facial nerve); carrying taste from anterior two-thirds of tongue and parasympathetic innervation to glands

Chorea A motor disorder characterized by abnormal, irregular, spasmodic, jerky, uncontrollable movements of the limbs or facial muscles, thought to be caused by degenerative changes in the basal ganglia

Choroid A delicate membrane; choroid plexuses are found in the ventricles of the brain

Choroid plexus Vascular structure consisting of pia with blood vessels, with a surface layer of ependymal cells; responsible for the production of CSF

Cingulum A bundle of association fibers in the white matter under the cortex of the cingulate gyrus; part of Papez (limbic) circuit

Circle of Willis Anastomosis between internal carotid and basilar arteries, located at the base of the brain, surrounding the pituitary gland

Cistern(a) Expanded portion of subarachnoid space containing CSF, e.g., cisterna magna (cerebello-medullary cistern), lumbar cistern

Claustrum A thin sheet of gray matter, of unknown function, situated between the lentiform nucleus and the insula

Clonus Abnormal sustained series of contractions and relaxations following stretch of the muscle; usually elicited in the ankle joint; present following lesions of the descending motor pathways, and associated with spasticity

Conjugate eye movement Coordinated movement of both eyes together, so that the image falls on corresponding points of both retinas

CNS Abbreviation for central nervous system

Colliculus A small elevation; superior and inferior colliculi comprising the tectum of the midbrain; also facial colliculus in the floor of the fourth ventricle

Commissure A group of nerve fibers in the CNS connecting structures on one side to the other across the midline (e.g., corpus callosum of the cerebral hemispheres; anterior commissure)

Consensual reflex Light reflex; refers to the bilateral response of the pupil after shining a light in one eye

Contralateral On the opposite side (e.g., contralateral to a lesion)

Corona radiata Fibers radiating from the internal capsule to various parts of the cerebral cortex — a term often used by neuroradiologists

Corpus callosum The main (largest) neocortical commissure of the cerebral hemispheres

Corpus striatum Caudate, putamen, and globus pallidus, nuclei inside cerebral hemisphere, with motor function; the basal ganglia

Cortex Layers of gray matter (neurons and neuropil) on the surface of the cerebral hemispheres (mostly six layers) and cerebellum (three layers)

Cortico-bulbar Descending fibers connecting motor cortex with motor cranial nerve nuclei and other nuclei of brainstem (including reticular formation)

Corticofugal fibers Axons carrying impulses away from the cerebral cortex

Corticopetal fibers Axons carrying impulses toward the cerebral cortex

Cortico-spinal tract Descending tract, from motor cortex to anterior (ventral) horn cells of the spinal cord (sometimes direct); also called pyramidal tract

Cranial nerve nuclei Collections of cells in brainstem giving rise to or receiving fibers from cranial nerves (CN III–XII); may be sensory, motor, or autonomic

Cranial nerves Twelve pairs of nerves arising from the brain and innervating structures of the head and neck (CN I is actually a CNS tract)

CSF Cerebrospinal fluid, in ventricles and subarachnoid space (and cisterns)

Cuneatus (cuneate) Sensory tract (fasciculus cuneatus) of the dorsal column of spinal cord, from the upper limbs and body; cuneate nucleus of medulla

Decerebrate posturing (rigidity) Characterized by extension of the upper and lower limbs; lesion at the brainstem level between the vestibular nuclei and the red nucleus

Decorticate posturing (rigidity) Characterized by extension of the lower limbs and flexion of the upper; lesion is located above the level of the red nucleus

Decussation The point of crossing of CNS tracts, e.g., decussations of the pyramidal (cortico-spinal) tract, medial lemnisci, and superior cerebellar peduncles

Dementia Progressive brain disorder that gradually destroys a person's memory, starting with short-term memory, and loss of intellectual ability, such as the ability to learn, reason, make judgments, and communicate, and finally, inability to carry out normal activities of daily living; usually affects people with advancing age

Dendrite Receptive process of a neuron; usually several processes emerge from the cell body, each of which branches in a characteristic pattern

Dendritic spine Cytoplasmic excrescence of a dendrite and the site of an excitatory synapse

Dentate (toothed or notched) Dentate nucleus of the cerebellum (intracerebellar nucleus); dentate gyrus of the hippocampal formation

Dermatone A patch of skin innervated by a single spinal cord segment (e.g., T1 supplies the skin of the inner aspect of the upper arm; T10 supplies umbilical region)

Descending tract Central motor pathway (e.g., from cortex to brainstem or spinal cord)

Diencephalon Consisting of the thalamus, epithalamus (pineal), subthalamus, and hypothalamus

Diplopia Double vision; a single object is seen as two objects

Dominant hemisphere The hemisphere responsible for language; this is the left hemisphere in about 85 to 90% of people (including left-handed individuals)

Dorsal column Fasciculus gracilis and fasciculus cuneatus of the spinal cord, pathways (tracts) for discriminative touch, conscious proprioception and vibration

Dorsal root Afferent sensory component of a spinal nerve, located in the subarachnoid (CSF) space

Dorsal root ganglion (DRG) A group of peripheral neurons along the dorsal root, whose axons carry afferent information from the periphery; their central process enters the spinal cord

Dura Dura mater, the thick external layer of the meninges (brain and spinal cord)

Dural venous sinuses Large venous channels for draining blood from the brain; located within dura of the meninges

Dysarthria Difficulty with the articulation of words

Dyskinesia Purposeless movements of the limbs or trunk, usually due to a lesion of the basal ganglia; also difficulty in performing voluntary movements

Dysmetria Disturbance of the ability to control the range of movement in muscular action, causing under- or overshooting of the target (usually associated with cerebellar lesions)

Dysphagia Difficulty with swallowing

Dyspraxia Impaired ability to perform a voluntary act previously well performed, with intact movement, coordination, and sensation

Efferent Away from the central nervous system; usually means motor to muscles

Emboliform Emboliform nucleus of the cerebellum, one of the intracerebellar (deep cerebellar) nuclei; with globose nucleus forms the interposed nucleus

Entorhinal Associated with olfaction (smell); the entorhinal area is the anterior part of the parahippocampal gyrus, adjacent to the uncus

Ependyma Epithelium lining of ventricles of the brain and central canal of spinal cord; specialized tight junctions at the site of the choroid plexus

Extrapyramidal system An older clinically used term, usually intended to include the basal ganglia portion of the motor systems and not the pyramidal (cortico-spinal) motor system

Facial nerve 7th cranial nerve (CN VII); motor to muscles of facial expression; carries taste from anterior two-thirds of tongue; also parasympathetic to two salivary glands, lacrimal and nasal glands (see also chorda tympani)

Falx Dural partition in the midline of the cranial cavity; the large falx cerebri between the cerebral hemispheres, and the small falx cerebelli

Fascicle A small bundle of nerve fibers

Fasciculus A large tract or bundle of nerve fibers

Fasciculus cuneatus Part of dorsal column of spinal cord; ascending tract for discriminative touch, conscious proprioception and vibration from upper body and upper limb

Fasciculus gracilis Part of dorsal column of spinal cord; ascending tract for discriminative touch, conscious proprioception and vibration from lower body and lower limb

Fastigial nucleus One of the deep cerebellar (intracerebellar) nuclei

Fiber Synonymous with an axon (either peripheral or central)

Flaccid paralysis Muscle paralysis with hypotonia due to a lower motor neuron lesion

Flocculus Lateral part of flocculonodular lobe of cerebellum (vestibulocerebellum)

Folium (plural folia) A flat leaf-like fold of the cerebellar cortex

Foramen An opening, aperture, between spaces containing CSF (e.g., Monro, between lateral ventricles and third ventricle; Magendie, between fourth ventricle and cisterna magna; Luschka, lateral foramen of fourth ventricle)

Forebrain Anterior division of embryonic brain; cerebrum and diencephalon

Fornix The efferent (noncortical) tract of the hippocampal formation, arching over the thalamus and terminating in the mammillary nucleus of the hypothalamus and in the septal region

Fourth (4th) ventricle Cavity between brainstem and cerebellum, containing CSF

Funiculus A large aggregation of white matter in the spinal cord, may contain several tracts

Ganglion (plural ganglia) A collection of nerve cells in the PNS — dorsal root ganglion (DRG) and sympathetic ganglion; also inappropriately used for certain regions of gray matter in the brain (i.e., basal ganglia)

Geniculate bodies Specific relay nuclei of thalamus — medial (auditory) and lateral (visual)

Genu Knee or bend; middle portion of internal capsule; genu of facial nerve

Glial cell Also called neuroglial cell; supporting cells in the central nervous system — astrocyte, oligodendrocyte, and ependymal — also microglia

Globus pallidus Efferent part of basal ganglia; part of the lentiform nucleus with the putamen; located medially

Glossopharyngeal nerve 9th cranial nerve (CN IX); motor to muscles of swallowing and carries taste from posterior one-third of tongue; nerve for the gag reflex

Gracilis (gracile) Sensory tract (fasciculus gracilis) of the dorsal column of spinal cord; nucleus gracilis of medulla

Gray matter Nervous tissue, mainly nerve cell bodies and adjacent neuropil; looks "grayish" after fixation in formalin

Gyrus (plural gyri) A convolution or fold of the cerebral hemisphere; includes cortex and white matter

Habenula A nucleus of the limbic system, adjacent to the posterior end of the roof of the 3rd ventricle (part of the epithalamus)

Hemiballismus Violent jerking or flinging movements of one limb, not under voluntary control, due to a lesion of subthalamic nucleus

Hemiparesis Muscular weakness affecting one side of the body

Hemiplegia Paralysis of one side of the body

Herniation Bulging or expansion of the tissue beyond its normal boundary

Heteronymous hemianopia Loss of different halves of the visual field of both eyes, as defined by projection to the visual cortex of both sides; bitemporal for the temporal halves and binasal for the nasal halves

Hindbrain Posterior division of the embryonic brain; includes pons, medulla, and cerebellum (located in the posterior cranial fossa)

Hippocampus or hippocampus "proper" Part of limbic system; a cortical area "buried" within the medial temporal lobe, consisting of phylogenetically old (three-layered) cortex; protrudes into floor of inferior horn of lateral ventricle

Homonymous hemianopia Loss of the same visual field in both eyes (i.e., left or right) as defined by the projection to the visual cortex on one side — involving the nasal half of the visual field in one eye and the temporal half in the other eye; also quadrantanopia

Horner's syndrome Miosis (constriction of the pupil), anhidrosis (dry skin with no sweat), and ptosis (drooping of the upper eyelid) due to a lesion of the sympathetic pathway to the head

Hydrocephalus Enlargement of the ventricles, usually due to excessive accumulation of cerebrospinal fluid within the ventricles (e.g., obstruction)

Hypoglossal nerve 12th cranial nerve (CN XII); motor to muscles of the tongue

Hypo/hyper reflexia Decrease (hypo) or increase (hyper) of the stretch (deep tendon) reflex

Hypo/hyper tonia Decrease or increase of the tone of muscles, manifested by decreased or increased resistance to passive movements

Hypokinesia Markedly diminished movements (spontaneous)

Hypothalamus A region of the diencephalon that serves as the main controlling center of the autonomic nervous system and is involved in several limbic circuits; also regulates the pituitary gland

Infarction Local death of an area of tissue due to loss of its blood supply

Infundibulum (funnel) Infundibular stem of the posterior pituitary (neurohypophysis)

Innervation Nerve supply, sensory and/or motor

Insula (island) Cerebral cortical area not visible from outside view and situated at the bottom of the lateral fissure (also called the island of Reil)

Internal capsule White matter between lentiform nucleus and head of caudate nucleus, and thalamus; consists of anterior limb, genu and posterior limb

Ipsilateral On the same side of the body (e.g., ipsilateral to a lesion)

Ischemia A condition in which an area is not receiving an adequate blood supply

Ischemic penumbra A region adjacent to or surrounding an area of infarcted brain tissue that is not receiving sufficient blood; the neurons may still be viable

Kinesthesia The conscious sense of position and movement

Lacune A pathological small "hole" remaining after an infarct in the internal capsule; also irregularly-shaped venous "lakes" or channels draining into the superior sagittal sinus

Lateral ventricle CSF cavity in each cerebral hemisphere; consists of anterior horn, body, atrium (or trigone), posterior horn, and inferior (temporal) horn

Lemniscus A specific pathway in CNS (medial lemniscus for discriminative touch, conscious proprioception, and vibration; lateral lemniscus for audition)

Lentiform Lens-shaped; lentiform nucleus, a part of the corpus striatum; also called lenticular nucleus; composed of putamen (laterally) and globus pallidus

Leptomeninges Arachnoid and pia mater, part of meninges

Lesion Any injury or damage to tissue (e.g., vascular, traumatic)

Limbic system Part of brain associated with emotional behavior

Locus ceruleus A small nucleus located in the uppermost pons on each side of the fourth ventricle; contains melanin-like pigment, visible as a dark-bluish area in freshly sectioned brain

Lower motor neuron Anterior horn cell of spinal cord and its axon; also the cells in the motor cranial nerve nuclei of the brainstem; called the alpha motor neuron; its loss leads to atrophy of the muscle and weakness, with hypotonia and hyporeflexia; also fascicluations are to be noted

Mammillary Mammillary bodies; nuclei of the hypothalamus that are seen as small swellings on the ventral surface of diencephalon (also spelled mamillary)

Massa intermedia A bridge of gray matter connecting the thalami of the two sides across third ventricle; present in 70% of human brains (also called the inter-thalamic adhesion)

Medial lemniscus Brainstem portion of sensory pathway for discriminative touch, conscious proprioception and vibration, formed after synapse (relay) in nucleus gracilis and nucleus cuneatus

Medial longitudinal fasciculus (MLF) A tract throughout the brainstem and upper cervical spinal cord that interconnects visual and vestibular input with other nuclei controlling movements of the eyes and the head and neck

Medulla Caudal portion of the brainstem; may also refer to the spinal cord as in a lesion within (intramedullary) or outside (extramedullary) the cord

Meninges Covering layers of the central nervous system (dura, arachnoid, and pia)

Mesencephalon The midbrain (upper part of the brainstem)

Microglia The "scavenger" cells of the CNS, i.e., macrophages; considered by some as one of the neuroglia

Midbrain Part of the brainstem; also known as mesencephalon (the middle division of the embryonic brain)

Motor Associated with movement or response

Motor unit A lower motor neuron, its axon, and the muscle fibers that it innervates

MRI/NMR Magnetic Resonance Imaging (nuclear magnetic resonance), a diagnostic imaging technique that uses an extremely strong magnet, not x-rays

Muscle spindle Specialized receptor within voluntary muscles that detects muscle length; necessary for the stretch/myotatic reflex (DTR); contains muscle fibers within itself capable of adjusting the sensitivity of the receptor

Myelin Proteolipid layers surrounding nerve fibers, formed in segments, which is important for rapid (saltatory) nerve conduction

Myelin sheath Covering of nerve fiber, formed and maintained by oligodendrocyte in CNS and Schwann cell in PNS; interrupted by nodes of Ranvier

Myelopathy Generic term for disease affecting the spinal cord

Myopathy Generic term for muscle disease

Myotatic reflex Stretch reflex, also called deep tendon reflex (DTR); elicited by stretching the muscle; causes a reflex contraction of the same muscle; monosynaptic (also spelled myotactic reflex)

Myotome Muscle groups innervated by a single spinal cord segment; in fact, usually two adjacent segments are involved (e.g., biceps, C5 and C6)

Neocerebellum Phylogenetically newest part of the cerebellum, present in mammals and especially well developed in humans; involved in coordinating precise voluntary movements and also in motor planning

Neocortex Phylogenetically newest part of the cerebral cortex, consisting of six layers (and sublayers) characteristic of mammals and constituting most of the cerebral cortex in humans

Neostriatum The phylogenetically newer part of the basal ganglia consisting of the caudate nucleus and putamen; also called the striatum

Nerve fiber Axonal cell process, plus myelin sheath, if present

Neuralgia Pain — severe, shooting, "electrical," along the distribution of a peripheral nerve (spinal or cranial)

Neuraxis The straight longitudinal axis of the embryonic or primitive neural tube, bent in later evolution and development

Neuroglia Accessory or interstitial cells of the central nervous system; includes astrocytes, oligodendrocytes, ependymal cells, and microglial cells

Neuron The basic structural unit of the nervous system, consisting of the nerve cell body and its processes — dendrites and axon

Neuropathy Disorder of one or more peripheral nerves

Neuropil An area between nerve cells consisting of a complex arrangement of nerve cell processes, including axon terminals, dendrites, and synapses

Nociception Refers to an injurious stimulus causing a neuronal response; may or may not be associated with the sensation of pain

Node of Ranvier Gap in myelin sheath between two successive internodes; necessary for saltatory (rapid) conduction

Nucleus (plural nuclei) An aggregation of neurons within the CNS; in histology, the nucleus of a cell

Nystagmus An involuntary oscillation of the eye(s), slow in one direction and rapid in the other; named for the direction of the quick movement

Oculomotor nerve 3rd cranial nerve (CN III); motor to most muscles of the eye

Olfactory nerve 1st cranial nerve (CN I); special sense of smell

Oligodendrocyte A neuroglial cell, forms and maintains the myelin sheath in the CNS; each cell is responsible for several internodes on different axons

Optic chiasm(a) Partial crossing of optic nerves — nasal half of retina representing the temporal visual fields — after which the optic tracts are formed

Optic disc Area of the retina where the optic nerve exits; also the site for the central retinal artery and vein; devoid of receptors, hence the blind spot

Optic nerve 2nd cranial nerve (CN II); special sense of vision; actually a tract of the CNS, from the ganglion cells of the retina until the optic chiasm

Paleocortex Phylogenetically older cerebral cortex consisting of three to five layers

Papilledema Edema of the optic disc, visualized with an ophthalmoscope (also called a choked disc); usually a sign of abnormal increased intracranial pressure

Paralysis Complete loss of muscular action

Paraplegia Paralysis of both legs and lower part of trunk

Paresis Muscle weakness or partial paralysis

Paresthesia Spontaneous abnormal sensation (e.g., tingling; pins and needles)

Pathway A chain of functionally related neurons (nuclei) and their axons, making a connection between one region of CNS and another; a tract (e.g., visual pathway, dorsal column-medial lemniscus sensory pathway)

Peduncle A thick stalk or stem; a bundle of nerve fibers (cerebral peduncle of the midbrain; also three cerebellar peduncles — superior, middle, and inferior)

Perikaryon The cytoplasm surrounding the nucleus of a cell; sometimes refers to the cell body of a neuron

Peripheral nervous system (PNS) Nerve roots, peripheral nerves and ganglia outside the CNS (motor, sensory, and autonomic)

PET Positron Emission Tomography; a technique used to visualize areas of the living brain that become "activated" under certain task conditions; uses very short-acting biologically active radioactive compounds

Pia (mater) The thin innermost layer of the meninges, attached to the surface of the brain and spinal cord; forms the inner boundary of the subarachnoid space

Plexus An interweaving arrangement of vessels or nerves

Pons (bridge) The middle section of the brainstem that lies between the medulla and the midbrain; appears to constitute a bridge between the two hemispheres of the cerebellum

Projection fibers Bidirectional fibers connecting the cerebral cortex with structures below, including basal ganglia, thalamus, brainstem, and spinal cord

Proprioception The sense of body position (conscious or unconscious)

Proprioceptor One of the specialized sensory endings in muscles, tendons, and joints; provides information concerning movement and position of body parts (proprioception)

Prosody Vocal tone, inflection, and melody accompanying speech

Ptosis Drooping of the upper eyelid

Pulvinar The posterior nucleus of the thalamus; functionally, involved with vision

Putamen The larger (lateral) part of the lentiform nucleus, with the globus pallidus; part of the neostriatum with the caudate nucleus

Pyramidal system Named because the cortico-spinal tracts occupy pyramid-shaped areas on the ventral aspect of the medulla; may include cortico-bulbar fibers; the term pyramidal tract refers specifically to the cortico-spinal tract

Quadrigeminal Referring to the four colliculi of the midbrain; also called the tectum

Quadriplegia Paralysis affecting the four limbs (also called tetraplegia)

Radicular Refers to a nerve root (motor or sensory)

Ramus (plural rami) The division of the mixed spinal nerve (containing sensory, motor, and autonomic fibers) into anterior and posterior

Raphe An anatomical structure in the midline; in the brainstem, several nuclei of the reticular formation are in the midline of the medulla, pons, and midbrain (these nuclei use serotonin as the neurotransmitter)

Red nucleus Nucleus in the midbrain (reddish color in a fresh specimen)

Reflex Involuntary movement of a fixed nature in response to a stimulus

Reflex arc Consisting of an afferent fiber, a central connection, a motor neuron, and its efferent axon leading to a muscle movement

Reticular Pertaining to or resembling a net — reticular formation of brainstem

Reticular formation Diffuse nervous tissue, nuclei and connections, in brainstem; quite old phylogenetically

Rhinencephalon In humans, refers to structures related to the olfactory system

Rigidity Abnormal muscle stiffness (increased tone) with increased resistance to passive movement of both agonists and antagonists (e.g., flexors and extensors), usually seen in Parkinson's disease; velocity independent

Root The peripheral nerves — sensory (afferent, dorsal) and motor (efferent, ventral) — as they emerge from the spinal cord and are found in the subarachnoid space

Rostral Toward the nose, or the most anterior end of the neuraxis

Rubro Red; pertaining to the red nucleus, as in rubro-spinal tract and cortico-rubral fibers

Saccadic To jerk; extremely quick movements, normally of both eyes together (conjugate movement), in changing the direction of gaze

Schwann cell Neuroglial cell of the PNS responsible for formation and maintenance of myelin; there is one Schwann cell for each internode of myelin

Secretomotor Parasympathetic motor nerve supply to a gland

Sensory Afferent; to do with receiving information, from the skin, the muscles, the external environment, or from internal organs

Septum pellucidum A double membrane of connective tissue separating the anterior horns of the lateral ventricles, situated in the median plane

Septal region An area below the anterior end of the corpus callosum on the medial aspect of the frontal lobe that includes cortex and the septal nuclei

Somatic Used in neurology to denote the body, exclusive of the viscera (as in somatic afferent neurons from the skin and body wall); the word soma is also used to refer to the cell body of a neuron

Somatic senses Touch (discriminative and crude), pain, temperature, proprioception, and the "sense of vibration"

Somatotopic The orderly representation of the body parts in CNS pathways, nuclei, thalamus, and cortex; topographical representation

Somesthetic Consciousness of having a body; somesthetic senses are the general senses of touch, pain, temperature, position, movement, and "vibration"

Spasticity Velocity-dependent increased tone and increased resistance to passive stretch of the antigravity muscles; in humans, flexors of the upper limb and extensors of the lower limb; usually accompanied by hyper-reflexia

Special senses Sight (vision), hearing (audition), balance (vestibular), taste (gustatory), and smell (olfactory)

Spinal accessory nerve 11th cranial nerve (CN XI); refers usually to the part of the nerve that originates in the upper spinal cord (C1–5) and innervates the muscles of the neck, the sternomastoid and trapezius muscles

Spinal shock Complete "shut down" of all spinal cord activity (in humans) following an acute complete lesion of the cord (e.g., severed cord after a diving or motor vehicle accident); usually up to two to three weeks in duration

Spino-cerebellar tracts Ascending tracts of the spinal cord, anterior and posterior, for "unconscious" proprioception to the cerebellum

Spino-thalamic tracts Ascending tracts of the spinal cord for pain and temperature (lateral) and nondiscriminative or light touch and pressure (anterior)

Split brain A brain in which the corpus callosum has been severed in the midline, usually as a therapeutic measure for intractable epilepsy

Stereognosis The recognition of an object using the tactile senses and also central processing, involving association areas especially in the parietal lobe

Strabismus A squint; lack of conjugate fixation of the eyes; may be constant or variable

Stria A slender strand of fibers (e.g., stria terminalis from amygdala)

Striatum The phylogenetically more recent part of the basal ganglia (neostriatum) consisting of the caudate nucleus and the putamen (lateral portion of the lentiform nucleus)

Stroke A sudden severe attack of the CNS; usually refers to a sudden focal loss of neurologic function due to death of neural tissue; mostly due to a vascular lesion, either infarct (embolus, occlusion) or hemorrhage

Subarachnoid space Space between arachnoid and pia mater, containing CSF (cerebrospinal fluid)

Subcortical Not in the cerebral cortex, i.e., at a functionally or evolutionary "lower" level in the CNS; usually refers to the white matter of the cerebral hemispheres, and also may include the basal ganglia

Subicular region Part of hippocampal formation; transitional cortex (three to five layers) between that of the hippocampus proper and the parahippocampal gyrus

Substantia gelatinosa A nucleus of the gray matter of the dorsal (sensory) horn of the spinal cord composed of small neurons; receives pain and temperature afferents

Substantia nigra A flattened nucleus in the midbrain with motor functions — consisting of two parts: the pars compacta with melanin pigment in the neurons (the dopamine neurons, which degenerate in Parkinson's disease), and the pars reticulata, which is an output nucleus of the basal ganglia

Subthalamus Region of the diencephalon beneath the thalamus, containing fiber tracts and the subthalamic nucleus; part of the functional basal ganglia

Sulcus (plural sulci) Groove between adjacent gyri of the cerebral cortex; a deep sulcus may be called a fissure

Synapse Area of structural and functional specialization between neurons where transmission occurs (excitatory, inhibitory, or modulation), using neurotransmitter substances (e.g., glutamate, GABA); similarly at the neuromuscular junction (using acetylcholine)

Syringomyelia A pathological condition characterized by expansion of the central canal of the spinal cord with destruction of nervous tissue around the cavity

Tectum The "roof" of the midbrain (behind the aqueduct) consisting of the paired superior and inferior colliculi; also called the quadrigeminal plate

Tegmentum The "core area" of the brainstem, between the ventricle (or aqueduct) and the cortico-spinal tract; contains the reticular formation, cranial nerve and other nuclei, and various tracts

Telencephalon Rostral part of embryonic forebrain; primarily cerebral hemispheres of the adult brain

Tentorium The tentorium cerebelli is a sheet of dura between the occipital lobes of the cerebral hemispheres and the cerebellum; its hiatus or notch is the opening for the brainstem — at the level of the midbrain

Thalamus A major portion of the diencephalon with sensory, motor, and integrative functions; consists of several nuclei with connections to areas of the cerebral cortex

Third (3ʳᵈ) ventricle Midline ventricle at the level of the diencephalon (between the thalamus of each side), containing CSF

Tic Brief, repeated, stereotyped, semipurposeful muscle contraction; not under voluntary control, although may be suppressed for a limited time

Tinnitus Persistent ringing or buzzing sound in one or both ears

Tomography Radiological images, done sectionally, including CT and MRI

Tone Referring to muscle, its firmness, and elasticity — normal, hyper, hypo — elicited by passive movement and also assessed by palpation

Tract A bundle of nerve fibers within the CNS, with a common origin and termination, (e.g., optic tract, cortico-spinal tract)

Transient ischemic attack (TIA) A nonpermanent focal deficit, caused by a vascular event; by definition, usually reversible within a few hours, with a maximum of 24 hours

Trapezoid body Transverse crossing fibers of the auditory pathway situated in the ventral portion of the tegmentum of the lower pons

Tremor Oscillating, "rhythmic" movements of the hands, limbs, head, or voice; intention (kinetic) tremor of the limb commonly seen with cerebellar lesions; tremor at rest commonly associated with Parkinson's disease

Trigeminal nerve 5th cranial nerve (CN V); major sensory nerve of the head (face, eye, tongue, nose, sinuses); also supplies muscles of mastication

Trochlear nerve 4th cranial nerve (CN IV); motor to the superior oblique eye muscle

Two-point discrimination Recognition of the simultaneous application of two points close together on the skin; distance varies with the area of the body (compare finger tip to back)

Uncus An area of cortex — the medial protrusion of the rostral (anterior) part of the parahippocampal gyrus of the temporal lobe; the amygdala is situated deep to this area; important clinically as in uncal herniation

Upper motor neuron Neuron located in the motor cortex or other motor areas of the cerebral cortex or in the brainstem — giving rise to a descending tract to lower motor neurons in the brainstem (for cranial nerves) or spinal cord (for body and limbs)

Upper motor neuron lesion A lesion of the brain (cortex, white matter of hemisphere), brainstem, or spinal cord interrupting descending motor influences to the lower motor neurons of the brainstem or spinal cord, characterized by weakness, spasticity, and hyperreflexia, and often clonus; usually accompanied by a Babinski response

Vagus 10th cranial nerve (CN X); supplies motor fibers to the larynx; the major parasympathetic nerve to organs of the thorax and abdomen

Velum A membranous structure; the superior medullary velum forms the roof of the fourth ventricle

Ventricles Cerebrospinal (CSF) fluid-filled cavities inside the brain

Vermis Unpaired midline portion of the cerebellum, between the hemispheres

Vertigo Abnormal sense of spinning, whirling, or motion, either of the self or of one's environment

Vestibulocochlear 8th cranial nerve (CN VIII); special senses of hearing and balance (acoustic nerve is not really correct)

White matter Nervous tissue of CNS made up of nerve fibers (axons), some of which are myelinated; appears "whitish" after fixation in formalin

INDEX